FRONTIERS OF FUNDAMENTAL PHYSICS

Frontiers of Fundamental Physics

Proceedings of the Sixth International Symposium "Frontiers of Fundamental and Computational Physics", Udine, Italy, 26–29 September 2004

Edited by

B.G. SIDHARTH
B.M. Birla Science Centre,
Adarsh Nagar, Hyderabad, India

F. HONSELL
University of Udine, Italy

and

A. DE ANGELIS
University of Udine, Italy

Springer

A C.I.P. Catalogue record for this book is available from the Library of Congress.

ISBN-10 1-4020-4151-9 (HB)
ISBN-13 978-1-4020-4151-8 (HB)
ISBN-10 1-4020-4339-2 (e-book)
ISBN-13 978-1-4020-4339-0 (e-book)

Published by Springer,
P.O. Box 17, 3300 AA Dordrecht, The Netherlands.

www.springer.com

Printed on acid-free paper

TABLE OF CONTENTS

INTRODUCTION

One of the first Computer Science sites in Italy, in recent years, the Friuli region has become a very active hub in Computational Physics and other applications of Informatics to Human and Natural Sciences. In particular the University of Udine has developed a tradition in innovative cross-disciplinary research areas involving Computer Science and Physics, providing digital tools for laboratories such as NASA and CERN.

The sixth International Symposium "Frontiers of Fundamental and Computational Physics" (FFP6) aimed at providing a platform for a wide range of physicists to meet and share thoughts on the latest trends in various research areas including High Energy Physics, Theoretical Physics, Gravitation and Cosmology, Astrophysics, Condensed Matter Physics, Fluid Mechanics. Such frontier lines were unified by the use of computers as an, often primary, research instrument, or dealing with issues related to information theory.

The present Sixth International Symposium in the series was organized at the University of Udine, Italy from 26th to 29th of September 2004. The University of Udine and the B.M. Birla Science Centre in Hyderabad have collaborated in the organization of this Symposium and the edition of these Proceedings, under the auspices of their joint initiative the International Institute of Applicable Mathematics and Information Sciences. The contributions in the Proceedings are grouped as follows:

- Field Theory, Relativity and Cosmology
- Foundations of Physics and of Information Sciences
- Nuclear and High-Energy Particle Physics and Astrophysics; Astroparticle Physics
- Complex Systems; Fluid Mechanics
- New Approaches to Physics Teaching

This Symposium had an attendance of over 100 participants. There were 63 papers/presentations, including 4 introductory invited lectures delivered by the Nobel Laureates L. Cooper and G. 't Hooft, and by the eminent physicists Y. Ne'eman and W. Greiner. As was the case for the previous editions, the conference was a great success. Full papers in these Proceeding underwent a refereeing process. When full papers have not been submitted, abstracts have been included.

The Editors
Udine, March 2005

Previous Symposia

- 5th International Symposium on Frontiers of Fundamental Physics
  06 - 08 January 2003
  B M Birla Science Center - Adarsh Nagar - HYDERABAD 500463 India
- 4th International Symposium on Frontiers of Fundamental Physics
  09 - 13 December 2000
  B M Birla Science Center - Adarsh Nagar - HYDERABAD 500463 India
- 3rd International Symposium on Frontiers of Fundamental Physics
  B M Birla Science Center - Adarsh Nagar - HYDERABAD 500463 India
- 2nd International Symposium on Frontiers of Fundamental Physics
  30 December 1998 - 01 January 1999
  B M Birla Science Center - Adarsh Nagar - HYDERABAD 500463 India
- 1st International Symposium on Frontiers of Fundamental Physics
  December 1997
  B M Birla Science Center - Adarsh Nagar - HYDERABAD 500463 India

Part I

Field Theory, Relativity and Cosmology

COSMOLOGICAL THEORIES OF SPECIAL AND GENERAL RELATIVITY - I

MOSHE CARMELI [a]

[a] *Department of Physics, Ben Gurion University of the Negev, Beer Sheva 84105, Israel*

Abstract

In the standard cosmological theory one uses the Einstein concepts of space and time as were originally introduced for the special theory of relativity and the general relativity theory. According to this approach all physical quantities are described in terms of the continuum spatial coordinates and time. Using general relativity theory a great progress has been made in understanding the evolution of the Universe. Cosmologists usually measure spatial distances and redshitfs of faraway galaxies as expressed by the Hubble expansion. In recent years this fact was undertaken to develop new theories in terms of distances and velocities (redshift). While in Einstein's relativity the propagation of light plays the major role, in the new theory it is the expansion of the Universe that takes that role and appears at the outset. The cosmic time becomes crucial in these recent theories, which in the standard theory is considered to be absolute but here it is relative. In this lecture this new approach to cosmology is presented.

1 Introduction

It is well known that both Einstein's theories are based on the fact that light propagates at a constant velocity. However, the Universe also expands at a constant rate when gravity is negligible. Moreover, cosmologists usually measure spatial distances and redshifts of faraway galaxies as expressed by the Hubble expansion. In recent years this fact was undertaken to develop new theories in terms of distances and velocities (redshift). While in Einstein's special relativity the propagation of

B.G. Sidharth, F. Honsell and A. de Angelis. (eds.), Frontiers of Fundamental Physics, 3–11.

light plays the major role, in the new theory it is the expansion of the Universe that takes that role. It is the concept of cosmic time that becomes crucial in these recent theories. In the standard theory the cosmic time is considered to be absolute. Thus we talk about the Big Bang time with respect to us here on Earth as an absolute quantity. Consider, for example, another galaxy that has, let us say, a relative cosmic time with respect to us of 1 billion year. Now one may ask what will be the Big Bang time with respect to this galaxy. Will it be the Big Bang time with respect to us minus 1 billion year? A person who lives in that galaxy will look at our galaxy and say that ours is far away from him by also 1 billion year. Will that mean, with respect to him, our galaxy is closer to the Big Bang time by 1 billion year? Or will we seem to him to be farther by 1 billion year? All this leads to the conclusion that there is no absolute cosmic time. Rather, it is a relative concept.

Based on this assumption, we present in this lecture a theory that relates distances between galaxies and their relative velocities. These are actually the dynamical variables that astronomers measure. In the first part the theory will be relating distances to velocities with very weak gravitational field (special relativity). We then, in the second part, extend the theory to include the gravitational field of the Universe. Before doing that, we present a brief review of Einstein's special relativity theory.

2 Einstein's special relativity: A review

We will not give a full detail of this very important theory. Rather, we outline some of its fundamentals [1,2]

1. The Michelson-Morley experiment for the constancy of the speed of light.
2. Einstein's two postulates: (a) Constancy of the speed of light; (b) Validity of the laws of physics in internal coordinate systems. (See Figure 1)
3. The Lorentz transformation. This is given by

$$ct' = (ct - \beta x) / \sqrt{1-\beta^2}, \qquad x' = (x - \beta ct) / \sqrt{1-\beta^2}, \tag{2.1}$$

where $\beta = v/c$, and $y' = y$, $z' = z$.
4. Minkowski's unification of space and time.
5. Invariance of the laws of physics under the Lorentz transformation.
6. Minkowski's line element

$$ds^2 = c^2dt^2 - \left(dx^2 + dy^2 + dz^2\right) \tag{2.2}$$

7. The light cone (see Figure 2).

3 Cosmological special relativity

We outline this theory very briefly with the following points [3, 4, 5].
1. Hubble's law $R = \tau v$, where R is the distance to a galaxy, v is the receding

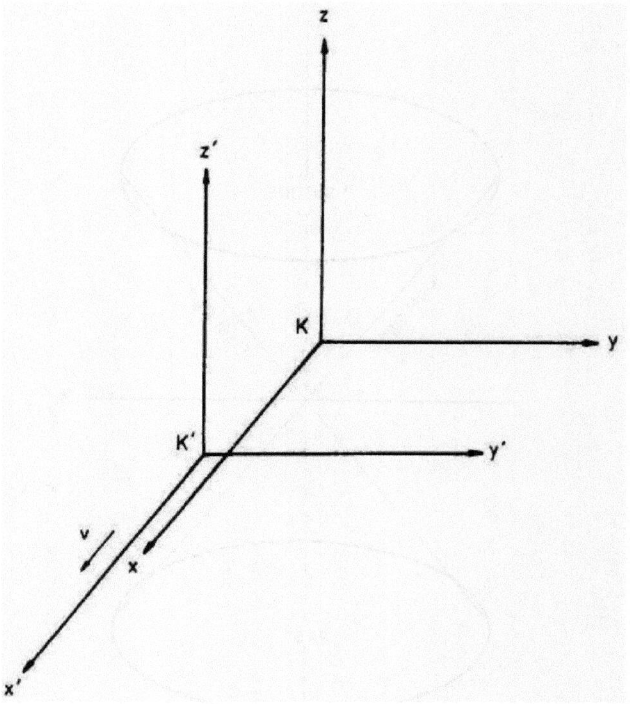

Figure 1. Two coordinate systems K and K', one moving with respect to the other with a velocity v in the x-direction.

velocity of the galaxy, τ is universal constant equal to 12.486 Gyr (Big Bang time).
2. Cosmic time is not absolute but a relative concept.
Example: Another galaxy is one billion year with respect to us. One may ask, what is the Big Bang time with respect to this galaxy. Will it be BB time with respect to us minus 1 billion year?From the point of view of that galaxy ours is far away from it by also 1 billion year. Does that mean, with respect to that galaxy, ours is closer to the BB time by 1 billion year?Or our galaxy will seem to be farther by 1 billion year? All this leads to the conclusion that there is no absolute cosmic time. Rather, it is a relative concept.
3. Universe with negligible gravity.
4. Line element of an expanding Universe with negligible gravity:

$$ds^2 = \tau^2 dv^2 - (dx^2 + dy^2 + dz^2). \tag{3.1}$$

It is equal to zero for the Hubble expansion but is not vanishing at cosmic times smaller than τ.

Notice the similarity to Minkowskian line element

$$ds^2 = c^2 dt^2 - (dx^2 + dy^2 + dz^2) \tag{3.2}$$

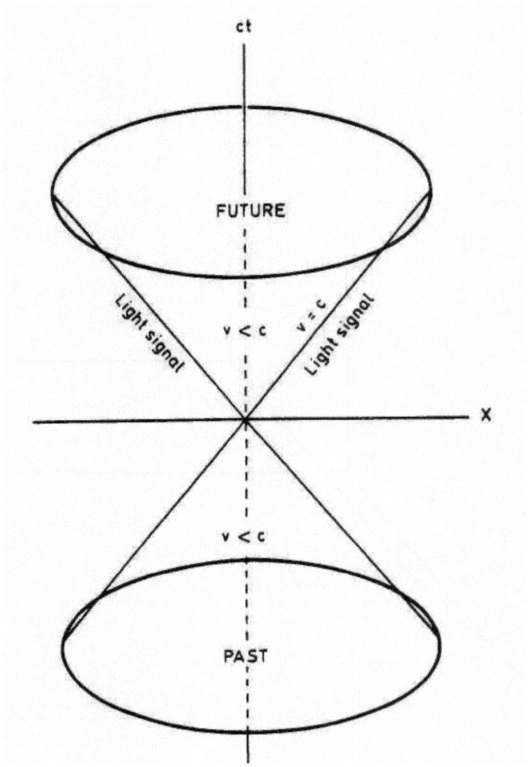

Figure 2. The light cone in two dimensions, $x^0(=ct)$ and $x^1(=x)$. The propagation of two light signals in opposite directions passing through $x = 0$ at time $t = 0$, is represented by the two diagonal straight lines. (Compare the galaxy cone given in the sequel.)

which vanishes for light propagation but is different from zero for particles of finite mass.

5. Postulates of cosmological special relativity: (1) The laws of physics are valid at all cosmic times; (2) τ has the same value at all cosmic times. (Similarly to Einstein's special relativity postulates.)

6. The cosmological transformation. This is the analog to the Lorentz transformation. It relates physical quantities at different cosmic times (similarly to the Lorentz transformation that relates quantities at different velocities):

$$x' = \frac{x - (\tau - t)\, v}{\sqrt{(t/\tau)\,(2 - t/\tau)}}, \qquad \tau v' = \frac{\tau v - x\,(\tau - t)\,/\tau}{\sqrt{(t/\tau)\,(2 - t/\tau)}}, \tag{3.3}$$

$$y' = y, \qquad z' = z,$$

$0 \le t \le \tau$, $t = 0$ at the Big Bang, $t = \tau$, now.

7. Example: Denote the temperature of the Universe at a cosmic time t by T and

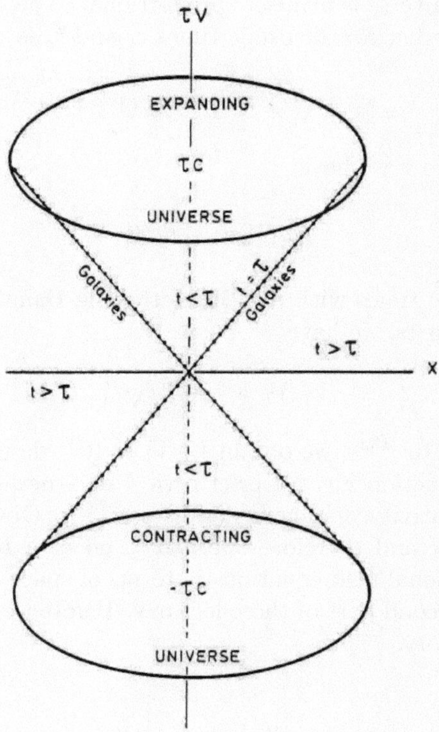

Figure 3. The galaxy cone in cosmological relativity, describing the cone in the $x - v$ space satisfying $\tau^2 v^2 - x^2 = 0$, where x represents the three-dimensional space. The heavy dots describe galaxies. The galaxy cone represents the locations of the galaxies at a given time rather than their path of motion in the real space.

that at present by T_0 (=2.73K), we then have

$$T = T_0 / \sqrt{(t/\tau)\,(2 - t/\tau)}. \tag{3.4}$$

At $t/\tau = 1/2$ we get $T = 3.15$K. (This result assumes negligible gravity and needs a correction by a factor of 13, and thus the temperature at $t/\tau = 1/2$ is 41K.)
8. The galaxy cone. This is the analog to the light cone in Einstein's special relativity. It represents the expansion of the Universe with negligible gravity. It is a four-dimensional cone. The coordinates are the velocity and the other three coordinates are the space coordinates. At the cone surface the Hubble expansion is represented, whereas the inner part of the cone represents events at cosmic times less than τ. The similarity to the light cone is remarkable.

9. Inflation at the early Universe. From the cosmological transformation we obtain the relationship between the mass density ρ_0 now to its value ρ at a backward time t:

$$\rho = \rho_0 / \sqrt{1 - t^2/\tau^2}. \tag{3.5}$$

The volume of the Universe is inversely proportional to its density, hence the ratio of the volumes at two backward cosmic times t_1 and t_2 is

$$V_2/V_1 = \sqrt{(1 - t_2^2/\tau^2) / (1 - t_1^2/\tau^2)}. \tag{3.6}$$

For t_1, t_2 very close to τ we have

$$V_2/V_1 = \sqrt{t_2'/t_1'}, \tag{3.7}$$

where primes indicate times with respect to the Big Bang. For $t_2' - t_1' \approx 10^{-32}$s, and t_2' much less than 1s, we have

$$V_2/V_1 = 10^{-16}/\sqrt{t_1'}. \tag{3.8}$$

For example, at $t_1' \approx 10^{-132}$s, we obtain $V_2/V_1 \approx 10^{50}$ describing inflation.

The above introduction gives a brief review of a new special relativity (cosmological special relativity, for more details see [5]). Obviously the Universe is filled up with gravity and therefore one has to go to a Riemannian space with the Einstein gravitational field equations in terms of space and redshift (velocity). This is done in the second part of these lectures. Before that we outline Einstein's general relativity theory.

4 General relativity theory: A brief outline

1. Postulates: (a) Principle of general covariance; (b) Principle of equivalence.
2. Riemannian curved space for the gravitational field. The metric tensor $g_{\mu\nu}$ as the gravitational potential [6].
3. The Einstein field equations:

$$G_{\mu\nu} = R_{\mu\nu} - \frac{1}{2} g_{\mu\nu} R = \kappa T_{\mu\nu} \tag{4.1}$$

4. The geodesic equation as the equation of motion:

$$\frac{d^2 x^\rho}{ds^2} + \Gamma^\rho_{\alpha\beta} \frac{dx^\alpha}{ds} \frac{dx^\beta}{ds} = 0. \tag{4.2}$$

5 Extension to curved space

The theory presented here, cosmological general relativity, uses a Riemannian four-dimensional presentation of gravitation in which the coordinates are those of Hubble, i.e. distances and velocity rather than the traditional space and time. We solve the field equations and show that there are three possibilities for the

Universe to expand. The theory describes the Universe as having a three-phase evolution with a decelerating expansion, followed by a constant and an accelerating expansion, and it predicts that the Universe is now in the latter phase. It is shown, assuming $\Omega_m = 0.245$, that the time at which the Universe goes over from a decelerating to an accelerating expansion, i.e., the constant-expansion phase, occurs at 8.5 Gyr ago. Also, at that time the cosmic radiation temperature was 146K. Recent observations of distant supernovae imply that the Universe's growth is accelerating, contrary to what has always been assumed, that the expansion is slowing down due to gravity. Our theory confirms these recent experimental results by showing that the Universe now is definitely in a stage of accelerating expansion. The theory predicts also that now there is a positive pressure, $p = 0.034 g/cm^2$, in the Universe. It is worthwhile mentioning that the theory has no cosmological constant. It is also shown that the three-dimensional space of the Universe is Euclidean, as the Boomerang, Maxima, DASI and CBI microwave telescopes have shown. Comparison with general relativity theory is finally made and it is pointed out that the classical experiments as well as the gravitational radiation prediction follow from the present theory, too.

6 Cosmology in spacevelocity

In the framework of cosmological general relativity (CGR) gravitation is described by a curved four-dimensional Riemannian spacevelocity. CGR incorporates the BB constant τ at the outset. The Hubble law is assumed in CGR as a fundamental law. CGR, in essence, extends Hubble's law so as to incorporate gravitation in it; it is actually a *distribution theory* that relates distances and velocities between galaxies. The theory involves measured quantities and it takes a picture of the Universe as it is at any moment. The following is a brief review of CGR as was originally given by the author in 1996 [7].

The foundations of any gravitational theory are based on the principles of equivalence and general covariance. These two principles lead immediately to the realization that gravitation should be described by a four-dimensional curved spacetime, in our theory spacevelocity, and that the field equations and the equations of motion should be generally covariant. Hence these principles were adopted in CGR also. Use is made in a four-dimensional Riemannian manifold with a metric $g_{\mu\nu}$ and a line element $ds^2 = g_{\mu\nu} dx^\mu dx^\nu$. The difference from Einstein's general relativity is that our coordinates are: x^0 is a velocitylike coordinate (rather than a timelike coordinate), thus $x^0 = \tau v$ where τ is the Big Bang time and v the velocity. The coordinate $x^0 = \tau v$ is the comparable to $x^0 = ct$ in ordinary general relativity. The other three coordinates x^k, $k = 1, 2, 3$, are spacelike, just as in general relativity theory.

An immediate consequence of the above choice of coordinates is that the null condition $ds = 0$ describes the expansion of the Universe in the curved spacevelocity (generalized Hubble's law with gravitation) as compared to the propagation of light in the curved spacetime in general relativity. This means one solves the field equations (to be given in the sequel) for the metric tensor, then from the null condition $ds = 0$ one obtains immedialety the dependence of the relative distances between the galaxies on their relative velocities.

As usual in gravitational theories, one equates geometry to physics. The first is expressed by means of the Einstein tensor. The physical part is expressed by the energy-momentum tensor which now has a different physical meaning from that in Einstein's theory. More important, the coupling constant that relates geometry to physics is now also *different.*

Accordingly the field equations are

$$R_{\mu\nu} - \frac{1}{2}g_{\mu\nu}R = \kappa T_{\mu\nu}, \tag{6.1}$$

exactly as in Einstein's theory, with κ given by $\kappa = 8\pi k/\tau^4$, (in general relativity it is given by $8\pi G/c^4$), where k is given by $k = G\tau^2/c^2$, with G being Newton's gravitational constant, and τ the Big Bang constant time. When the equations of motion are written in terms of velocity instead of time, the constant k will replace G. Using the above equations one then has $\kappa = 8\pi G/c^2\tau^2$.

The energy-momentum tensor $T^{\mu\nu}$ is constructed, along the lines of general relativity theory, with the speed of light being replaced by τ. If ρ is the average mass density of the Universe, then it will be assumed that $T^{\mu\nu} = \rho u^\mu u^\nu$, where $u^\mu = dx^\mu/ds$ is the four-velocity. In general relativity theory one takes $T_0^0 = \rho$. In Newtonian gravity one has the Poisson equation $\nabla^2\phi = 4\pi G\rho$. At points where $\rho = 0$ one solves the vacuum Einstein field equations in general relativity and the Laplace equation $\nabla^2\phi = 0$ in Newtonian gravity. In both theories a null (zero) solution is allowed as a trivial case. In cosmology, however, there exists no situation at which ρ can be zero because the Universe is filled with matter. In order to be able to have zero on the right-hand side of (6.1) one takes T_0^0 not as equal to ρ, but to $\rho_{eff} = \rho - \rho_c$, where ρ_c is the critical mass density, a *constant* in CGR given by $\rho_c = 3/8\pi G\tau^2$, whose value is $\rho_c \approx 10^{-29} g/cm^3$, a few hydrogen atoms per cubic meter. Accordingly one takes $T^{\mu\nu} = \rho_{eff} u^\mu u^\nu$; $\rho_{eff} = \rho - \rho_c$ for the energy-momentum tensor. Moreover, the above choice of the energy-momentum tensor is the only possibility that yields a constant expansion when $\rho = \rho_c$ as it should be.

In Part II we apply CGR to obtain the accelerating expanding Universe and related subjects.

References

[1] A. Einstein, *Autobiographical Notes*, P.A. Schlipp, Ed., Open Court (1979)

[2] A. Einstein, *The Meaning of Relativity*, 5th Edition, Princeton Univ. Press, Princeton (1955)

[3] M. Carmeli, *Found. Phys.* **25**, 1029 (1995)

[4] M. Carmeli, *Found. Phys.* **26**, 413 (1996)

[5] M. Carmeli, *Cosmological Special Relativity*, 2nd Edition, World Scientific (2002)

[6] M. Carmeli, *Classical Fields: General Relativity and Gauge Theory*, Wiley, New York (1982) [reprinted by World Scientific (2001)]

[7] M. Carmeli, *Commun. Theor. Phys.* **5**, 159 (1996)

COSMOLOGICAL THEORIES OF SPECIAL AND GENERAL RELATIVITY - II

MOSHE CARMELI [a]

[a] *Department of Physics, Ben Gurion University of the Negev, Beer Sheva 84105, Israel*

Abstract

Astronomers measure distances to faraway galaxies and their velocities. They do that in order to determine the expansion rate of the Universe. In Part I of these lectures the foundations of the theory of the expansion of the Universe was given. In this part we present the theory. A formula for the distance of the galaxy in terms of its velocity is given. It is very simple: $r(v) = c\tau/\beta \sinh \beta v/c$, where τ is the Big Bang time, $\beta = \sqrt{1-\Omega_m}$, and Ω_m is the mass density of the Universe. For $\Omega_m < 1$ this formula clearly indicates that the Universe is expanding with acceleration, as experiments clearly show.

1 Gravitational field equations

In the four-dimensional spacevelocity the spherically symmetric metric is given by

$$ds^2 = \tau^2 dv^2 - e^{\mu} dr^2 - R^2 \left(d\theta^2 + \sin^2\theta d\phi^2\right), \tag{1.1}$$

where μ and R are functions of v and r alone, and comoving coordinates $x^\mu = (x^0, x^1, x^2, x^3) = (\tau v, r, \theta, \phi)$ have been used. With the above choice of coordinates, the zero-component of the geodesic equation becomes an identity, and since r, θ and ϕ are constants along the geodesics, one has $dx^0 = ds$ and therefore $u^\alpha = u_\alpha = (1, 0, 0, 0)$. The metric (1.1) shows that the area of the sphere $r = constant$ is given by $4\pi R^2$ and that R should satisfy $R' = \partial R/\partial r > 0$. The possibility that $R' = 0$ at a point r_0 is excluded since it would allow the lines $r = constants$ at the

B.G. Sidharth, F. Honsell and A. de Angelis. (eds.), Frontiers of Fundamental Physics, 13–19.

neighboring points r_0 and $r_0 + dr$ to coincide at r_0, thus creating a caustic surface at which the comoving coordinates break down.

As has been shown in Part I the Universe expands by the null condition $ds = 0$, and if the expansion is spherically symmetric one has $d\theta = d\phi = 0$. The metric (1.1) then yields $\tau^2 dv^2 - e^{\mu} dr^2 = 0$, thus

$$\frac{dr}{dv} = \tau e^{-\mu/2}. \tag{1.2}$$

This is the differential equation that determines the Universe expansion. In the following we solve the gravitational field equations in order to find out the function $\mu(r.v)$.

The gravitational field equations, written in the form

$$R_{\mu\nu} = \kappa \left(T_{\mu\nu} - g_{\mu\nu} T/2\right), \tag{1.3}$$

where

$$T_{\mu\nu} = \rho_{eff} u_\mu u_\nu + p \left(u_\mu u_\nu - g_{\mu\nu}\right), \tag{1.4}$$

with $\rho_{eff} = \rho - \rho_c$ and $T = T_{\mu\nu} g^{\mu\nu}$, are now solved. One finds that the only nonvanishing components of $T_{\mu\nu}$ are $T_{00} = \tau^2 \rho_{eff}$, $T_{11} = c^{-1}\tau p e^{\mu}$, $T_{22} = c^{-1}\tau p R^2$ and $T_{33} = c^{-1}\tau p R^2 \sin^2\theta$, and that $T = \tau^2 \rho_{eff} - 3c^{-1}\tau p$.

One obtains three independent field equations (dot and prime denote derivatives with v and r)

$$e^{\mu}\left(2R\ddot{R} + \dot{R}^2 + 1\right) - R'^2 = -\kappa\tau c^{-1} e^{\mu} R^2 p, \tag{1.5}$$

$$2\dot{R}' - R'\dot{\mu} = 0, \tag{1.6}$$

$$e^{-\mu}\left[\frac{1}{R}R'\mu' - \left(\frac{R'}{R}\right)^2 - \frac{2}{R}R''\right] + \frac{1}{R}\dot{R}\dot{\mu} + \left(\frac{\dot{R}}{R}\right)^2 + \frac{1}{R^2} = \kappa\tau^2\rho_{eff}. \tag{1.7}$$

2 Solution of the field equations

The solution of (1.6) satisfying the condition $R' > 0$ is given by

$$e^{\mu} = R'^2 / \left(1 + f(r)\right), \tag{2.1}$$

where $f(r)$ is an arbitrary function of the coordinate r and satisfies the condition $f(r) + 1 > 0$. Substituting (2.1) in the other two field equations (1.5) and (1.7) then gives

$$2R\ddot{R} + \dot{R}^2 - f = -\kappa c^{-1}\tau R^2 p, \tag{2.2}$$

$$\frac{1}{RR'}\left(2\dot{R}\dot{R}' - f'\right) + \frac{1}{R^2}\left(\dot{R}^2 - f\right) = \kappa\tau^2\rho_{eff}, \tag{2.3}$$

respectively.

The simplest solution of the above two equations, which satisfies the condition $R' = 1 > 0$, is given by $R = r$. Using this in Eqs. (2.2) and (2.3) gives $f(r) = \kappa c^{-1}\tau p r^2$, and $f' + f/r = -\kappa\tau^2\rho_{eff} r$, respectively. Using the values of $\kappa = 8\pi G/c^2\tau^2$ and $\rho_c = 3/8\pi G\tau^2$, we obtain

$$f(r) = (1 - \Omega_m)\, r^2/c^2\tau^2, \tag{2.4}$$

where $\Omega_m = \rho/\rho_c$. We also obtain

$$p = \frac{1 - \Omega_m}{\kappa c\tau^3} = \frac{c}{\tau}\frac{1 - \Omega_m}{8\pi G} = 4.544\,(1 - \Omega_m) \times 10^{-2} g/cm^2, \tag{2.5}$$

$$e^{-\mu} = 1 + f(r) = 1 + \tau c^{-1}\kappa p r^2 = 1 + (1 - \Omega_m)\, r^2/c^2\tau^2. \tag{2.6}$$

Accordingly, the line element of the Universe is given by

$$ds^2 = \tau^2 dv^2 - \frac{dr^2}{1 + (1 - \Omega)\, r^2/c^2\tau^2} - r^2\left(d\theta^2 + \sin^2\theta d\phi^2\right), \tag{2.7}$$

or,

$$ds^2 = \tau^2 dv^2 - \frac{dr^2}{1 + (\kappa\tau/c)\, p r^2} - r^2\left(d\theta^2 + \sin^2\theta d\phi^2\right). \tag{2.8}$$

This line element is the comparable to the FRW line element in the standard theory.

It will be recalled that the Universe expansion is determined by Eq. (1.2), $dr/dv = \tau e^{-\mu/2}$. The only thing that is left to be determined is the sign of $(1 - \Omega_m)$ or the pressure p. Thus we have

$$\frac{dr}{dv} = \tau\sqrt{1 + \kappa\tau c^{-1} p r^2} = \tau\sqrt{1 + \frac{1 - \Omega_m}{c^2\tau^2} r^2}. \tag{2.9}$$

3 Physical meaning

For $\Omega_m > 1$ one obtains

$$r(v) = \frac{c\tau}{\alpha}\sin\alpha\frac{v}{c}, \qquad \alpha = \sqrt{\Omega_m - 1}. \tag{3.1}$$

This is obviously a closed Universe, and presents a decelerating expansion.

For $\Omega_m < 1$ one obtains

$$r(v) = \frac{c\tau}{\beta}\sinh\beta\frac{v}{c}, \qquad \beta = \sqrt{1 - \Omega_m}. \tag{3.2}$$

This is now an open accelerating Universe.

For $\Omega_m = 1$ we have, of course, $r = \tau v$.

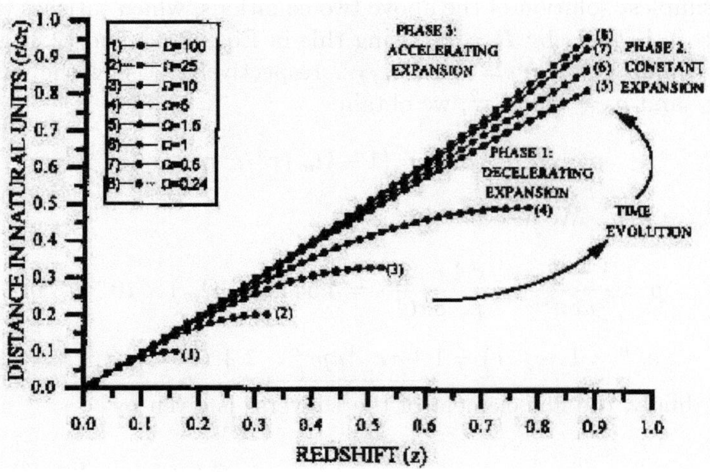

Figure 1. Hubble's diagram describing the three-phase evolution of the Universe according to cosmological general relativity theory.

4 The accelerating Universe

From the above one can write the expansion of the Universe in the standard Hubble form $v = H_0 r$ with

$$H_0 = h\left[1 - (1 - \Omega_m)\, v^2/6c^2\right], \tag{4.1}$$

where $h = \tau^{-1}$. Thus H_0 depends on the distance it is being measured [12]. It is well-known that the farther the distance, the lower the value for H_0 is measured. This is possible only for $\Omega_m < 1$, i.e. when the Universe is accelerating. In that case the pressure is positive.

Figure 1 describes the Hubble diagram of the above solutions for the three types of expansion for values of Ω_m from 100 to 0.245. The figure describes the three-phase evolution of the Universe. Curves (1)-(5) represent the stages of *decelerating expansion* according to Eq. (3.1). As the density of matter ρ decreases, the Universe goes over from the lower curves to the upper ones, but it does not have enough time to close up to a Big Crunch. The Universe subsequently goes over to curve (6) with $\Omega_m = 1$, at which time it has a constant expansion for a fraction of a second. This then followed by going to the upper curves (7) and (8) with $\Omega_m < 1$, where the Universe expands with *acceleration* according to Eq. (3.2). Curve no. 8 fits the present situation of the Universe. For curves (1)-(4) in the diagram we use the cutoff when the curves were at their maximum.

5 Theory versus experiment

To find out the numerical value of τ we use the relationship between $h = \tau^{-1}$ and H_0 given by Eq. (4.1)(CR denote values according to Cosmological Relativity):

$$H_0 = h\left[1 - \left(1 - \Omega_m^{CR}\right) z^2/6\right], \tag{5.1}$$

where $z = v/c$ is the redshift and $\Omega_m^{CR} = \rho_m/\rho_c$ with $\rho_c = 3h^2/8\pi G$. (Notice that our $\rho_c = 1.194 \times 10^{-29} g/cm^3$ is different from the standard ρ_c defined with H_0.) The redshift parameter z determines the distance at which H_0 is measured. We choose $z = 1$ and take for $\Omega_m^{CR} = 0.245$, its value at the present time (corresponds to 0.32 in the standard theory), Eq. (5.1) then gives $H_0 = 0.874h$. At $z = 1$ the corresponding Hubble parameter H_0 according to the latest results from HST can be taken [20] as $H_0 =$ 70km/s-Mpc, thus $h =$ (70/0.874)km/s-Mpc, or $h =$ 80.092km/s-Mpc, and $\tau = 12.486 Gyr = 3.938 \times 10^{17} s$.

What is left is to find the value of Ω_Λ^{CR}. We have $\Omega_\Lambda^{CR} = \rho_c^{ST}/\rho_c$, where $\rho_c^{ST} = 3H_0^2/8\pi G$ and $\rho_c = 3h^2/8\pi G$. Thus $\Omega_\Lambda^{CR} = (H_0/h)^2 = 0.874^2$, or $\Omega_\Lambda^{CR} = 0.764$. As is seen from the above equations one has

$$\Omega_T = \Omega_m^{CR} + \Omega_\Lambda^{CR} = 0.245 + 0.764 = 1.009 \approx 1, \tag{5.2}$$

which means the Universe is Euclidean.

Our results confirm those of the supernovae experiments and indicate on the existance of the dark energy as has recently received confirmation from the Boomerang cosmic microwave background experiment [21, 22], which showed that the Universe is Euclidean.

6 Comparison with general relativity

One has to add the time coordinate and the result is a five-dimensional theory of space-time-velocity. One can show that all the classical experiments predicted by general relativity are also predicted by CGR. Also predicted a wave equation for gravitational radiation. In the linear approximation one obtains

$$\left(\frac{1}{c^2}\frac{\partial^2}{\partial t^2} - \nabla^2 + \frac{1}{\tau^2}\frac{\partial^2}{\partial v^2}\right)\gamma_{\mu\nu} = -2\kappa T_{\mu\nu}, \tag{6.1}$$

where $\gamma_{\mu\nu}$ is a first approximation term,

$$g_{\mu\nu} \approx \eta_{\mu\nu} + h_{\mu\nu} = \eta_{\mu\nu} + \gamma_{\mu\nu} - \eta_{\mu\nu}\gamma/2, \tag{6.2}$$

$$\gamma = \eta^{\alpha\beta}\gamma_{\alpha\beta}. \tag{6.3}$$

Hence CGR predicts that gravitational waves depend not only on space and time but also on the redshift of the emitting source.

Table 1. Cosmological parameters in cosmological general relativity and in standard theory.

	COSMOLOGICAL RELATIVITY	STANDARD THEORY
Theory type	Spacevelocity	Spacetime
Expansion type	Tri-phase: decelerating, constant, accelerating	One phase
Present expansion	Accelerating (predicted)	One of three possibilities
Pressure	$0.034g/cm^2$	Negative
Cosmological constant	$1.934 \times 10^{-35}s^{-2}$ (predicted)	Depends
$\Omega_T = \Omega_m + \Omega_\Lambda$	1.009	Depends
Constant-expansion occurs at	8.5Gyr ago (Gravity is included)	No prediction
Constant-expansion duration	Fraction of second	Not known
Temperature at constant expansion	146K (Gravity is included)	No prediction

7 New developments on dark matter

Using the theory presented here, John Hartnett has recently shown that there is no need for the existence of dark matter in spiral galaxies. We only give the references to this work by Hartnett [25, 26, 27].

References

[1] M. Carmeli, *Cosmological Special Relativity*, 2nd Edition, World Scientific (2002)

[2] M. Carmeli and T. Kuzmenko, "Value of the cosmological constant", astro-ph/0110590

[3] M. Carmeli, Proc. 20th Texas Symp., AIP Conf. Proc., 586 (2001), astro-ph/012033

[4] M. Carmeli, "Accelerating Universe: Theory vs experiment", astro-ph/0205396

[5] M. Carmeli, "Five-dimensional cosmological theory of unified space, time and velocity", *Nuclear Phys B***124**, 258 (2003); asrto-ph/0205113

[6] M. Carmeli, "The line element in the Hubble expansion", In: *Gravitation and cosmology*, A. Lobo *et al.* Eds., Universitat de Barcelona, 2003, pp. 113-130 (Proceedings of the Spanish Relativity Meeting, Menorca, Spain, 22-24 September 2002); astro-ph/0211043

[7] M. Carmeli, *Commun. Theor. Phys.* **5**, 159 (1996)

[8] M. Carmeli, *Classical Fields: General Relativity and Gauge Theory*, Wiley, New York (1982) [reprinted by World Scientific (2001)]

[9] S. Behar and M. Carmeli, *Intern. J. Theor. Phys.* **39**, 1375 (2000), astro-ph/0008352

[10] M. Carmeli and S. Behar, "Cosmological general relativity", pp. 5-26, in: *Quest for Mathematical Physics*, T.M. Karade *et al.* Eds., New Delhi (2000)

[11] M. Carmeli and S. Behar, "Cosmological relativity: a general relativistic theory for the accelerating Universe", Talk given at Dark Matter 2000, Los Angeles, February 2000, pp.182–191, in: *Sources and Detection of Dark Matter/Energy in the Universe*, D. Cline, Ed., Springer (2001)

[12] P.J.E. Peebles, Status of the big bang cosmology, in: *Texas/Pascos 92: Relativistic Astrophysics and Particle Cosmology*, C.W. Akerlof and M.A. Srednicki, Editors, p. 84, New York Academy of Sciences, New York (1993)

[13] P.M. Garnavich *et al.*, *Astrophys. J.* **493**, L53 (1998) [Hi-Z Supernova Team Collaboration (astro-ph/9710123)]

[14] B.P. Schmidt *et al.*, *Astrophys. J.* **507**, 46 (1998) [Hi-Z Supernova Team Collaboration (astro-ph/9805200)]

[15] A.G. Riess *et al.*, *Astronom. J.* **116**, 1009 (1998) [Hi-Z Supernova Team Collaboration (astro-ph/9805201)]

[16] P.M. Garnavich *et al.*, *Astrophys. J.* **509**, 74 (1998) [Hi-Z Supernova Team Collaboration (astro-ph/9806396)]

[17] S. Perlmutter *et al.*, *Astrophys. J.* **483**, 565 (1997) [Supernova Cosmology Project Collaboration (astro-ph/9608192)].

[18] S. Perlmutter *et al.*, *Nature* **391**, 51 (1998) [Supernova Cosmology Project Collaboration (astro-ph/9712212)]

[19] S. Perlmutter *et al.*, *Astrophys. J.* **517**, 565 (1999) [Supernova Cosmology Project Collaboration (astro-ph/9812133)]

[20] W.L. Freedman *et al.*, "Final results from the Hubble Space Telescope", Talk given at the 20th Texas Symposium on Relativistic Astrophysics, Austin, Texas 10-15 December 2000, (astro-ph/0012376)

[21] P. de Bernardis *et al.*, *Nature* **404**, 955 (2000), (astro-ph/0004404)

[22] J.R. Bond *et al.*, in *Proc. IAU Symposium* 201 (2000), (astro-ph/0011378)

[23] A.V. Filippenko and A.G. Riess, p.227 in: *Particle Physics and Cosmology: Second Tropical Workshop*, J.F. Nieves, Editor, AIP, New York (2000)

[24] A.G. Riess *et al.*, *Astrophys. J.*, in press, astro-ph/0104455

[25] J. Hartnett, "Carmeli's accelerating Universe is spatially flat without dark matter", *Intern. J. Theor. Phys.*, in press (accepted for publication) (2004), gr-qc/0407082

[26] J. Hartnett, "Can the Carmeli metric correctly describe spiral galaxy rotation curves?" 2004, gr-qc/0407083

[27] J. Hartnett, "Carmeli's cosmology: The Universe is spatially flat without dark matter", invited talk given at the international conference "Frontiers of Fundamental Physics 6", held in Udine, Italy, September 26-29, 2004

CARMELI'S COSMOLOGY: THE UNIVERSE IS SPATIALLY FLAT WITHOUT DARK MATTER

JOHN G. HARTNETT [a]

[a] *School of Physics, University of Western Australia, 35 Strirling Hwy, Crawley, WA 6009 Australia*

Abstract

Carmeli's 5D brane cosmology has been applied to the expanding accelerating universe and it has been found that the distance redshift relation will fit the data of the high-z supernova teams without the need for dark matter. Also the vacuum energy contribution to gravity, Ω_Λ indicates that the universe is asymptotically expanding towards a spatially flat state, where the total mass energy density $\Omega + \Omega_\Lambda \rightarrow 1$.

1 Introduction

The Carmeli cosmology [3] [1] is as revolutionary in its implementation as it is in its interpretation. The metric used by Carmeli is unique in that it extends the number of dimensions of the universe by either one dimension if we consider only the radial velocity of the galaxies in the Hubble flow or by three if we consider all three velocity components. We will confine the discussion in this paper to only one extra dimension as does Carmeli. In that case the line element in five dimensions becomes

$$ds^2 = (1 + \Phi)c^2dt^2 - dr^2 + (1 + \Psi)\tau^2dv^2 \tag{1}$$

where $dr^2 = (dx^1)^2 + (dx^2)^2 + (dx^3)^2$ and Φ are Ψ potential functions to be determined. The time t is measured in the observer's frame. The new dimension (v) is the radial velocity of the galaxies in the expanding universe, in accordance with Hubble flow. The parameter τ is a constant at any epoch and its reciprocal (designated h) is approximately the Hubble constant H_0.

B.G. Sidharth, F. Honsell and A. de Angelis. (eds.), Frontiers of Fundamental Physics, 21–28.

The line element represents a spherically symmetric isotropic universe, and the expansion is the result of spacetime expansion. The expansion is observed at a definite time and thus dt = 0. Taking into account $d\theta = d\phi = 0$ (isotropy condition) and equation (1) becomes

$$-dr^2 + (1+\Psi)\tau^2 dv^2 = 0 \tag{2}$$

2 Phase space equation

The solution of equation (2)(given by equation B.38 and solved in section B.10 in [3] is reproduced here.

$$\frac{dr}{dv} = \tau\sqrt{1+(1-\Omega)\frac{r^2}{c^2\tau^2}} \tag{3}$$

The parameter Ω is the mass/energy density of the universe expressed as a fraction of the critical or "closure" density, i.e. $\Omega = \rho_m/\rho_c$ where ρ_m is the averaged matter/energy density of the universe. In this model,

$$\rho_c = \frac{3}{8\pi G\tau^2} = 10^{-29} gcm^{-3}$$

Then (3) may be integrated exactly to get

$$r(v) = \frac{c\tau}{\sqrt{1-\Omega}} \sinh\left(\frac{v}{c}\sqrt{1-\Omega}\right) \qquad \forall\Omega \tag{4}$$

Carmeli has expanded (4) in the limit of small $z = v/c$ and small Ω to get

$$r = \tau v\left(1+(1-\Omega)\frac{v^2}{6c^2}\right) \tag{5}$$

$$\Rightarrow \frac{r}{c\tau} = z\left(1+(1-\Omega)\frac{z^2}{6}\right) \qquad \forall\Omega < 1, z < 1 \tag{6}$$

Thus we can write the expansion in terms of normalized or natural units $r/c\tau$. Equation (6) is plotted in fig. 1 for various values of $\Omega = 1, 0.24$ and 0.03. Let us now re-write (4) in terms of natural units and for z small but arbitrary Ω.

$$\frac{r}{c\tau} = \frac{\sinh(z\sqrt{1-\Omega})}{\sqrt{1-\Omega}} \tag{7}$$

Equation (7) produces curves almost indistinguishable from (6) so this verifies that the approximations work well for $z < 1$.

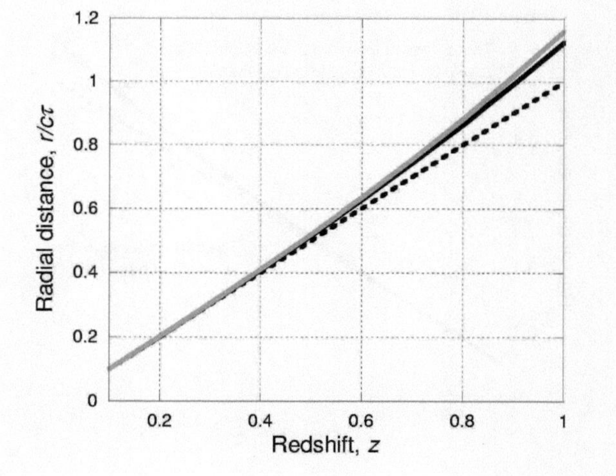

Figure 1. Plot of (6), $r/c\tau$ vs redshift (z) for $\Omega = 1$ (broken line), $\Omega = 0.245$ (solid black line) and $\Omega = 0.03$ (solid grey line).

3 Density verses redshift

Now let us consider what happens to the density of matter as we look back in the cosmos with redshift, z. It was assumed in fig. 1 that the value of Ω is fixed for each curve. Carmeli does this also in figure A4, page 134 in ref [3]. However, more correctly Ω varies as a function of z. For flat space we assume the following relation to hold,

$$\frac{\rho_m}{\rho_0} = (1+z)^3 = \frac{\Omega}{\Omega_0} \tag{8}$$

where ρ_m is a function of the redshift z, and ρ_0 is the averaged matter density of the universe locally. The parameter Ω_0 is then the local averaged matter density expressed as a fraction of "closure" density. Equation (8) results from the fact that as the redshift increases the volume decreases as $(1+z)^3$. Notice at $z = 1$ that the universe is 8 times smaller in volume and therefore 8 times more dense, that is, at $z = 1$, $\Omega = 8\Omega_0$.

Substituting (8) into (7) we get

$$\frac{r}{c\tau} = \frac{\sinh\left(z\sqrt{1-\Omega_0(1+z)^3}\right)}{\sqrt{1-\Omega_0(1+z)^3}} \tag{9}$$

Carmeli was able to simulate the form of the $0.1 < z < 1$ redshift data of [6] published in 1998 which announced an accelerating universe following the observations of [4] and [5]. See figure A4, page 134 in [3]. But in fact he had predicted this

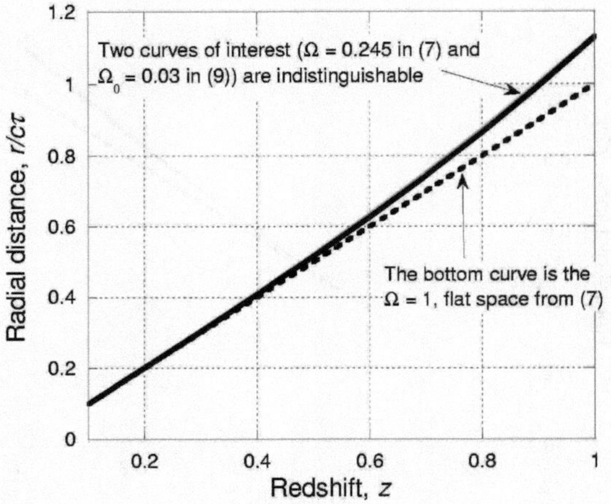

Figure 2. Plot of (7) with $\Omega = 1$ (broken line) and $\Omega = 0.245$ (solid black line) and (9) with $\Omega_0 = 0.03$ (solid grey line). Note: the top two curves lay on top of each other .

in 1996 [2]. So this means that Carmeli assumed a value of total matter (normal + dark matter) density $\Omega = 0.245$, which was the accepted value in 1998.

Now let's plot (7) with $\Omega = 0.245$ and (9) with $\Omega_0 = 0.03$. See fig. 2. This means that my modified equation (9) with $\Omega_0 = 0.03$ gives the same result as Carmeli's unapproximated equation (7) with his assumed value of $\Omega = 0.245$, but this includes dark matter. In fact, comparing (7) and (9), a local matter density of only $\Omega_0 = 0.03$–0.04 is necessary to have agreement. This effectively *eliminates the need for the existence of dark matter* on the cosmic scale.

Table I shows the critical data from the comparison at redshifts between $z = 0.25$ and $z = 1$. It can be seen that the difference between the two equations over the domain of the measurements is much less significant than the fit to the data. If we assume $\Omega_0 = 0.04$ instead of $\Omega_0 = 0.03$, since both are within measured parameters, we get closer agreement at smaller redshifts but worse near $z = 1$.

Table 1. Comparison of equations (7) and (9)

Redshift z	0.25	0.5	0.75	1.0
$r/c\tau$ from (7) with $\Omega = 0.245$	0.251984	0.515984	0.804591	1.13157
$r/c\tau$ from (9) with $\Omega_0 = 0.03$	0.252459	0.518935	0.810416	1.13157
% difference with $\Omega_0 = 0.03$	0.19	0.57	0.72	0.00
% difference with $\Omega_0 = 0.04$	0.17	0.43	0.23	1.28

In any case (7) and (9) really must be modified as $z \to 1$ to allow for relativistic

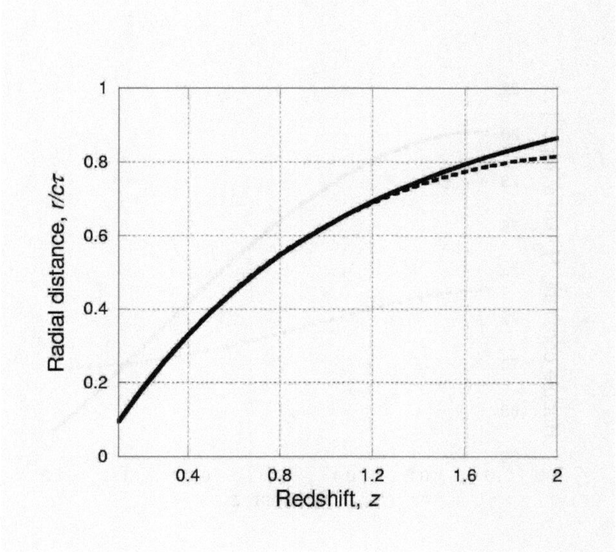

Figure 3. Plot of (10) with $\Omega = 0.245$ (solid curve) and (11) with $\Omega_0 = 0.03$ (broken curve). Note: the two curves separate for $z > 1.2$.

effects, by replacing v/c with the relativistic form $v/c = ((1+z)^2-1)/((1+z)^2+1)$. Therefore we can re-write (7) and (9) respectively as

$$\frac{r}{c\tau} = \frac{1}{\sqrt{1-\Omega}} \sinh\left(\frac{(1+z)^2-1}{(1+z)^2+1}\sqrt{1-\Omega}\right) \tag{10}$$

and

$$\frac{r}{c\tau} = \frac{1}{\sqrt{1-\Omega_0(1+z)^3}} \sinh\left(\frac{(1+z)^2-1}{(1+z)^2+1}\sqrt{1-\Omega_0(1+z)^3}\right) \tag{11}$$

where the varying matter density has been taken into account. In fig. 3, (10) and (11) are compared. The density approximation may be no longer valid past $z = 1$, because it is shown below that the vacuum energy term dominates and the universe is far from flat.

4 Hubble parameter

Based on the above analysis we can rewrite equation A.54 from [3] for H_0 as

$$H_0 = h\left(1 - (1 - \Omega_0(1+z)^3)\frac{z^2}{6}\right) \qquad \forall\Omega < 1, z < 1 \tag{12}$$

which according to equation A.51 from [3] may be further generalized without approximation, and using the relativistic form of the redshift. Still that equation

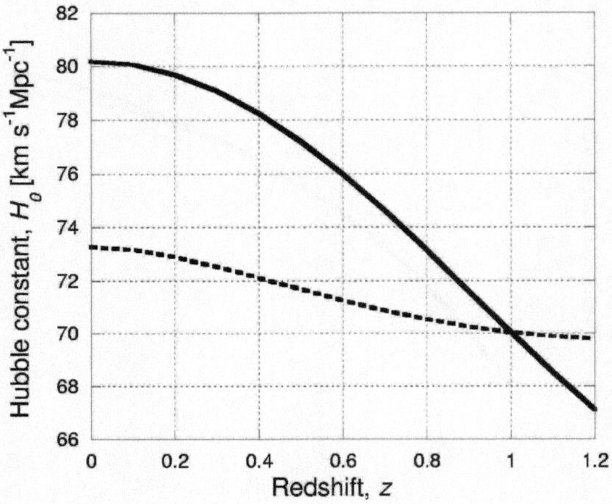

Figure 4. Plot of (12) (solid curve) and (13) (broken curve). Note: the two curves intersect at $z = 1$ where $H_0 = 70kms^{-1}Mpc^{-1}$.

may only be approximate for $z > 1$ because of the density assumptions. However it becomes

$$H_0 = h\frac{\xi}{\sinh \xi} \tag{13}$$

where $\xi = \frac{(1+z)^2-1}{(1+z)^2+1}\sqrt{1-\Omega_0(1+z)^3}$.

Both (12) and (13) have been plotted in fig. 4, and for Carmeli's chosen value of $H_0 \approx 70kms^{-1}Mpc^{-1}$ at $z = 1$ in (12) yields $h \approx 80.2kms^{-1}Mpc^{-1}$ (very close to Carmeli's value) but (13) yields $h \approx 73.27kms^{-1}Mpc^{-1}$. This means without the small z approximation the value of h is reduced when compared to that in [3].

5 Dark energy

The vacuum or dark energy parameter Ω_Λ does not appear explicitly in Carmeli's model. It is only by a comparison with F-L models can an assignment be made. On page 138 of [3] by comparing with the standard model it is shown that $\Omega_\Lambda = (H_0/h)^2$, therefore we can write

$$\Omega_\Lambda = \left(\frac{\xi}{\sinh \xi}\right)^2. \tag{14}$$

From (14) it is expected that using the unapproximated equation (13) for H_0

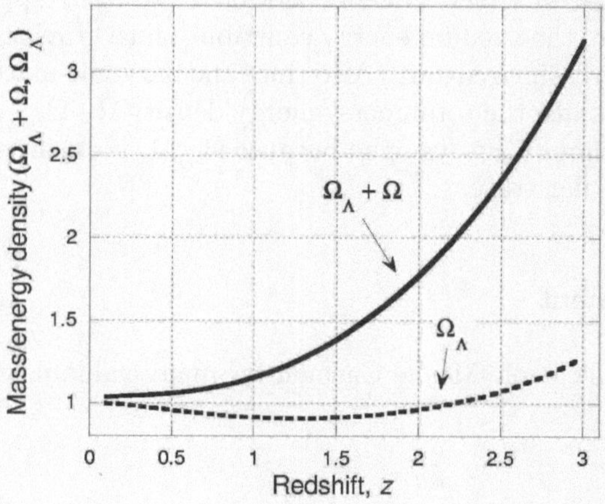

Figure 5. Plot of Ω_Λ (broken curve) and total density $\Omega + \Omega_\Lambda$ (solid curve) as a function of redshift z. Notice that Ω_Λ tends to unity as z tends to zero and the total density tends to the local matter density Ω_0 plus the vacuum energy density Ω_Λ.

the value of Ω_Λ will be larger than Carmeli's value using the form of (6). Fig. 5 shows the values for the vacuum energy density Ω_Λ (broken curve) and for the total energy density $\Omega + \Omega_\Lambda$ (solid curve) as a function of redshift, z. From (14) it follows that as the universe expands the total density tends to the vacuum energy density $\Omega_\Lambda \to 1$ (since $\Omega_0 \to 0$). This means a totally 3D spatially flat universe in a totally relaxed state. For small z the total density becomes

$$\Omega + \Omega_\Lambda \approx (1 + \Omega_0) + 3z\Omega_0. \tag{15}$$

It follows from (15) that for $\Omega_0 = 0.03$ at z = 0 the total density $\Omega + \Omega_\Lambda \approx 1.03$. This value is consistent with Carmeli's result of 1.009. However, it follows from (8) and (15) that the universe will always be open, $\Omega < 1$ as it expands. From fig. 5 the total density $\Omega + \Omega_\Lambda$ is always greater than unity and as the universe expands, it asymptotically approaches unity—therefore a spatially flat universe devoid of dark matter.

6 Conclusion

The 5D brane world of Moshe Carmeli has been has been applied to the expanding accelerating universe and the redshift distance relation has been generalised for redshifts up to at least $z = 1.2$. It has been found that if a certain form of the dependence of baryonic matter density on redshift is assumed then the resulting distance-redshift relation will fit the data of the high-z supernova teams without the need for dark matter. Even though it does not explicitly appear in the Carmeli *spacevelocity* metric, the vacuum energy contribution to gravity, Ω_Λ tends to unity as a function of decreasing redshift. Also since the baryonic matter density $\Omega_0 \to 0$ as the universe expands, the total mass/energy density $\Omega + \Omega_\Lambda \to 1$. This indicates that the universe, though always open because $\Omega < 1$, is asymptotically expanding towards a spatially flat state.

7 Acknowledgment

I would like to thank Prof. Moshe Carmeli for many valuable discussions.

References

[1] S. Behar, M. Carmeli, *Cosmological Relativity: A New Theory of Cosmology*, Int. J. Theor. Phys. 39(5) (2000)

[2] M. Carmeli, Commun. Theor. Phys. 5:159 (1996)

[3] M. Carmeli, *Cosmological Special Relativity*, Singapore, World Scientific. (2002).

[4] P.M. Garnavich et al, *Constraints on Cosmological Models from Hubble Space Telescope Observations of High-z Supernovae*, Bulletin of the American Astronomical Society 29(7) (1997).

[5] S. Perlmutter et al, emphCosmology From Type Ia Supernovae: Measurements, Calibration Techniques, and Implications., Bulletin of the American Astronomical Society 29(5): 1351 (1997).

[6] A.G. Riess et al, *Observational evidence from supernovae for an accelerating universe and a cosmological constant*, Astron. J. 116(Sept): 1009-1038 (1998)

BLACK HOLES AND THE INFORMATION PARADOX

GERARD 'T HOOFT [a]

[a] *Institute for Theoretical Physics, University of Utrecht, The Nederlands*

Abstract

In electromagnetism, like charges repel, opposite charges attract. A remarkable feature of the gravitational force is that like masses attract. This gives rise to an instability: the more mass you have, the stronger the attractive force, until an inevitable implosion follows, leading to a "black hole". It is in the black hole where an apparent conflict between Einstein's General Relativity and the laws of Quantum Mechanics becomes manifest. Most physicists now agree that a black hole should be described by a Schrödinger equation, with a Hermitean Hamiltonian, but this requires a modification of general relativity. Both General Relativity and Quantum mechanics are shaking on their foundations.

B.G. Sidharth, F. Honsell and A. de Angelis. (eds.), Frontiers of Fundamental Physics, 29–29.

A QUANTUM APPROACH TO COSMOLOGY

ANTONIO ALFONSO-FAUS [a]

[a] *E.U.I.T. Aeronáutica, Plaza Cardenal Cisneros s/n, 28040 Madrid, Spain*

Abstract

We present a theory based upon the treatment of the gravitational field as a sea of gravity quanta, as defined elsewhere. The resultant model for the Universe is a static one, like Einstein first saw, with a new feature: a local shrinking quantum world that completely explains the Hubble red shift under a new point of view. The presently accepted expansion of the Universe is interpreted here as an apparent effect, as seen from the Lab system of reference. The static Universe has immersed in it a local shrinking atomic world: a fundamental change in the interpretation of the Hubble's observations. The conservation principles (momentum, angular momentum and energy) can be dealt with under 2 different points of view: local (apparent) and COSMOLOGICAL ("real"). The 2 are in complete agreement with observation. They are also free of well known contradictions or paradoxes/incoherencies (i.e. in the Big Bang model). By dealing now with very well known first principles (Heisenberg, Mach, de Broglie, Weinberg's relation) under the same 2 points of view, we arrive at the conclusion that our new approach is in accordance with the Einstein's field equations of General Relativity, and Quantum Mechanics. We consider this to be a promising first step towards the way of dealing with the gravitational field coherently both from the General Relativity and from the Quantum Mechanical theories. The agreement with the present values of the cosmological parameters is very satisfactory.

B.G. Sidharth, F. Honsell and A. de Angelis. (eds.), Frontiers of Fundamental Physics, 31–35.

1 Introduction

The point of view of considering the gravitational field as a sea of gravity quanta [1] has been dealt with elsewhere. The published result there for the mass mg of this quantum is given by the relation $m_g = \hbar/(c2t)$, where t is the age of the Universe (then today mg is of the order of 2×10^{-66} grams).

The first important consequence of the above approach is the need to introduce a new concept, that we call the Mass Boom, [2], [3], [4] and [5]. In essence it expresses the property of any gravitational mass that, due to the emission of these gravity quanta, having a negative energy, its mass increases linearly with cosmological time. This linear dependence between mass and time makes it possible to identify the mass of the Universe M with the cosmological time t. A philosophical statement like *we are made of time* obviously follows and merits a deep reflexion. Clearly this approach is of the Machean type.

The second important consequence of this approach is that the speed of light must decrease with time. In fact it can be equated to the inverse of t, $c = 1/t$ [4] and [5] Then the resultant model for the Universe is a static one, as Einstein first proposed, and mathematically stated as $a(t) = ct =$ constant, i.e. a constant cosmological scale factor. The expansion of the Universe, a generally accepted interpretation of the red shift, is interpreted here as an apparent effect seen from the laboratory system of reference. The reinterpretation of the Hubble red shift is that the quantum world is shrinking, an effect coming directly from the time variation of Planck's *constant* [3], proportional to $1/t2$ or equivalently to $c2$.

The work we present here is based upon the above results. We analyze, from this new point of view, the conservation principles, and solve both: the Schrödinger equation together with the Einstein cosmological equations, which represents a first step in the harmonization of Quantum Mechanics and Relativity. The conclusion is that the whole approach is very promising and liberates present theories (like the Big Bang) from contradictions ands paradoxes. The agreement with the known numerical values for the cosmological parameters, as accepted today, is very satisfactory.

2 New concepts in the conservation principles

A summary of the new concepts is as follows:

- The mass of the gravity quanta, $m_g = \hbar/(c2t)$.
- The Mass Boom effect: any gravitational mass m has a time dependence as $m =$ constant t (t the age of the Universe).
- The decrease of the speed of light with time, $c = 1/t$.
- The $G = c3$ relation, following the Action Principle.

- Heisenberg and the De Broglie wavelength (perhaps the Compton wavelength as an alternative), $\hbar/mc$ = constant in the Lab.

- The v/c constancy as seen from the LAB system in order to conserve the constancy of the relativistic relations at any time.

- The Mass Boom is always present (as long as gravity is present).

- The decrease of the speed of light with time as $c = 1/t$, always present as a consequence of the constancy of momentum (in the absence of mechanical perturbations).

- The apparent interpretation of the cosmological expansion, following the Hubble's observations.

- The $h = c2$ relation (which explains the contraction of the quantum world).

- Weinberg's relation under a new point of view: not only explains the quantum of mass at the local Lab. It explains the Universe as a quantum black hole whose mass increases linearly with time.

- The introduction of $H = 1$, the cosmological Planck's constant, given by the relation $H = \hbar t2 = 1$, which is the essence of the quantum approach to cosmology.

- The Cosmological Planck's units using H, instead of $\hbar$, defining the cosmological quantum given by the whole Universe (mass, size and time $M = t$, and size $ct = 1 = 10^{28}$ cm with the constant homogeneous tic given by Planck's time).

- The fluctuation of the whole Universe, seen as a quantum black hole (corroborated by the Weinberg's relation using H).

- The determination of the age of the Universe as $t = 10^{61}$ units of time (the age of the Universe today) and given by the ratio of Planck's length at $t = 1$ (the constant length 1028 cm) and the present value of 10^{-33} cm).

- The solution to the cosmological Schrödinger equation coupled with the Einstein's cosmological equations (harmonization of General Relativity and Quantum Mechanics).

- The new entropy concept, that includes the gravitational entropy: $S = kM/m_g = M = t$ (for the Universe, [6]).

3 The solution to the Schrödinger cosmological equation coupled to the Einstein cosmological equations

The Schrödinger equation can be formulated from a cosmological point of view by using the "cosmological" Planck's constant $H = \hbar t2 = 1$ (a real constant). The resultant equation is then:

$$\frac{H2}{2m} \frac{\partial 2\Psi}{\partial x2} + V\Psi = iH \frac{\partial \Psi}{\partial t}$$

We see that all the terms in this equation vary as $1/t$. Then multiplying by t we have both members of the equation constant, as in the normal quantum mechanical treatment. The solution is then, assuming the wave function to be represented by a product of two functions: one depending on space and the other depending on time only, in the usual way one has:

$$\Psi(x,t) = const\ (sinx)\ t2$$

On the other hand the Einstein cosmological equations have the solution $a(t) = t2$ [5] which coincides with the above time dependence, as seen from the Lab reference system. Hence we have the same time dependent solution for both: General Relativity and Quantum Mechanics.

4 Conclusions

The main implications of this work are: a change in basic paradigms, perhaps the most important one is the new explanation for gravitation, in quantum mechanical terms, and coherent with general relativity. Also a new approach to the entropy concepts, in particular to the Hawking-Bekenstein treatments, where the entropy of a black hole is here defined as linear with mass [7]. A return to the Einstein initial cosmological model is the most significant change in the formulation of cosmological models.

References

[1] A. Alfonso-Faus, "Gravity Quanta, Entropy and Black Holes", *Physics Essays* **12, n. 4** (1999). Also in arXiv: gr-qc/0002065 v1, February 18, 2000.

[2] A. Alfonso-Faus, "Mass Boom versus Big Bang: Einstein was right". International Symposium n° V "Frontiers in Fundamental Physics". Hyderabad, India 2003. arXiv:physics 0302058

[3] A. Alfonso-Faus, "Mass Boom versus Big Bang: the role of Planck's constant", *Journal of Theoretics* **June 12** (2003) (arXiv: physics/0309108)

[4] A. Alfonso-Faus, "Quantum Gravity and General Relativity Consistent with a Decreasing Speed of Light and Mach's principle". *Spacetime and Substance* **3** (13) (2002) 130

[5] A. Alfonso-Faus, "Laboratory Physics and Cosmology", *Physics Essays*, July 2004 and arXiv: physics/0407150 v1

[6] A. Alfonso-Faus, "Entropy in the Universe: A new Approach", *online Entropy*, **2** (2000) 168 (at www.mdpi.orgg/entropy/)

[7] A. Alfonso-Faus, "Black Hole Entropy: Linear or Quadratic?" arXiv:astro-ph/0211053 v1, November 4, 2002

MY FOCUS ON THE QUANTUM SOURCE OF GRAVITY

SHANTILAL G. GORADIA

Gravity Research Institute, Inc., 3003 Royal Huntsman Court,
South Bend, Indiana 46637, USA,
sg@gravityresearchinstitute.org

Abstract

If Newtonian gravitation is modified to use surface-to-surface separation between particles, it can have the strength of nuclear force between nucleons. This may be justified by possible existence of quantum wormholes in particles. All gravitational interactions would be between coupled wormholes, emitting graviton flux in proportional to particle size, allowing for the point-like treatment above. When the wormholes are 1 Planck length apart, the resultant force is 10^{40} times the normal gravitational strength for nucleons. I explain the stability of the nucleus by predicting that the force mediated by pions is repulsive, not attractive.

1 Introduction

Newtonian gravity encounters issues for microscopic dimensions and cannot explain the nuclear binding force. Experimentalists and string theorists face a yet incomplete task of detecting and incorporating the spin 2 graviton into a fully quantized and renormalized theory. If we use the surface-to-surface separation between these particles to quantify the gravitational attraction instead of the center-to-center separation, at small separations relative to the particle radii the force between these particles grows much larger than classical gravity, and may resolve the above issues. The first step in the road map suggested by Richard Feynman for consistency in our physical theories is to see if Newton's law can be modified to be consistent with Einstein's law and can be further modified to be consistent with the uncertainty principle [4].

B.G. Sidharth, F. Honsell and A. de Angelis. (eds.), Frontiers of Fundamental Physics, 37–44.

2 Modification of the Inverse Square Law

As an example, for two coupled nucleons (Fig. 1a), I chose the Planck length $L = (Gh/c^3)^{0.5}$ as the surface separation, as it is the minimum possible spatial distance that makes any sense in physics. Assuming zero separation distance would imply that the two particles are joined to form one particle, losing their distinctions as separate particles. The diameter of the nucleon is about 1 fm (10^{-15} meters). The Newtonian gravitational force is then

$$F_N = Gm^2/D^2, \tag{1}$$

where D is the center-to-center distance, $\sim$ 1 fm. If we select the surface-to-surface separation instead, the force would become

$$F_P = Gm^2/d^2, \tag{2}$$

with $d = L = 10^{-20}$ fm. The ratio of these two forces is

$$\frac{F_P}{F_N} = \frac{D^2}{d^2} = 10^{40}, \tag{3}$$

which is also the strength of the proposed gravity relative to Newtonian gravity. As the nucleons are separated, D/d shrinks, and F_P rapidly approaches F_N. Mathematically,

$$\lim_{D\to\infty} \frac{D}{d} = 1. \tag{4}$$

A similar analysis can be made of the quark-lepton interaction (Fig. 1b).

Nucleons are responsible for over 99 percent of gravity, therefore they are the primary focus of this paper. For nucleons, I recover Newtonian gravity at practically 1000 fm. This modification yields a force with high intensity at short range, rapidly falling off to a very low intensity at long range. The values of a field and its rate of change with time are like the position and velocity of a particle. This modification meets the uncertainty principle requirement that the field can never be measured to be precisely zero.

"Einstein, in a paper written in 1919, attempted to demonstrate that his gravitational fields play an important role in the structure and stability of elementary particles. His hypothesis was not accepted because of gravity's extreme weakness" [10]. While Einstein's attempt is worth mentioning, it is not the foundation of my theory. Einstein could be wrong, but it seems he may not be. "It has been proposed that the gravitational constant inside a hadron is very large, $\sim 10^{38}$ times the Newtonian G" [10]. This "strong gravity" inside the hadron is similar to my proposed modification, but in my modification, instead of needing to change G itself, I change the distance measurement and get the same result. My theory does not create a conflict with the color force theory either. Strong gravity is consistent with string theory. The short range forces are weakened at long range by a high order of magnitude. This makes other attributes of the short range forces, infinitesimal at long range.

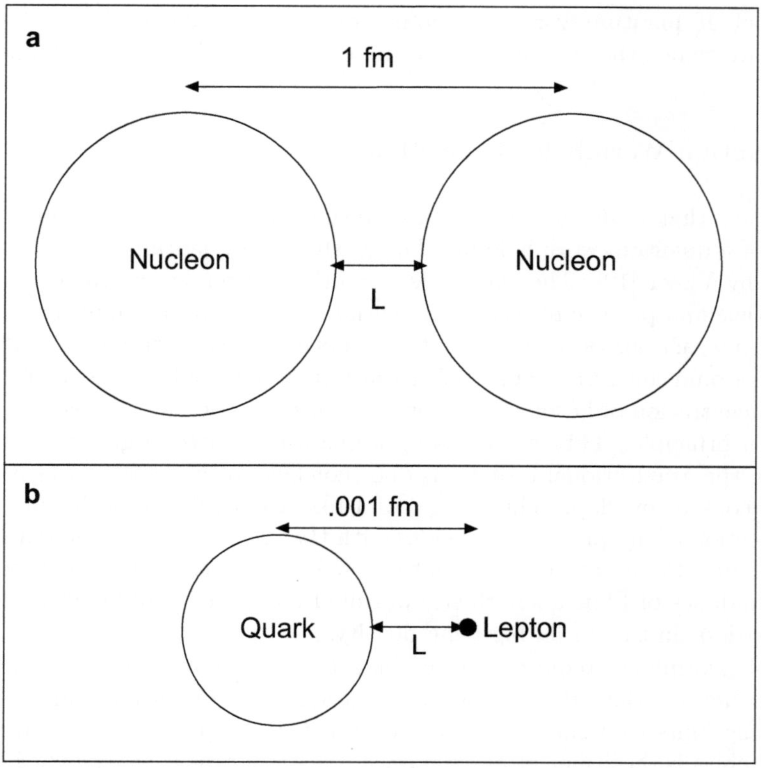

Figure 1. Pictorial view of gravitational interaction showing surface and center separations (not to scale). L is the Planck length, 10^{-20} fm. **a,** Two nucleons at minimum separation; **b,** A quark and a lepton, also at minimum separation. The standard inverse-square law would use the center-to-center distances to calculate the force between the particles; using the surface-to-surface distance yields a much stronger force for these separations, equal to the relative strengths of the strong and weak nuclear forces, respectively.

One may question the mathematically simple application of the Planck scale to a problem where the relevant distances seem to be fm. After I first published my findings in early 1999, Frank Wilczek wrote a series of articles [13], explaining how these scales can be reconciled and provided responses. While this may seem simplistic, it seems to be mathematically valid, and frequently significant problems can be solved simply in the end, as also illustrated by Morris and Thorne [8]. Complexity in physics lies in the abstraction of simplicity. Classical centers of shapes and therefore surfaces, though used here only for intuitive reasoning are invoked in nuclear coupling constants by implicit comparison to Newtonian gravity and in other descriptions in modern physics. My model is very consistent and therefore suggestive, however it does not reconcile the fact that nucleons overlap. Thanks are due to Gerald 't Hooft for this comment. Quantum wormholes, as currently theorized, may resolve this issue and give a mathematical foundation to my model. If quantum wormholes do not resolve the issue, we face a challenge to investigate some other verifiable quantum entity to explain this phenomenon.

3 Quantum Wormhole Connection

I postulate that each nucleon has a quantum mouth, potentially matching the mouth of a quantum wormhole [3]. The existence of quantum wormholes was examined by Visser [12]. The wormhole's mouth then represents the entire mass of the particle and propagates its $1/r$ potential to the rest of the universe. All gravitational interactions become interactions between these wormholes. Radiation by nucleons would consist of energy being emitted by the mouth of the wormhole. The high concentration of bosons, whether gravitons or photons, is consistent with the exclusion principle. This would justify a quantum source of gravity. The mouth emitting the gravitational radiations does not have to be at the surface, allowing the nucleons to overlap. This may sound like a radical approach, but it is not. The direction of my proposal coincides with that in the particle related article by Einstein and Rosen, introducing what is now known as Einstein–Rosen bridges [1]. The abundance of Planck-length size wormholes required could have evolved from perturbations in the initial big-bang density.

Stable wormholes require "exotic", negative energy matter. "... it is not possible to rule out the existence of such material; and quantum field theory gives tantalizing hints that such material might, if fact, be possible [8]." The stability of wormholes is on firmer grounds now. "...the theoretical analysis of Lorentzian wormholes is "merely" an extension of *known physics*-no new physical principle or fundamentally new physical theories are involved [11]." Literature search reveals no detection of any central force within nucleons, raising a question about the existence of gravitons within nucleons. Fig. 2 shows the mental picture of the graviton flux from nucleons with some background data. Richard Feynman seems to have investigated transfusion of two particles into gravitons [2], but not in this context. The structure of the quantum space-time is *foamy* [6]. The potential conversion of two gluons into one graviton and vice versa would be debatable. However, such *foamy* structure may give a green light for some other form of a particle mechanism. Since the spin-dependent nuclear force could be positive or negative, my theory maybe suggestive of photon-like particles as mediators of gravity.

Some long range forces are potentially simple, cumulative long range manifestations of their short range counter parts and vice versa with their intermediate range immeasurable by microscopic or macroscopic means. My model showing the strong gravity as a function of D^2 instead of particle mass (logical function of D^3) is consistent with the holographic principle. This is analogous to entropy's proportionality to the horizon area or D^2. Mach principle may imply that the universe spinning in the reference frame of nucleons may subject the nucleons to some form of gravity, not residual color force. As long as the observable characteristics of the proposed wormholes are stable, their stability and types are of secondary importance because the coupling constants are averages of observations.

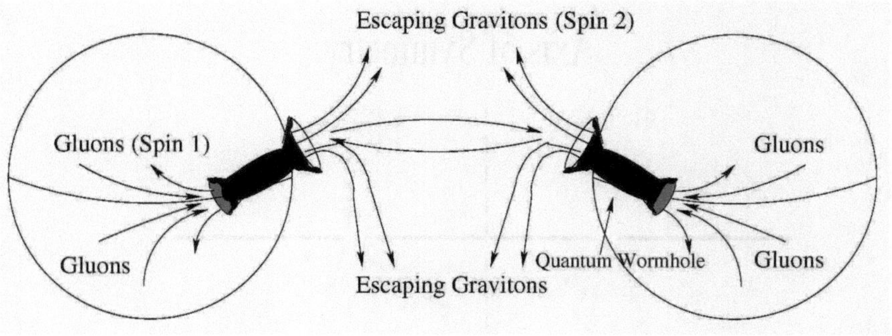

Figure 2. Mental image of nuclear interactions via quantum wormholes. The graviton flux would be proportional to the mass of the interacting particle, yielding couplings of 10^{40} for nucleons, 10^{34} for lighter quark-lepton pairs and ~ 1 for point-like leptons.

The understanding of the coupling constants lies at the heart of our understanding other important issues. "Using the concept of strong gravity, one can show the stability and structure of elementary particles, which could not be achieved by weak gravity" [10]. The sudden decrease in nuclear potential near the surfaces of nucleons may be a result of pion intervention pushing the nucleons apart as needed to stabilize the nucleus against the potential collapse by strong gravity. Pions might not be pulling the nucleons together as originally theorized. Since pions are observed to be spin-zero and their range matches the size of nuclei, this possibility cannot be ruled out.

A toy bird thrown in the air follows deterministic path, but a real bird has a soul. Therefore, it does not follow a deterministic path. A particle does likewise. This raises a question: does a particle have a soul? Per ancient Hindu Vedas (*Katha* Upanishad 1.2.20), a particle has a particle-soul (*anu-atma*), and it is also connected to the Supreme, which is omnipresent (*param-atma*). I see some qualitative consistency between this Vedic doctrine and my proposal based on a potentially justifiable speculation that *anu-atma* and *param-atma* are analogous to particle and normal space-time respectively, connected by some quantum entity.

4 Double Slit Experiment

Per my theory, in a two-slit experiment (Fig. 3), the network of geodesics downstream of the slits would depend upon whether both slits are open or only one of them is open, not upon the number of slits used for shooting the photons at the same time. Strong gravity at the edges of the slits would strongly impact the network of geodesics downstream geodesics. It does not matter whether the experiment shoots the photons through one slit or both. The screen pattern is a function of the network of geodesics.

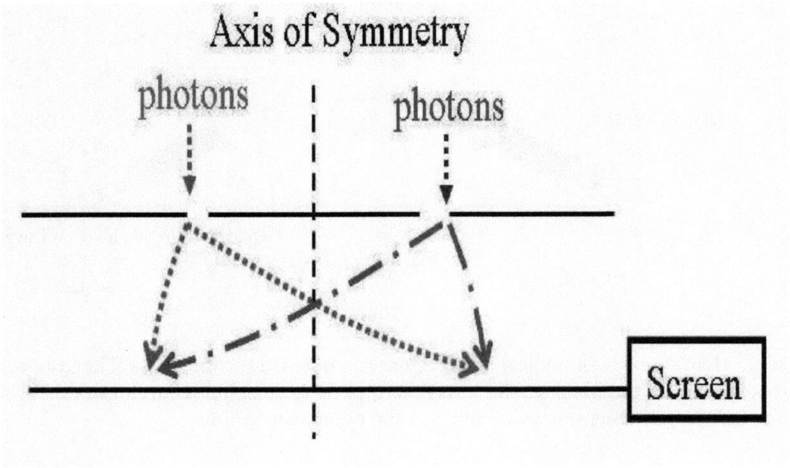

Figure 3. Depiction of double slit experiment. As shown, the screen pattern is independent of whether the left, right, or both slits are used, as long as the slits are open.

5 The Uncertainty Principle

If one assumes the uncertainty created for the information crossing the quantum wormholes as the Planck time multiplied by the energy of the wormhole (Planck energy in this specific case), the product of the two closely equals the Planck constant used in the uncertainty principle as shown below, potentially supporting my postulation.

$$\Delta E = (10^{19} GeV \times 10^{9} eV/GeV)/(1.6 \times 10^{19} eV/J) = 0.6 \times 10^{9} J \tag{5}$$

$$\Delta t = 10^{-43} s \tag{6}$$

Multiplying the above equations

$$\Delta E \times \Delta t = 0.6 \times 10^{19} J \times 10^{-43} s = 0.6 \times 10^{-34} \cdot s. \tag{7}$$

This yields Heisenburg's Uncertainty, which is

$$\Delta E \times \Delta t \geq 0.5 \times 10^{-34} J \cdot s \simeq \hbar/2 \tag{8}$$

In my theory, I do not have to express the range of nuclear force as "short" with an unanswered question as to precisely how short. The difference between the two large dynamic numbers of proposed strong gravity and the repulsive nuclear force manifests into observed short range, short enough to fix the size of the nucleus.

6 The Early Universe

If God created the universe from nothing, my theory shows that mass energy on one side of the throat of the quantum wormhole is equal to the gravitational field energy on the other side of its throat, both canceling each other. This would imply that the gravitational field is negative energy considering that mass is positive energy. The appearance of binding energy as mass defect in nuclear engineering calculations may lead to justify the binding energy (result of strong gravity) as negative energy. This is consistent with inflationary universe, no matter how big is the universe. "There is nothing known that places any limit on the amount of inflation that can occur while the total energy remains exactly zero [5]."

7 Prediction

My model provides a consistent, intuitive and simplistic, but mathematical explanation of the observed relative values of coupling constants, something no other theory has done. If a theory explains observations, it need not predict. Experimentally, my theory may be explored by a careful examination of the nuclear force at distances above 10 fm. Recently published test results verified the gravitational inverse square law down to 218μm [7]. The test results do not verify the higher dimensional theories that motivated the test, but they are not in conflict with my theory, as at these separations my modified force should be indistinguishable from Newtonian gravity. The generalized equation in the conclusion predicts a string coupling constant of $(10^{-35})^2 = 10^{-70}$.

My theory does away with the need to renormalize gravity, since the value of "r" is never zero.

8 Conclusion

In summary, in the early part of last century, when the nuclear force was declared to be a separate force, the Planck length and its implications were not well understood. Planck's system of fundamental units was considered heretical until came the proposal by Peres and Rosen [9]. The weakness of gravity was unquestioned. Therefore, it was impossible to explain strong gravity force in terms of Newtonian gravity and Einstein's view was undermined. In light of my article this issue needs to be revisited. My consistent results show that strong gravity creates an illusion of a different force between nucleons. Mathematically, the strong force coupling constant $C_s = D^2$, where D = nucleon diameter in Planck lengths.

9 Acknowledgments

I thank the staff at the Physics Department, University of Notre Dame for comments.

References

[1] Einstein, A. and Rosen, N., *Phys. Rev.* **48**, 73 (1935).

[2] Feynman, R., *Acta Phys. Pol.* **24**, 697 (1963).

[3] Goradia, S. G., "Potential Link: Gravity and Nuclear Force," physics/0210040 (2002).

[4] Feynman, R., *Six Easy Pieces*, Perseus Books Group, Cambridge, Massachusetts (1995).

[5] Guth, A. H., *Phys. Rev. D* **23**, 347 (1981).

[6] Hawking, S. W., *Phys. Rev. D* **46**, 603 (1992).

[7] Hoyle, C. D., *et al.*, *Phys. Rev. Lett.* **86**, 1418 (2001).

[8] Morris, M. S. and Thorne, K. S., *Am. J. Phys.* **56**, 395 (1988).

[9] Peres, A. and Rosen, N., *Phys. Rev.* **118**, 335 (1960).

[10] Shrivastava, S. K. *Aspects of Gravitational Interactions*, Nova Science Publishers, Commack, New York (1998).

[11] Visser, M., *Lorentzian Wormholes, From Einstein to Hawking*, Springer-Verlag, New York, New York (1996).

[12] Visser, M., "Quantum Wormholes," *Physical Review D* **43**, 402 (1991).

[13] Wilczek, F., *Physics Today* **54**, 12 (2001).

VACUUM DECAY BY P-BRANES PRODUCTION

LORENZO SINDONI [a] AND STEFANO ANSOLDI [b,c]

[a] *Dipartimento di Fisica, Università degli Studi di Trieste, via Valerio 2, I-34127 Trieste (TS), Italy*

[b] *Dipartimento di Matematica ed Informatica, Università degli Studi di Udine, via delle Scienze 206, I-33100 Udine (UD), Italy*

[c] *INFN, Sezione di Trieste, Italy*

Abstract

We present a generalization to the N-dimensional case for the nucleation coefficient of a spherical p-brane, separating two (anti-)de Sitter spacetimes. We use a semiclassical approximation based on the analytical continuation to the Euclidean sector of a suitable effective action describing a p-brane in General Relativity.

1 Introduction

Vacuum decay can be seen as a phase transition in spacetime and a long time ago the relevance of gravity for the process was studied [1]. The standard treatment of this process makes use of a scalar field, known as the *inflaton*, that drives the transition between the false and true vacuum states. This situation can be described by instanton calculations as, for instance, the Coleman-de Luccia and the Hawking-Moss instantons.

Here we present a different approach, generalizing past works of one of the authors [2, 3]. In particular we are going to use (anti-)de Sitter solutions in N spacetime dimensions. In this background we put a spherically symmetric $(N-1)$-brane that splits spacetime into two domains. The system can be described by Israel junction conditions [4], which provide the equations of motion for the

B.G. Sidharth, F. Honsell and A. de Angelis. (eds.), Frontiers of Fundamental Physics, 45–49.

timelike brane. The associated solutions are of two kinds: the first one consists of a degenerate brane of zero radius, while the second one consists of a *bounce* brane collapsing from infinity towards a finite nonzero turning point, and then re-expanding. To model vacuum decay we consider the tunnelling from the zero radius solution to the bounce solution. The corresponding physical picture is the following: a very small brane[1] inside a de Sitter geometry with cosmological constant Λ_+, due to quantum effects, has a non-vanishing probability to *tunnel* into a brane, containing a de Sitter spacetime with a different cosmological constant Λ_-. This represents the formation of a bubble of a different vacuum phase that then expands to infinity, realizing a transition of the whole spacetime geometry. We can obtain an expression for the probability of such a process using an effective action for this system.

2 Classical Dynamics

The stress-energy tensor for a distribution of matter localized on an hypersurface Σ (the p-brane we mentioned above) can be written in the form $S_{\mu\nu}\delta(\eta)$, where δ is a Dirac delta, and η can be thought as a transverse coordinate to Σ. In N-dimensional General Relativity it is possible to write down the equations of motion for this infinitesimally thin distribution of matter by splitting Einstein equations in the tangential and transverse part (see [4] for the 4-dimensional case; it can be extended to higher dimensions). Israel junction conditions, then, are

$$[\mathcal{K}_{ij} - h_{ij}\mathcal{K}] \propto S_{ij},$$

where $\mathcal{K}_{ij}$ and $\mathcal{K}$ are, respectively, the extrinsic curvature tensor and its trace and h_{ij} is the induced metric on Σ. Here we introduced the standard notation $[A] = \lim_{\eta\to 0^+}\{A(\eta) - A(-\eta)\}$. Israel junction conditions describe how the $(N-1)$-brane is embedded in the (in principle different) geometries of the two spacetime domains that it separates. For our purposes, we are going to write down these equations for a spherical brane with surface stress energy tensor $S_{ij} = kh_{ij}$ separating two de Sitter spacetimes. This can be done explicitly in terms of the radius R of the brane[2]. Then Israel junction conditions reduce to the single differential equation

$$\mathcal{H}(R,\dot{R}) = \left[\epsilon\sqrt{\dot{R}^2 + f(R)}\right] R^{(N-3)} - kR^{N-2} = 0; \tag{1}$$

k is the constant *tension* of the brane, ϵ are signs to be determined by the equation itself [5], and $f(R) = 1 - \Lambda R^2$ is the metric function appearing in the static line

[1] In the mathematical treatment of the classical situation the brane has, in fact, zero radius; from the physical point of view, with quantum gravity in mind, we can imagine this brane as a result of zero point quantum fluctuations.

[2] We are going to consider R as a function $R(\tau)$ of the proper time τ of an observer sitting on the brane and denote by an overdot the derivative with respect to τ.

element adapted to the spherical symmetry for the (anti-)de Sitter spacetime[3]. For suitable values of the cosmological constants equation (1) has two types of solutions: the first is $R \equiv 0$, while the second represents a brane collapsing and re-expanding from and to infinity. For our purpose it is also important to note that equation (1) can also be obtained by an effective action, which in the N-dimensional case can be written as

$$S_{\text{eff.}} = \int \left\{ R^{N-3}\dot{R} \left[\chi\text{th}^{-1} \left(\frac{\dot{R}}{\sqrt{\dot{R}^2 + f(R)}} \right) \right] - \mathcal{H}(R, \dot{R}) \right\} d\tau, \tag{2}$$

with the additional *constraint* $\mathcal{H} = 0$ that has to be imposed on the solutions of the corresponding Euler-Lagrange equation [5].

3 Tunnelling

The action (2) is crucial in our semiclassical quantization program, since it can be used to quantize the system *via* a path integral approach. Here we are going to consider the tunnelling process from the $R \equiv 0$ solution to the bounce solution, within the saddle-point approximation. This gives the possibility to estimate the following approximated amplitude

$$A_{\text{s.p.}} \propto \exp\left(-S^{(\text{e})}_{\text{eff.}}\right), \tag{3}$$

where $S^{(\text{e})}_{\text{eff.}}$ is the Euclidean effective action obtained by analytically continuing the action (2) to the Euclidean sector. In order to simplify some expressions, we introduce the following adimensional quantities:

$$x = kR \quad , \quad t = k\tau \quad , \quad \alpha = \frac{\Lambda_- + \Lambda_+}{k^2} \quad , \quad \beta = \frac{\Lambda_- - \Lambda_+}{k^2}.$$

Moreover it is a well known result that the adimensional version of the equation of motion (1) can be cast in the following form

$$(x')^2 + V(x) = 0,$$

where the prime now denotes the derivative with respect the adimensional time t. The potential $V(x)$ is given by

$$V(x) = 1 - \frac{x^2}{x_0^2}, \tag{4}$$

[3]Since the compact notation could be misleading, we remember that we have two spacetimes with different cosmological constants $\Lambda_\pm$ and, thus, two metric functions $f_\pm(R)$; please, also remember the meaning of the square brackets defined above.

where $x_0 = 2/\sqrt{(1 + 2\alpha + \beta^2)}$ is the adimensional turning radius, provided that the argument of the square root is positive. If this condition holds, there is a bounce trajectory, otherwise we have only the $x = 0$ solution. Starting from (2), (3) and (4) it is possible to evaluate the Euclidean action on the tunnelling trajectory, i.e on the segment $[0, x_0]$. The final result can be expressed as

$$S_{\mathrm{eff}}^{(\mathrm{e})} = \frac{x_0^{(N-3)/2}}{2(N-2)k^{(N-2)}} \left[(\beta - \epsilon)J(N, C_\sigma)\right], \tag{5}$$

where $J(N,p)$ is a function of the dimension of spacetime, N, and of

$$C_\sigma = 1 - \left(\frac{\beta - \sigma}{2}\right)^2 x_0^2;$$

with

$$\sigma_\pm = \pm 1.$$

A detailed description of the functional form of $J(N,p)$ is beyond the scope of this contribution and can be found elsewhere [7]. We just remark one important feature of it, namely that $J(N,p)$ is defined only when $p < 1$, a condition that, according to the form of C_σ, is always satisfied in our case.

4 Conclusions

In this contribution we have summarized how it is possible to compute the tunnelling amplitude for a spherical p-brane: the process, in view of our short discussion in the introduction, can be used to model the transition between two different vacuum phases, one being the de Sitter spacetime with cosmological constant Λ_+ and the other being the inflating inside of the p-brane, which is characterized by the cosmological constant Λ_-. This computation was already performed in four dimensions in the seminal work by Coleman and de Luccia [1] and that result was reproduced in [3]. Here we have carried out its generalization to arbitrary dimensions (referring to [7] for a more detailed analysis).

ACKNOWLEDGMENTS One of us (L.S.) wants to thank the Physics Department of University of Udine for financial support.

References

[1] S. Coleman and F. de Luccia, *Phys. Rev. D* **21** (1980) 3305

[2] V. A. Berezin, V. A. Kuzmin and I. I. Tkachev, *Phys. Rev. D* **36** (1987) 2919; E. Fahri, A. H. Guth and J. Guven, *Nucl. Phys. B* **339** (1990) 417; S. Ansoldi, *Class. Quantum Grav.* **19** (2002) 6321.

[3] S. Ansoldi, A. Aurilia, R. Balbinot and E. Spallucci, *Class. Quantum Grav.* **14** (1997) 2727;

[4] W. Israel, *Nuovo Cimento* **1** (1966) 1; W. Israel, *Nuovo Cimento* **2** (1967) 463 (errata); C. Barrabes and W. Israel, *Phys. Rev. D* **43** (1991) 1129.

[5] L. Sindoni, *MSc Thesis*, Università di Trieste (in italian).

[6] S. Ansoldi and L. Sindoni, see this *conference proceedings*.

[7] S. Ansoldi and L. Sindoni, *in preparation*.

SUPER-STRONG INTERACTING GRAVITONS AS A MAIN ENGINE OF THE UNIVERSE WITHOUT EXPANSION OR DARK ENERGY

MICHAEL A. IVANOV [a]

[a] *Physics Dept.,*
Belarus State University of Informatics and Radioelectronics,
6 P. Brovka Street, BY 220027, Minsk, Republic of Belarus.
E-mail: ivanovma@gw.bsuir.unibel.by

Abstract

The basic cosmological conjecture about the Dopplerian nature of redshifts may be false if gravitons are super-strong interacting particles. A quantum mechanism of classical gravity and the main features of a new cosmological paradigm based on it are described here.

If we assume that the background of super-strong interacting gravitons exists, then the classical gravitational attraction between any pair of bodies arises due to a Le Sage's kind mechanism. A net force of attraction and repulsion will be non-zero if one suggests that graviton pairs exist and these pairs are destructed by collisions. This pairing is like to the one having place in a case of superconductivity. The portion of pairing gravitons, $2\bar{n}_2/\bar{n}$, a spectrum of single gravitons, $f(x)$, and a spectrum of subsystem of pairing gravitons, $f_2(2x)$, are shown on Fig. 1 as functions of the dimensionless parameter $x \equiv \hbar\omega/kT$ (for more details, see [1]).

By the Planckian spectra of gravitons we find for the Newtonian constant [1]:

$$G = \frac{2}{3} \frac{\cdot D^2 c(kT)^6}{\pi^3 \hbar^3} \cdot I_2 \qquad (1)$$

where $I_2 = 2.3184 \cdot 10^{-6}$, T is an effective temperature of the background, and D is some new dimensional constant. It is necessary to accept for a value of this constant: $D = 1.124 \cdot 10^{-27} m^2/eV^2$.

B.G. Sidharth, F. Honsell and A. de Angelis. (eds.), Frontiers of Fundamental Physics, 51–54.

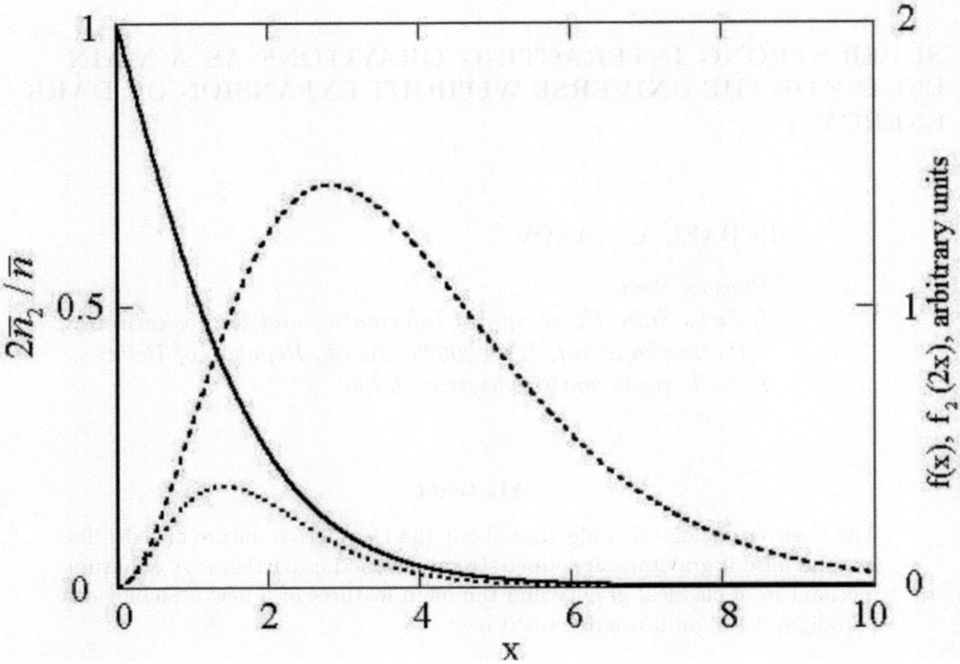

Figure 1. The portion of pairing gravitons, $2\bar{n}_2/\bar{n}$, (solid line), a spectrum of single gravitons, $f(x)$, (dashed line), and a spectrum of graviton pairs, $f_2(2x)$, (dotted line) versus the dimensionless parameter x.

In a presence of the graviton background, which is considered in a flat space-time, an energy of any photon should decrease with a distance r, so we have for a redshift z [2]: $z = \exp(ar) - 1$, where $a = H/c$ with the Hubble constant:

$$H = \frac{1}{2\pi} D \cdot \bar{\epsilon} \cdot (\sigma T^4), \tag{2}$$

where $\bar{\epsilon}$ is an average graviton energy, σ is the Stephan-Boltzmann constant.

It means that in this approach the two fundamental constants, G and H, are connected between themselves:

$$H = (G \frac{45}{32\pi^5} \frac{\sigma T^4 I_4^2}{c^3 I_2})^{1/2}, \tag{3}$$

with $I_4 = 24.866$. Using the known value of G, one can compute:$H = 3.026 \cdot 10^{-18} s^{-1} = 94.576\ km \cdot s^{-1} \cdot Mpc^{-1}$ by $T = 2.7K$.

From another side, an additional relaxation of any photonic flux due to non-forehead collisions of gravitons with photons leads to a luminosity distance D_L :

$$D_L = a^{-1} \ln(1+z) \cdot (1+z)^{(1+b)/2} \equiv a^{-1} f_1(z), \tag{4}$$

where $b = 3/2 + 2/\pi = 2.137$.

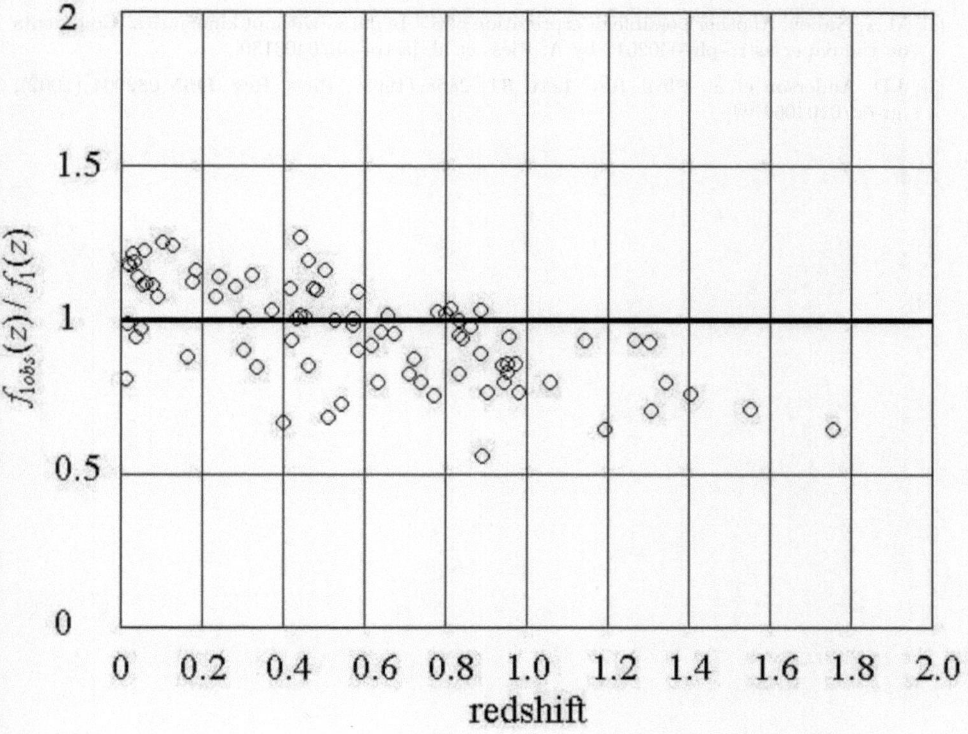

Figure 2. The ratio of observed to theoretical functions $f_{1obs}(z)/f_1(z)$ (dots); observational data are taken from Table 5 of [3]. If this model is true, the ratio should be equal to 1 for any z (solid line).

This function fits supernovae data well for $z < 0.5$ [4]. It excludes a need of any dark energy to explain supernovae dimming. If one introduces distance moduli $\mu_0 = 5 \log D_L + 25 = 5 \log f_{1obs} + c_1$, where c_1 is a constant, $f_{1obs}(z)$ is an observed analog of $f_1(z)$, we can compute the ratio $f_{1obs}(z)/f_1(z)$ using recent supernovae observational data from [3] (see Fig. 2).

In this approach, every massive body would be decelerated due to collisions with gravitons [2] that may be connected with the Pioneer 10 anomaly [5].

References

[1] M.A. Ivanov. Screening the graviton background, graviton pairing, and Newtonian gravity [gr-qc/0207006].

[2] M.A. Ivanov, General Relativity and Gravitation, **33**, 479 (2001); Erratum: **35**, 939 (2003); [astro-ph/0005084 v2].

[3] A.G. Riess et al. Type Ia Supernova Discoveries at $z > 1$ From the Hubble Space Telescope ..., [astro-ph/0402512] (to appear in ApJ, 2004).

[4] M.A. Ivanov. Another possible interpretation of SN 1a data - without kinematics: Comments on the paper astro-ph/0402512 by A. Riess et al. [astro-ph/0403130].

[5] J.D. Anderson et al. Phys. Rev. Lett. **81**, 2858 (1998); Phys. Rev. **D65** 082004 (2002); [gr-qc/0104064 v4]

GENERALIZED COSMOLOGICAL CONSTRAINTS ON NEUTRINO OSCILLATIONS - RELAXED OR STRENGTHENED

DANIELA P. KIRILOVA [a]

[a] *Institute of Astronomy, Bulgarian Academy of Sciences, Sofia, Bulgaria*

Abstract

I discuss cosmological constraints, based on BBN production of He-4, on electron-sterile neutrino oscillations proceeding after electron neutrino freeze-out. The general case of sterile neutrino state partially filled at the onset of oscillations is discussed $0 < \delta N_s < 1$. It is shown that $\delta N_s \neq 0$ has two-fold effect on He-4: (a) it enhances the energy density, increasing the over-production of He-4 and strengthening BBN constraints on oscillations and (b) it suppresses the kinetic effects of oscillations on BBN, especially neutrino spectrum distortion, thus decreasing He-4 overproduction and relaxing BBN constraints.

1 Introduction

Evidences for neutrino oscillations were obtained by atmospheric, solar and terrestrial experiments. Solar and atmospheric neutrino anomalies were resolved in terms of neutrino oscillations, flavour oscillations being the dominant channel. Experimental constraints on the impact of the sterile neutrino ν_s in the neutrino anomalies were obtained.

Stringent constraints on active-sterile oscillations were obtained also from Big Bang Nucleosynthesis (BBN) considerations [1, 2, 3, 4, 5, 6]. For more details concerning BBN constraints see refs. [7, 8]. *LMA and LOW active-sterile solar oscillation solutions and atmospheric active-sterile solutions were excluded by BBN* many years before global analysis of experimental neutrino data pointed to the preference of flavour oscillations for solving the neutrino anomalies.

B.G. Sidharth, F. Honsell and A. de Angelis. (eds.), Frontiers of Fundamental Physics, 55–59.

Update of BBN constraints on oscillations with $\delta m^2 > 10^{-6}$ eV^2, assuming equilibrium neutrino energy spectrum, is available in refs. [9, 10]. In the nonresonant case they can be approximated by:

$$(\delta m^2_{\nu_e\nu_s}/\mathrm{eV}^2)\sin^4 2\theta^{\nu_e\nu_s} = 3.16 \times 10^{-5}(\delta N_\nu)^2$$

$$(\delta m^2_{\nu_\mu\nu_s}/\mathrm{eV}^2)\sin^4 2\theta^{\nu_\mu\nu_s} = 1.74 \times 10^{-5}(\delta N_\nu)^2$$

In case of oscillations effective after ν_e decoupling, i.e. $(\delta m^2/\mathrm{eV}^2)\sin^4 2\theta < 10^{-7}$, the re-population of active neutrino becomes slow and the spectrum distortion of ν_e is considerable. For that nonequilibrium oscillations case BBN constraints were discussed in [5, 6]. The analytical fits to the exact constraints, corresponding to 3% He-4 overproduction, are:

$$\delta m^2(\sin^2 2\vartheta)^4 \leq 1.5 \times 10^{-9}\mathrm{eV}^2 \quad \delta m^2 > 0$$
$$|\delta m^2| < 8.2 \times 10^{-10}\mathrm{eV}^2 \quad \delta m^2 < 0, \ \textit{large} \ \vartheta,$$

All these BBN constraints assumed initially zero sterile state $\delta N_s = 0$ before oscillations. $\delta N_s = n_{\nu_s}/n_{\nu_e}$ and n_{ν_e} is the equilibrium number density of n_{ν_e}. However, ν_s may be present at BBN epoch as is predicted by different models.

Here we discuss BBN with neutrino oscillations in the case $\delta N_s \neq 0$. We present the results of the analysis of $\delta N_s \neq 0$ effects on oscillations and on BBN. We provide generalized BBN constraints on $\nu_e \leftrightarrow \nu_s$ oscillations for $0 < \delta N_s < 1$.

2 BBN with $\nu_e \leftrightarrow \nu_s$ and non-empty initially ν_s

^{4}He is abundantly produced (25% by mass), precisely measured ($\sim$ 3% uncertainty) and calculated ($\sim$ 0.1% uncertainty) and has simple post-BBN chemical evolution. Besides it is very sensitive to the kinetics of the nucleon transitions before the nucleosynthesis epoch, and therefore, as far as nucleons freezing depends strongly on the expansion rate, it is considered the best speedometer.

Hence, ^{4}He is the preferred primordial element for probing non-standard physics.

^{4}He yield essentially depends on the freezing of the reactions: $\nu_e + n \leftrightarrow p + e^-$, $e^+ + n \leftrightarrow p + \tilde{\nu}_e$, defined by the equilibration of expansion rate $H \sim g_{eff}T^4$ and the weak rate $\Gamma_w \sim G_F E^2 N_\nu^3$: $\Gamma_w \sim H$. The mass fraction of He-4 is roughly estimated as:

$$Y_p \sim 2(n/p)_f/(1 + n/p)_f \exp(-t/\tau_n)$$

where the frozen neutron-to-proton ratio is to a good approximation given by $(n/p)_f \sim \exp\{-(m_n - m_p)/T_f\}$. It depends on relativistic degrees of freedom at BBN, $g_{eff} = 10.75 + 7/4\ \delta N_s$, which enter through H, on the electron neutrino spectrum and on the neutrino-antineutrino asymmetry through Γ_w.

Active-sterile oscillations exert two types of effects on BBN:

- **dynamical effect**

Through bringing an additional neutrino type into equilibrium they increase the expansion rate [1] $H(t) \sim g_{eff}^{1/2}$, leading to earlier n/p-freezing, $T_f \sim (g_{eff})^{1/6}$, and ^{4}He overproduction. This effect gives up to 5% ^{4}He overproduction.

Hence, initially present $\delta N_s \neq 0$ increases the dynamical effect of oscillations, as far as it further increases g_{eff} and exaggerates He overproduction.

- **kinetic effect due to distortion of the electron neutrino spectrum**

Oscillations between non-equilibrium ν_s, $\delta N_s \neq 1$, and active neutrinos lead to deviations from the equilibrium ν spectrum because oscillation rate is energy dependent $\Gamma \sim \delta m^2/E$ [2].

The distortion leads both to a depletion of the active neutrino number densities N_ν and a decrease of the Γ_w, finally overproducing ^{4}He. This kinetic effect in case of oscillations effective after ν decoupling can be as strong as $\delta N_s = 6$! [11]

Larger δN_s decreases the spectrum distortion, and hence, it decreases the oscillations kinetic effect.

In conclusion, $\delta N_s \neq 0$ has two-fold effect on He-4: (a) it enhances the energy density, increasing the overproduction of He-4 and *strengthening BBN constraints* on oscillations and (b) it suppresses the kinetic effects of oscillations on BBN, especially neutrino spectrum distortion, thus decreasing He-4 overproduction and *relaxing BBN constraints.*

The figure shows the influence of different effects on helium-4 production Y_p. The dotted curve present the kinetic effect, the dashed curve – the dynamical effect of initially present ν_s. There is an interplay between the two effects on He due to the dependence of the kinetic effect on the initial population of ν_s.

3 BBN constraints on oscillations

Numerical analysis of He-4 production in the presence of $\nu_e \leftrightarrow \nu_s$, effective after neutrino decoupling, in the general case of partially filled initially sterile neutrino state $Y_p(\delta N_s, \delta m^2, \sin^2 2\vartheta)$ was provided in ref. [12].

Cosmological constraints on oscillation parameters corresponding to $\delta Y_p/Y_p = 3\%$ overproduction and different initial δN_s were calculated in ref. [13]. They strengthen with the increase of δN_s.

BBN constraints corresponding to $\delta Y_p/Y_p = 5\%$ and different ν_s populations were calculated in ref. [14]. These constraints relax with the increase of δN_s. The constraints for $\delta N_s = 0.5$ and $\delta N_s = 0$ and different He overproduction are illustrated in the figure 2. The two uppermost contours present the constraints for $\delta Y_p/Y_p = 5\%$ overproduction, the two lower curves – for $\delta Y_p/Y_p = 3\%$.

In conclusion BBN constraints may be either strengthened or relaxed with the increase of the initial population of the sterile state, depending on the level of the

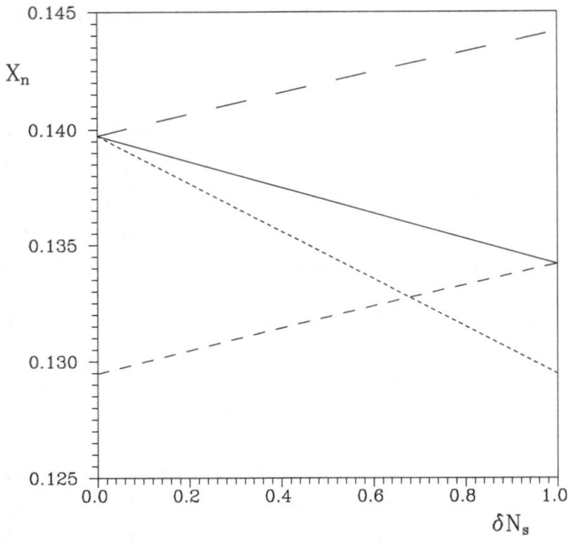

Figure 1. The solid curve presents frozen neutron number density relative to nucleons $X_n^f = N_n^f/N_{nuc}$ as a function of the sterile neutrino initial population, at $\delta m = \pm 10^{-8}$ eV^2, $\sin^2 2\theta = 1$. The dotted curve presents the kinetic effect, while the lower dashed curve presents energy density increase effect. The uppermost long dashed curve corresponds to the total effect when the decrease of the kinetic effect is not accounted for, i.e. in case the effects were additive.

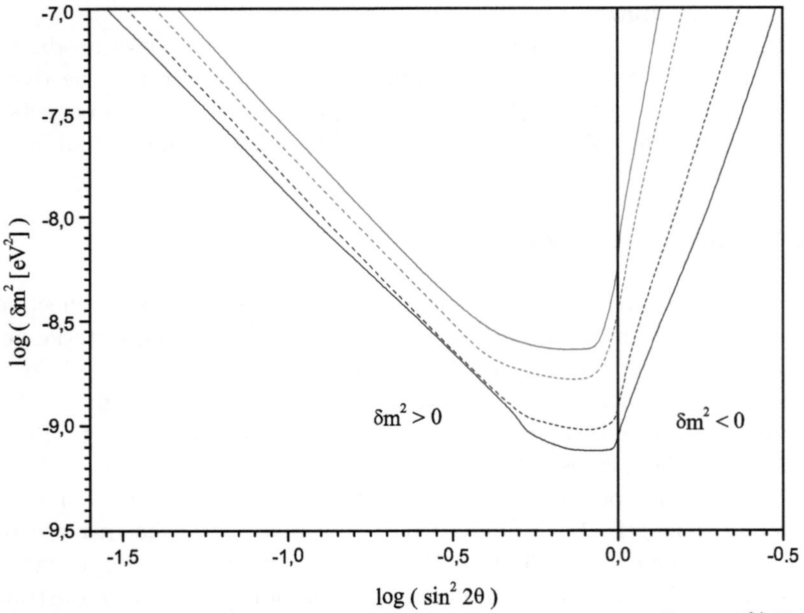

Figure 2. The lower dashed curve presents BBN constraints corresponding to 3% He overproduction and $\delta N_s = 0$, while the lowest curve presents the strengthened constraints due to higher δN_s, namely $\delta N_s = 0.5$. The upper curve gives the relaxed 5% He overproduction contour corresponding to $\delta N_s = 0.5$, while the upper dashed curve corresponds to 5% He and $\delta N_s = 0$.

He overproduction. The results are important for studying neutrino properties, for defining the role of the sterile neutrino in resolving the solar and atmospheric neutrino anomalies, for constraining models the presence of sterile neutrinos during BBN epoch.

References

[1] A. Dolgov : *Sov. J. Nucl. Phys.* **33**, 700 (1981).

[2] D. Kirilova : JINR E2-88-301 (1988).

[3] R. Barbieri, and A. Dolgov : *Phys. Lett.* **B 237**, 440 (1990); *Nucl. Phys.* **B 349**, 743 (1991).

[4] K. Enqvist, K. Kainulainen, J. Maalampi : *Phys. Lett.* **B 244**, 186 (1990); K. Enqvist, K. Kainulainen, M. Thompson : *Nucl. Phys.* **B 373**, 498 (1992).

[5] M. Chizhov, D. Kirilova : *in Neutrino96*, 478 (1996); *Phys. Lett.* **B 393**, 375 (1997); D. Kirilova, and M. Chizhov : *Phys. Rev.* **D 58**, 073004 (1998); *Nucl. Phys. B* **591**, 457 (2000).

[6] D. Kirilova, M. Chizhov : *Nucl. Phys. Suppl.* **B 100**, 360 (2001).

[7] A. Dolgov : *Phys. Rept.* **370**, 333 (2002).

[8] D. Kirilova : *Central Eur.J.Phys* **2**, 467 (2004).

[9] A. Dolgov : hep-ph/0306154 (2003).

[10] A. Dolgov, F. Villante : hep-ph/0308083 (2003).

[11] D. Kirilova : *Astropart. Phys* **19**, 409 (2003).

[12] D. Kirilova : *Int. J. Mod. Phys.* **D 13**, 831 (2004).

[13] D. Kirilova : ULB-TH/04-01, ULB (2004).

[14] D. Kirilova, M. Panayotova : IC/IR2004/13 (2004).

ATTRACTIONS OF AFFINE QUANTUM GRAVITY

JOHN R. KLAUDER [a]

[a] *Departments of Physics and Mathematics University of Florida Gainesville, FL 32611. Email: klauder@phys.ufl.edu*

Abstract

All attempts to quantize gravity face several difficult problems. Among these problems are: (i) metric positivity (positivity of the spatial distance between distinct points), (ii) the presence of anomalies (partial second-class nature of the quantum constraints), and (iii) perturbative nonrenormalizability (the need for infinitely many distinct counterterms). In this report, a relatively nontechnical discussion is presented about how the program of affine quantum gravity proposes to deal with these problems.

1 Introduction and Survey

The program of affine quantum gravity differs from that of string theory or loop quantum gravity: specifically, it differs in the insistence on a spatial metric tensor that is strictly positive definite; in the simultaneous and uniform treatment of both first- and second-class operator constraints; in dealing with nonperturbative renormalizability; and in maintaining a close connection with the motivating classical (Einstein) gravity theory. A suitable realization of these principles is most readily presented within a formalism that is, for the most part, generally unfamiliar to many readers. The purpose of this paper is to provide a relatively simple introduction to several concepts used to study quantum gravity from this new perspective.

B.G. Sidharth, F. Honsell and A. de Angelis. (eds.), Frontiers of Fundamental Physics, 61–67.

2 Reproducing Kernel Hilbert Spaces

This important idea can be readily explained. Let $|l\rangle \in \mathfrak{H}$, for all $l \in \mathcal{L}$, denote a set of states (chosen to be normalized, but that is not a requirement) that span the separable Hilbert space $\mathfrak{H}$ of interest. In addition we assume that (when it is finite dimensional) the space $\mathcal{L}$ is locally equivalent to a Euclidean space, and that the states $|l\rangle$ are continuously labeled by the (multi-dimensional) labels l. We will refer to the set of states $\{|l\rangle\}$ as *coherent states.*

Since the coherent states span $\mathfrak{H}$, it follows that two elements of a dense set of states may be given by

$$|\psi\rangle = \Sigma_{j=1}^{J}\alpha_j\,|l_j\rangle\,, \qquad J<\infty\,,$$
$$|\phi\rangle = \Sigma_{k=1}^{K}\beta_k\,|l_{(k)}\rangle\,, \qquad K<\infty\,.$$

As *functional representatives* of these abstract vectors we introduce

$$\psi(l) \equiv \langle l|\psi\rangle = \Sigma_{j=1}^{J}\alpha_j\,\langle l|l_j\rangle\,,$$
$$\phi(l) \equiv \langle l|\phi\rangle = \Sigma_{k=1}^{K}\beta_k\,\langle l|l_{(k)}\rangle\,.$$

Finally, as the *inner product* for these two elements we choose

$$(\psi,\phi) \equiv \Sigma_{j,k=1}^{J,K}\alpha_j^*\beta_k\,\langle l_j|l_{(k)}\rangle = \langle\psi|\phi\rangle\,.$$

This pre-Hilbert space is completed by adding the limit points of Cauchy sequences. The result is a functional representation composed entirely of continuous functions representing the separable Hilbert space $\mathfrak{H}$.

Such spaces are called reproducing kernel Hilbert spaces because if $J=1$ and $\alpha_1 = 1$, then it follows that $\psi(l) = \langle l|l_1\rangle$ and so

$$(\psi,\phi) = \Sigma_{k=1}^{K}\beta_k\,\langle l_1|l_{(k)}\rangle = \phi(l_1)\,,$$

a result which "reproduces" the vector $\phi(l)$. Thus the coherent state overlap $\langle l|l'\rangle$ serves as the "reproducing kernel" for this space.

Observe that all properties of this representation are determined by the jointly continuous coherent state overlap function $\langle l|l'\rangle$. Indeed, *any* continuous function of two (sets of) variables $K(l;l')$ serves to define a reproducing kernel Hilbert space provided K satisfies the condition

$$\Sigma_{j,k=1}^{J,J}\alpha_j^*\alpha_k K(l_j;l_k) \geq 0$$

for all possible complex choices of $\{\alpha_j\}$ and finite J.

3 Metric Positivity

Distinct points in a space-like 3-dimensional manifold have a positive separation distance. For a small coordinate separation $dx^a \not\equiv 0$, that distance, as usual, is given by $ds^2 = g_{ab}(x)\,dx^a\,dx^b > 0$. We require that the associated quantum operator $\hat{g}_{ab}(x)$ also satisfy metric positivity such that $\hat{g}_{ab}(x)\,dx^a\,dx^b > 0$ in the sense of operators for all nonvanishing dx^a. Moreover, we insist that $\hat{g}_{ab}(x)$ becomes self adjoint when smeared with a suitable real test function. In canonical quantization one chooses the canonical (ADM) momentum $\pi^{ab}(x)$ as the field to promote to an operator, $\hat{\pi}^{ab}(x)$. However, since the momentum acts to translate the metric, such a choice is inconsistent with the preservation of metric positivity. Instead, it is appropriate to choose the mixed valence momentum field $\pi^a_c(x) \equiv \pi^{ab}(x)\,g_{bc}(x)$ to promote to an operator $\hat{\pi}^a_c(x)$. This choice is dictated by the relation

$$e^{i\int \gamma^a_b(y)\,\hat{\pi}^b_a(y)\,d^3y}\,\hat{g}_{cd}(x)\,e^{-i\int \gamma^a_b(y)\,\hat{\pi}^b_a(y)\,d^3y} = (e^{\gamma(x)/2})^e_c\,\hat{g}_{ef}(x)\,(e^{\gamma(x)/2})^f_d\ ,$$

a relation that manifestly preserves metric positivity.

The full set of kinematical commutation relations is given by [1]

$$\begin{aligned}
[\hat{\pi}^a_b(x),\, \hat{\pi}^c_d(y)] &= \tfrac{1}{2}\,i\,[\delta^c_b\hat{\pi}^a_d(x) - \delta^a_d\hat{\pi}^c_b(x)]\,\delta(x,y)\ ,\\
[\hat{g}_{ab}(x),\, \hat{\pi}^c_d(y)] &= \tfrac{1}{2}\,i\,[\delta^c_a\hat{g}_{bd}(x) + \delta^c_b\hat{g}_{ad}(x)]\,\delta(x,y)\ ,\\
[\hat{g}_{ab}(x),\, \hat{g}_{cd}(y)] &= 0\ .
\end{aligned}$$

These are the so-called affine commutation relations appropriate to the affine fields $\hat{\pi}^a_c(x)$ and $\hat{g}_{ab}(x)$, both of which may be taken as self adjoint when smeared with real test functions. These commutation relations provide a realization of the group $\mathrm{IGL}(3,\mathbb{R})$, and as such they are more in the spirit of a current algebra than traditional canonical commutation relations.

It is important to add that by choosing $\hat{\pi}^a_c(x)$ as the partner field to go with $\hat{g}_{ab}(x)$, it follows that the momentum field $\hat{\pi}^{ab}(x)$ does *not* make an operator when smeared but only a form.

4 Quantization of Constraints

There are several schemes in common usage to quantize canonical systems with constraints. Traditionally, these schemes treat first- and second-class constraints differently. Gravity is not a traditional gauge theory since the set of classical constraints form an *open* first class system, which means that the Poisson brackets among the constraints have the form of a Lie algebra except that the structure constants are actually structure functions depending on the canonical variables. On quantization, these structure functions become operators that do not commute with the constraints, and as a consequence, the quantum constraints are partially second class in character. As noted above this usually entails a separate procedure for their analysis.

However, the recently introduced *projection operator method* [2] to incorporate quantum constraints treats first- and second-class constraints in an identical fashion and thereby it seems ideal to apply to gravity. Here, we content ourselves with a sketch of how this procedure is applied to simple systems.

We start by assuming that $\phi_\alpha(p,q)=0$, $\alpha=1,...,A$, represent a set of real classical constraints for some multi-dimensional system. We choose some quantization procedure and identify $\{\Phi_\alpha(P,Q)\}$ as a set of self-adjoint operators representing the constraints. Ideally, following Dirac, we would identify the physical Hilbert space $\mathfrak{H}_{phys}\subset\mathfrak{H}$ as composed of vectors $|\psi_{phys}\rangle$ that have the property that

$$\Phi_\alpha(P,Q)\,|\psi_{phys}\rangle=0$$

for all α. Consistency of this procedure requires that (i) $\langle\psi_{phys}|\psi_{phys}\rangle<\infty$, and (ii) $[\Phi_\alpha(P,Q),\Phi_\beta(P,Q)]\,|\psi_{phys}\rangle=0$. Unfortunately, for certain constrained systems, either one or both of these consistency conditions is violated. In that case it is useful to propose another scheme.

One alternative procedure, known as the projection operator method, involves a projection operator

$$\mathbb{E}=\mathbb{E}(\Sigma_\alpha\Phi_\alpha^2(P,Q)\leq\delta(\hbar)^2)\ ,$$

an expression which means that

$$0\leq\mathbb{E}\Sigma_\alpha\Phi_\alpha^2(P,Q)\mathbb{E}\leq\delta(\hbar)^2 I\ .$$

In these expressions, $\delta(\hbar)$ denotes a small, positive cutoff, generally dependent on $\hbar$, that can be reduced to a suitable level. In this approach the (regularized) physical Hilbert space is taken as $\mathfrak{H}_{phys}=\mathbb{E}\mathfrak{H}$.

It is pedagogically useful to illustrate this procedure with three simple examples:

(1) Let $\Phi_\alpha=J_1,\ J_2,\ J_3$ denote the generators of SU(2), and the desired physical Hilbert space satisfies $J_k\,|\psi_{phys}\rangle=0$ for $k=1,2,3$. We can secure the physical Hilbert space of interest by choosing

$$\mathbb{E}(\Sigma_k J_k^2\leq\tfrac{1}{2}\hbar^2)\ .$$

This example represents a first-class constrained system.

(2) Let $\Phi_\alpha=P,\ Q$, a pair of canonical operators. In this case we choose

$$\mathbb{E}(P^2+Q^2\leq\hbar)\ ,$$

which projects onto states $|\psi_{phys}\rangle$ that satisfy $(Q+iP)\,|\psi_{phys}\rangle=0$. This example represents a second-class constrained system.

(3) Let $\Phi_\alpha=Q$, a single operator with zero in the continuous spectrum and for which $Q\,|\psi_{phys}\rangle=0$ has no normalizable solution. here we choose

$$\mathbb{E}(Q^2\leq\delta^2)\ ,$$

with no $\hbar$ dependence necessary. As $\delta\to 0$, this projection operator passes strongly (hence weakly) to the zero operator. To overcome this fact, we *rescale* the projection operator and take a suitable limit as δ goes to zero. As one example, we introduce coherent states

$$|p,q\rangle\equiv e^{-iqP}\,e^{ipQ}\,|0\rangle\ ,$$

and we consider

$$\langle\langle p'', q'' | p', q' \rangle\rangle \equiv \lim_{\delta \to 0} \frac{\langle p'', q'' | \mathbb{E}(Q^2 \leq \delta^2) | p', q' \rangle}{\langle 0 | \mathbb{E}(Q^2 \leq \delta^2) | 0 \rangle} .$$

The resultant expression forms a suitable reduced reproducing kernel which can be used to characterize the physical Hilbert space as a reproducing kernel Hilbert space.

If P and Q form an irreducible pair, and for the sake of illustration we choose $|0\rangle$ as a normalized solution of $(Q + iP)|0\rangle = 0$, i.e., $|0\rangle$ is the oscillator ground state, then

$$\langle\langle p'', q'' | p', q' \rangle\rangle = e^{-\frac{1}{2}(q''^2 + q'^2)} ,$$

a reproducing kernel which characterizes a *one*-dimensional Hilbert space. Different choices of the fiducial vector $|0\rangle$ may lead to different functional representatives, but they nevertheless still describe one-dimensional Hilbert spaces.

It is noteworthy that path integral expressions exist that directly generate matrix elements of any desired projection operator. For example, staying with elementary examples, coherent state path integrals that generate expressions such as $\langle p'', q'' | \mathbb{E} | p', q' \rangle$ may formally be written as conventional phase-space path integrals save for one change, namely, the choice of the integration measure for the Lagrange multipliers [2].

How these general ideas may be applied to quantum gravity can be found in [1].

5 Perturbative Nonrenormalizability

One of the most challenging aspects of conventional approaches to quantum gravity is its perturbative nonrenormalizability. Divergences can be regularized by the introduction of cutoffs, as usual, and then counterterms developed on the basis of perturbation theory can be identified and included in the formalism. For renormalizable theories there are only a finite number of distinct types of counterterms, while for nonrenormalizable theories – such as gravity – an infinite set of qualitatively distinct counterterms is mandated by perturbation theory. It is no wonder that the morass created by renormalized perturbation theory has driven many workers to alternative approaches such as string theory. On the other hand, perhaps we are deceiving ourselves; could it be that perturbatively suggested counterterms to nonrenormalizable models are in fact *irrelevant*? This heretical viewpoint is indeed suggested by the *hard-core picture* of nonrenormalizable interactions which we now outline [3].

To present the essential ideas as simply as possible let us initially examine certain singular potentials in quantum mechanics. In particular, consider the Euclidean-space path integral for a free particle in the presence of an additional potential $V(x) \geq 0$. In symbols, let us study

$$W_\lambda \equiv \mathcal{N} \int_{x(0)=x'}^{x(T)=x''} e^{-\frac{1}{2}\int \dot{x}(t)^2\,dt \,-\, \lambda \int V(x(t))\,dt}\, \mathcal{D}x\,.$$

As $\lambda \to 0^+$, it appears self evident that W_λ passes to the expression

$$W_0 \equiv \mathcal{N} \int_{x(0)=x'}^{x(T)=x''} e^{-\frac{1}{2}\int \dot{x}^2(t)\,dt}\, \mathcal{D}x = \frac{1}{\sqrt{2\pi T}}\, e^{-(x''-x')^2/2T}$$

appropriate to a free particle. Whatever the analytic dependence of $W_\lambda - W_0$ for small λ (e.g., $O(\lambda)$, $O(\lambda^{1/3})$, $O(e^{-1/\lambda})$, etc.), it is tacitly assumed that as $\lambda \to 0^+$, $W_\lambda \to W_0$, i.e., that W_λ is *continuously connected* to W_0. However, this limiting behavior is *not* always true.

Consider the example $V(x) = x^{-4}$. In this case the singularity at $x = 0$ is so strong that the contribution from all paths that reach or cross the origin is *completely suppressed* since $\int x(t)^{-4}\,dt = \infty$ for such paths, no matter how small $\lambda > 0$ is chosen. As a consequence, as $\lambda \to 0^+$ for $V(x) = x^{-4}$, it follows that

$$\lim_{\lambda \to 0^+} W_\lambda = W_0' \equiv \frac{\theta(x''x')}{\sqrt{2\pi T}} \left[e^{-(x''-x')^2/2T} - e^{-(x''+x')^2/2T} \right].$$

Stated otherwise, when $V(x) = x^{-4}$, W_λ is *decidedly not* continuously connected to the free theory W_0, but is instead continuously connected to an alternative theory – called a pseudofree theory – that accounts for the *hard-core* effects of the interaction. The interacting theory may well possess a perturbation expansion about the pseudofree theory (to which it is continuously connected), but the interacting theory will *not* possess any perturbation expansion about the free theory (to which it is not even continuously connected).

Let us next pass to scalar field theory and the Euclidean-space functional integral

$$S_\lambda(h) \equiv \mathcal{N} \int \exp\{\int h\phi\,d^nx - \tfrac{1}{2}\int [(\nabla\phi)^2 + m^2\phi^2]\,d^nx - \lambda \int \phi^4\,d^nx\}\, \mathcal{D}\phi$$

appropriate to the ϕ^4_n model in n spacetime dimensions. We recall for such expressions that there is a Sobolev-type inequality to the effect that

$$\{\int \phi(x)^4\,d^nx\}^{1/2} \leq K \int [(\nabla\phi(x))^2 + m^2\phi(x)^2]\,d^nx$$

holds for *finite* K (e.g., $K = 4/3$) whenever $n \leq 4$, but which *fails* to hold (i.e., $K = \infty$) whenever $n \geq 5$. Thus for nonrenormalizable interactions ϕ^4_n, for which $n \geq 5$, it follows that there are fields ϕ for which the free action is finite while the interaction action is infinite. Just as in the elementary example, there is no reason to believe that counterterms suggested by a regularized perturbation analysis (the underlying premise of which is to maintain a continuous connection with the free theory!) should have any relevance in defining the pseudofree theory $S_0'(h)$.

It is noteworthy that proposals have been advanced to define $S'_0(h)$ and thereby to develop a meaningful and nontrivial theory of nonrenormalizable scalar fields [4]. Monte Carlo studies of such proposals are currently under way.

Lastly we observe that gravity is also a theory for which the free action (limited to quadratic terms) does not dominate the interaction action (remaining terms), and consequently gravity would seem to be a candidate theory to be understood on the basis of a hard-core interaction, which, when regularized, leads to its perturbatively nonrenormalizable behavior. As plausible as this scenario seems, it will involve a considerable effort to establish it convincingly.

References

[1] J.R. Klauder, "The Affine Quantum Gravity Programme", Class. Quant. Grav. **19**, 817-826 (2002). See also J.R. Klauder, "Noncanonical Quantization of Gravity. I. Foundations of Affine Quantum Gravity", J. Math. Phys. **40**, 5860-5882 (1999); "Ultralocal Fields and their Relevance for Reparametrization Invariant Quantum Field Theory", J. Phys.A: Math. Gen. **34** 3277-3288 (2001); "Noncanonical Quantization of Gravity. II. Constraints and the Physical Hilbert Space", J. Math. Phys. **42**, 4440-4464 (2001).

[2] J.R. Klauder, "Quantization of Constrained Systems", Lect. Notes Phys. **572**, 143-182 (2001); "Coherent State Quantization of Constraint Systems", Ann. Phys. **254**, 419-453 (1997).

[3] J.R. Klauder, *Beyond Conventional Quantization*, Cambridge University Press, Cambridge, 2000.

[4] J.R. Klauder, "Overcoming Nonrenormalizability", Letters in Math. Phys. **63**, 229-239 (2003); "Overcoming Nonrenormalizability. Part 2", J. Stat. Phys. **116**, 1181-1187 (2004).

GRAVITATIONAL TUNNELLING OF RELATIVISTIC SHELLS

STEFANO ANSOLDI [a] AND LORENZO SINDONI [b]

[a] *Dipartimento di Matematica e Informatica, Università degli Studi di Udine, via delle Scienze, 206 - I-33100 Udine (UD), ITALY and I.N.F.N. - Sezione di Trieste*
[b] *Dipartimento di Fisica, Università degli Studi di Trieste, via A. Valerio, 2 - I-34127 Trieste (TS), ITALY*

Abstract

Thin shells in general relativity have been used in the past as keystones to obtain realistic models of cosmological and astrophysical situations. A crucial role for these developments was played by the compact description of their dynamics in terms of Israel's junction conditions. Starting from this geometrical formulation we present a problem related to the WKB regime of shell dynamics and suggest a possible solution.

General relativistic shells are an interesting system in general relativity and because of the simple geometrical description of their dynamics provided by Israel's junction conditions [1] they became preferred models for many crucial aspects of astrophysical and cosmological situations (see [2] for a more complete bibliography on the subject). Many of these models have been developed under the assumption of spherical symmetry, but (as it happens for instance in the case of gravitational collapse [3]) this does not seem a severe restriction and it is likely that the obtained results can be extended to more general situations. On the other hand, the reduction in the number of degrees of freedom that it is possible to obtain in the spherically symmetric case makes simpler the development of effective models and more transparent the discussion of the interesting subtleties that often appears in the geometrodynamics of shells. Here we are, indeed, going to discuss one of these

B.G. Sidharth, F. Honsell and A. de Angelis. (eds.), Frontiers of Fundamental Physics, 69–73.

subtleties that already manifests itself in the spherically symmetric case, where the junction conditions reduce to just one equation[1]

$$[\epsilon(\dot{R}^2 + f(R))^{1/2}] \equiv \epsilon_-(\dot{R}^2 + f_-(R))^{1/2} - \epsilon_+(\dot{R}^2 + f_+(R))^{1/2} = M(R)/R, \quad (1)$$

a first order integral of the second order equation of motion for the shell. In (1) R is the radius of the shell (a function of the proper time τ of an observer comoving with the shell); $M(R)$ describes the matter content of the shell (i.e. it is related to its stress-energy tensor); $f_\pm(R)$ are the metric functions in the two domains of spacetime separated by the shell when the line element is written in the static form adapted to the spherical symmetry; $\epsilon_\pm$ are signs (i.e. $0, \pm 1$). Much of the discussion that follows is centered on these last quantities, $\epsilon_\pm$, but, before embarking this program, we also remember that, starting from an effective Lagrangian (the particular form of which is not our concern here), we can compute the second order equation of motion that has (1) as a first integral and also obtain the effective momentum [4] conjugated to the only surviving degree of freedom R,

$$P(R, \dot{R}) = R\left[\tanh^{-1}\left(\epsilon\dot{R}/(\dot{R}^2 + f(R))^{1/2}\right)^{\mathrm{Sgn}(f(R))}\right]. \quad (2)$$

Moreover, equation (1) can be cast in the form of an effective equation [5, 4] for the motion of a unitary mass particle with zero energy in a potential $V(R)$, $\dot{R}^2 + V(R) = 0$. Then all the solutions of (1) are solutions of this effective equation and *viceversa*. This solves the problem of obtaining a qualitative description of how the radius R changes as a function of the proper time τ. Of course this is not the full story, since we still have to build up the global structure of the spacetime in which the shell leaves. It is in this process that we need also the information provided by the functions $f_\pm(R)$ and by the two signs $\epsilon_\pm$. In particular when *cutting and pasting* the Penrose diagrams to build up the complete spacetime, $\epsilon_\pm$ select the sides of the Penrose diagram crossed by the trajectory [5]. Expressions for $\epsilon_\pm$ can be obtained with little algebra, $\epsilon_\pm = \mathrm{Sign}\left(M(R)\left(f_- - f_+ \mp M^2(R)/R^2\right)\right)$, and the points where $\epsilon_\pm$ change from ± 1 to ∓ 1 are the points in which $f_\pm(R)$ are tangent to $V(R)$, if they exist. Since $f_\pm(R) \geq V(R)$ always, the signs can change i) when the shell is crossing a region with $f_\pm(R) \leq 0$ or ii) *along a classically forbidden trajectory*[2], where $V(R) > 0$. It is shown in [4] that integrating the analytic continuation of (2) on the classically forbidden trajectory we can compute WKB transition amplitudes for the tunnelling process through the potential barrier; these amplitudes agree with those already computed by other means[3] in [7].

[1]Following a standard notation, quantities in the two spacetime regions separated by the shell are identified by $\pm$ subscripts. We use square brackets "[...]" to denote their jump in going from the "$-$" to the "$+$" side of the shell and an overdot, " $\dot{}$ ", to indicate the derivative with respect to the proper time measured by a shell-comoving observer.

[2]Our formulation here is far too synthetic and we refer the reader to the literature on the subject (for example [5, 6]) for extended background material.

[3]This is a strong argument in favor of an expression for the effective momentum that, when evaluated along a classically forbidden trajectory, differs from (2), also evaluated on a classically forbidden trajectory, by a total derivative of a function of R, at most.

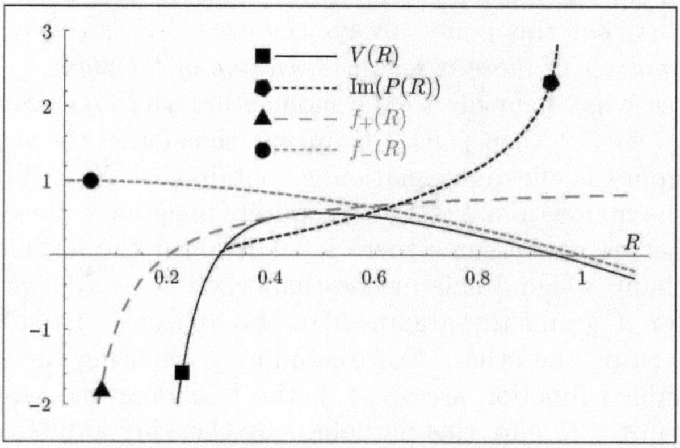

Figure 1. Graphical representation of the apparent problem with the behavior of the Euclidean momentum when one of the signs $\epsilon_\pm$ changes along the forbidden trajectory (here $-$ is de Sitter and $+$ is Schwarzschild). It seems that *or* the momentum is vanishing at both extrema (top) *or* it is continuous ...

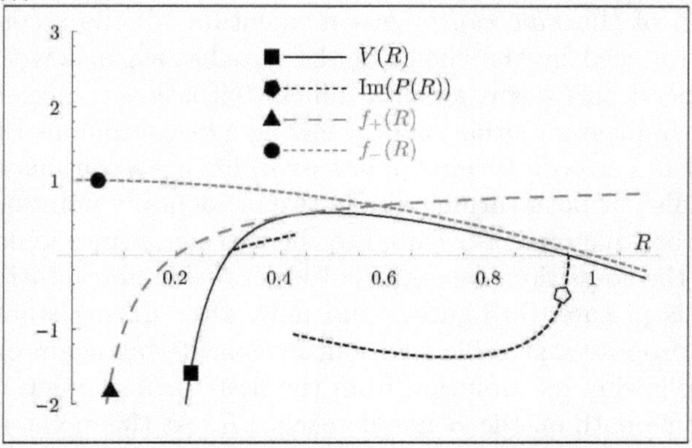

... (bottom). The two behaviors correspond to the choice of different branches for the arctan function that emerges from the $\tanh^{-1}$ dependence in the conjugated momentum of equation (2). Although we are not discussing this problem here, this situation also requires a deeper analysis in the Euclidean sector [8].

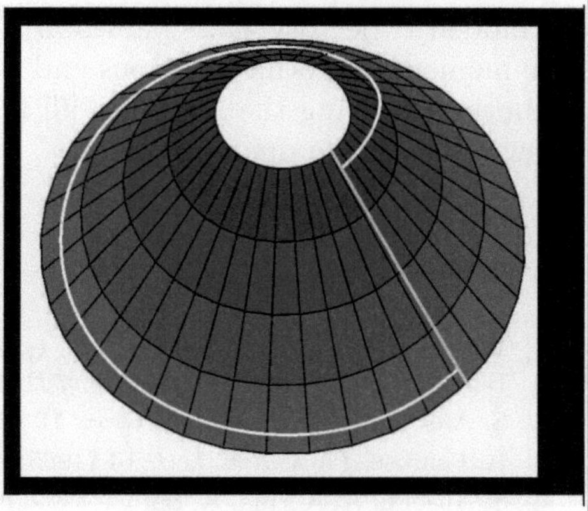

Figure 2. Representation of the Euclidean momentum wrapped around a $(R, \mathrm{Im}(P))$-cone that replaces the standard $(R, \mathrm{Im}(P))$-plane (Euclidean phase space). At each value of R the momentum takes values on an $\mathbb{S}^1$ with radius proportional to the current value of the radius R. In this way the momentum is *both* continuous and vanishing at the extrema. The gray straight segment in the picture corresponds to a segment of the horizontal R axis in figure 1.

In the cases discussed in [4] the signs $\epsilon_\pm$ are constant along the forbidden trajectory, but this is not always the case. *We are here interested in a more detailed analysis of those cases in which one of the signs, $\epsilon_\pm$, indeed changes.* Let us then see what happens to the momentum $P(R, \dot{R})$. Since on a forbidden trajectory $V(R) > 0$, then $f(R) > 0$: we can thus forget the weird exponent in (2). Moreover from the effective equation we obtain that $\dot{R}^2 < 0$ i.e. $\dot{R}$ is purely imaginary and the momentum $P(R)$ also is purely imaginary, since $\tanh^{-1}(\imath \dots) = \imath \arctan(\dots)$. Let us now assume there is an $\bar{R}$ along the forbidden trajectory where, say, ϵ_- changes sign. This means that when $R \to \bar{R}^\pm$ we have $(\dot{R}^2 + f_-(\bar{R}))^{1/2} \to 0^\pm$ (or $0^\mp$) and the argument of the $\arctan(\dots)$ tends to $-\infty$ on one side and to $+\infty$ on the other. Correspondingly, choosing the standard branch of the multi-valued function $\arctan(\dots)$, the Euclidean momentum has a discontinuity. We can try to cure this pathology by choosing a different branch of $\arctan(\dots)$: but then, following the evolution of the *now continuous* momentum till the second turning point, the offset introduced by the choice of the new branch makes the momentum non-vanishing there; this seems again a difficult situation to accept. Apparently, we thus face the unpleasant situation of i) having a discontinuous Euclidean momentum that vanishes at both turning points **or** ii) having a continuous momentum that does not vanish at both turning points (we incidentally point out that if we construct the Penrose diagrams associated to the two spacetimes joined by the shell before and after the transition, some difficulties in their interpretation also occur). This situation is pictured in figure 1 and now, after having stated the problem, we proceed to propose a possible solution, by considering again our Euclidean momentum and following its evolution from the first turning point. It starts from zero and after some path on the R line it reaches $\bar{R}$. At this point we enforce its continuity and keep following it until the second turning point, where we impose that **it is** zero. We said above that this cannot happen, but we implicitly made an assumption, namely that *the Euclidean momentum is a function taking values in the real line.* Relaxing this assumption we are going to see that not so much remains of the above problem. Figure 2 shows indeed that if we consider the Euclidean momentum as a function that at each point R along the forbidden trajectory takes values in a circle ($\mathbb{S}^1$) of radius R, then we can make the momentum both continuous and vanishing at both extrema! We end this contribution referring the reader to [9] for an extended discussion from the point of view of Euclidean quantum gravity.

References

[1] W. Israel, *Nuovo Cimento* **1** (1966) 1; W. Israel, *Nuovo Cimento* **2** (1967) 463 (errata); C. Barrabes and W. Israel, *Phys. Rev. D*, **43** (1991) 1129.

[2] S. Ansoldi, *Class. Quantum Grav.* **19** (2002) 6321.

[3] R. Penrose, *Phys. Rev. Lett.* **14** (1965) 57.

[4] S. Ansoldi, A. Aurilia, R. Balbinot and E. Spallucci, *Class. Quantum Grav.* **14** (1997) 2727;

[5] see for instance S. K. Blau, E. I. Guendelman and A. H. Guth, *Phys. Rev. D* **35** (1987) 1747; A. Aurilia, M. Palmer and E. Spallucci, *Phys. Rev. D* **40** (1989) 2511.

[6] V. A. Berezin, V. A. Kuzmin and I. I. Tkachev, *Phys. Rev. D* **36** (1987) 2919; V. A. Berezin and M. Okhrimenko, *Class Quantum Grav.* **18** (2001) 2195; P. Hajicek and J. Kijowski, *Phys. Rev. D* **62** (2000) 044025; L. Sindoni, *Graduation Thesis (in Italian)* (2004) University of Trieste.

[7] S. Coleman and F. de Luccia, *Phys. Rev. D* **21** (1980) 3305; S. Parke, *Phys. Lett.* **121B** (1983) 313.

[8] H. Ishihara, *private communication.*

[9] S. Ansoldi, *in preparation.*

THREE-DIMENSIONAL RELATIVISTIC SIMULATIONS OF ROTATING NEUTRON-STAR COLLAPSE TO A KERR BLACK HOLE

LUCA BAIOTTI [a], IAN HAWKE [b], PEDRO J. MONTERO [a]
FRANK LÖFFLER [b], LUCIANO REZZOLLA [a,c],
NIKOLAOS STERGIOULAS [d], JOSÉ A. FONT [e],
ED SEIDEL [b,c,f]

[a] *SISSA, International School for Advanced Studies and INFN, Trieste, Italy*

[b] *Max-Planck-Institut für Gravitationsphysik, Albert-Einstein-Institut, Golm, Germany*

[c] *Department of Physics, Louisiana State University, Baton Rouge, LA USA*

[d] *Department of Physics, Aristotle University of Thessaloniki, Thessaloniki, Greece*

[e] *Departamento de Astronomía y Astrofísica, Universidad de Valencia, Burjassot, Spain*

[f] *Center for Computation and Technology, LSU, Baton Rouge, LA USA*

Abstract

We discuss the application of our new three-dimensional fully general relativistic hydrodynamics code which uses high-resolution shock-capturing techniques and a conformal traceless formulation of the Einstein equations, to the study of the gravitational collapse of uniformly rotating neutron stars to Kerr black holes. We investigate the dynamics of the matter and of the trapped surfaces. We provide precise measurements of the black-hole mass and spin.

B.G. Sidharth, F. Honsell and A. de Angelis. (eds.), Frontiers of Fundamental Physics, 75–82.

1 Introduction

The numerical investigation of gravitational collapse of rotating stellar configurations leading to black-hole formation is a long standing problem in numerical relativity. However, it is through numerical simulations in general relativity that one can hope to improve our knowledge of fundamental aspects of Einstein's theory such as the cosmic censorship hypothesis and black-hole no-hair theorems, along with that of current open issues in relativistic astrophysics research, such as the mechanism responsible for gamma-ray bursts. Furthermore, numerical simulations of stellar gravitational collapse to black holes provide a unique mean of computing the gravitational waveforms emitted in such events, believed to be among the most important sources of detectable gravitational radiation.

However, the modelling of black-hole spacetimes with collapsing matter sources in multidimensions is one of the most formidable efforts of numerical relativity. This is due, on one hand, to the inherent difficulties and complexities of the system of equations which is to be integrated, the Einstein field equations coupled to the general-relativistic hydrodynamics equations, and, on the other hand, to the immense computational resources needed to integrate the equations in the case of three-dimensional (3D) evolutions. In addition to the practical difficulties encountered in the accurate treatment of the hydrodynamics involved in the gravitational collapse of a rotating neutron star to a black hole, the precise numerical computation of the gravitational radiation emitted in the process is particularly challenging as the energy released in gravitational waves is much smaller than the total rest-mass energy of the system.

The details of the formulation we use for the hydrodynamics and Einstein equations can be found in [3]. We stress that an important feature of this formulation is that it extends to a general relativistic context the powerful numerical methods developed in classical hydrodynamics, in particular high-resolution shock-capturing (HRSC) schemes based on linearized Riemann solvers. Such schemes are essential for a correct representation of shocks, whose presence is expected in several astrophysical scenarios.

Another fundamental improvement we implemented is the ability to excise from the evolved grid regions of spacetime within horizons. Such regions are causally disconnected from the rest of the spacetime and so do not have any influence on the exterior evolution. The field and hydrodynamical variables in these regions, though, would get extreme values and so compromise the whole evolution. More details on how the hydrodynamical excision is applied in practice, as well as tests showing that this method is stable, consistent and converges to the expected order are published in [9, 3].

Also for all the details about each of the following sections, which contain only brief summaries of our work on this subject, the reader is referred to [3].

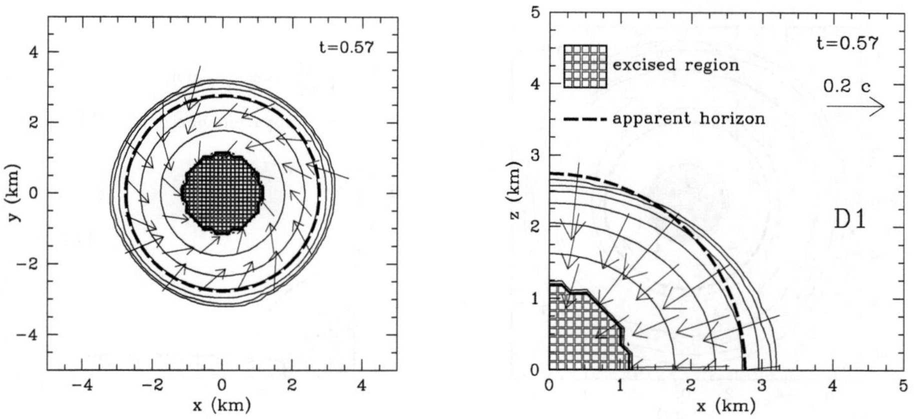

Figure 1. View of the final stages of the collapse of model D1.

2 Initial stellar models

The initial data for our simulations are constructed using a 2D numerical code, that computes accurate stationary equilibrium solutions for axisymmetric and rapidly rotating relativistic stars in polar coordinates [11, 12]. For simplicity, we have focused on initial models constructed with the polytropic EOS, choosing the polytropic exponent $\Gamma = 2$ and polytropic constant $K = 100$ to produce stellar models that are, at least qualitatively, representative of what is expected from observations of neutron stars. In the following we report in particular about two of the models we have studied, one slowly rotating (which we refer to as D1) and one rapidly rotating (D4).

3 Dynamics of the matter

We show in Fig. 1 a representative snapshot of the final stages of the evolution of a slowly rotating initial model. Soon after an apparent horizon is found and when this has grown to a sufficiently large area, the portion of the computational domain containing the singularity is excised (The determination of the apparent horizon is obtained using the fast finder of Thornburg [13]). This is indicated as an area filled with squares. Also shown with a thick dashed line is the coordinate location of the apparent horizon and it should be remarked that, because of rotation, this surface is not a coordinate two-sphere, although the departures are not significant and cannot be appreciated in Fig. 1 (see also Section 5). At $t = 0.57$ ms, the time which Fig. 1 refers to, most of the matter has already fallen within the apparent horizon and has assumed an oblate shape.

Overall, confirming what was already discussed by several authors in the past, the gravitational collapse of the slowly rotating stellar model D1 takes place in an

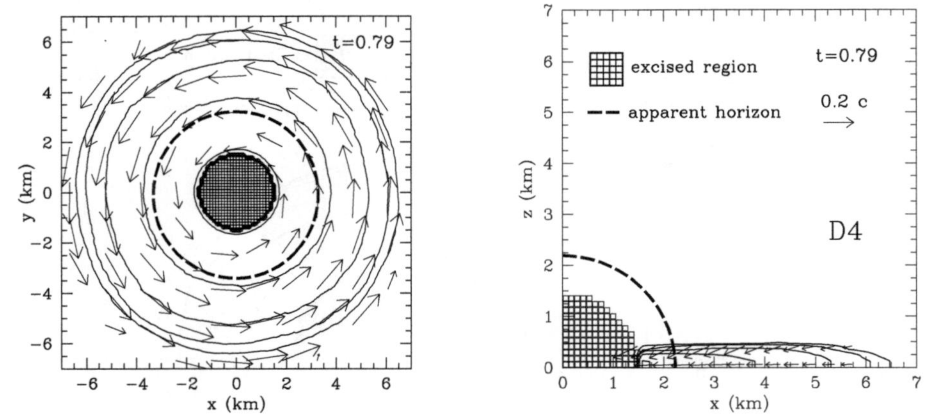

Figure 2. view of the final stages of the collapse of model D4.

almost spherical manner and we have found no evidence of shock formation which could prevent the prompt collapse to a black hole, nor appreciable deviations from axisymmetry.

As for the slowly rotating star D1, we show in Fig. 2 a representative snapshot of the evolution of the rapidly rotating model D4. The isocontour levels shown for the rest-mass density are the same used in Fig. 1. The dynamics is very similar to the one discussed for model D1 up to a time $t \sim 0.49$ ms. However, as the collapse proceeds, significant differences between the two models start to emerge and in the case of model D4 the large angular velocity of the progenitor stellar model produces significant deviations from a spherical infall. Indeed, the parts of the star around the rotation axis that are experiencing smaller centrifugal forces collapse more promptly and, as a result, the configuration increases its oblateness.

In this case the star flattens considerably, all of the matter near the rotation axis has fallen inside the apparent horizon, but a disc of low-density matter remains near the equatorial plane, orbiting at very high velocities $\gtrsim 0.2\,c$. The appearance of an effective barrier preventing a purely radial infall of matter far from the rotational axis may be the consequence of the larger initial angular momentum of the collapsing matter and of the pressure wave originating from the faster collapse along the rotational axis.

Note that the disc formed outside the apparent horizon is *not* dynamically stable and, in fact, it rapidly accretes onto the newly formed black hole. This is shown in Fig. 2, where one can notice that the disc is considerably flattened but also has large radial inward velocities which induce it to be accreted rapidly onto the black hole. Note that as the area of the apparent horizon increases, so does the excised region, which is allowed to grow accordingly.

4 Dynamics of the horizons

In order to investigate the formation of a black hole in our simulations, we have used horizon finders, available through the `Cactus` framework, which compute both the *apparent* horizon and the *event* horizon.

We also adopted several different methods for measuring the black-hole mass and spin. The first and simplest method of approximating the black-hole mass is to use the formula $M = \frac{C_{\rm eq}}{4\pi}$, where $C_{\rm eq} \equiv \int_0^{2\pi} \sqrt{g_{\phi\phi}} d\phi$ is the proper equatorial circumference. But this would only apply exactly to the Kerr (or Schwarzschild) solution; which is not the case here.

A second method for estimating the horizon mass is to look at the distortion of the horizon using the ratio of polar and equatorial proper circumferences, $C_r \equiv C_{\rm pol}/C_{\rm eq}$ [6]. For a perturbed Kerr black hole this is expected to oscillate around the asymptotic Kerr value with the form of a quasi-normal mode. By fitting to this mode we extract an estimate of the angular momentum parameter a/M_{hor} from the relation [5] $\frac{a}{M_{hor}} = \sqrt{1-(-1.55+2.55C_r)^2}$, where M_{hor} coincides with the black-hole mass M only if the spacetime has become axisymmetric and stationary.

A third method of approximating J and hence measuring M is to use the *isolated* and *dynamical*-horizon frameworks of Ashtekar and collaborators [1, 8]. This assumes the existence of an axisymmetric Killing vector field intrinsic to a marginally trapped surface such as an apparent horizon. This gives an unambiguous definition of the spin of the black hole and hence of its total mass. If there is an energy flux across the horizon, the isolated-horizon framework needs to be extended to the *dynamical*-horizon formalism [2].

In practice, our approach to the dynamical-horizon framework has been through the use of a code by Schnetter which implements the algorithm of [8] to calculate the horizon quantities. The advantage of the dynamical-horizon framework is that it gives a measure of mass and angular momentum which is accurately computed locally, without a global reconstruction of the spacetime. One possible disadvantage is that the horizon itself is required to be (close to) axisymmetric; so in case it deviates largely from axial symmetry, no information can be found. However, because arbitrarily large distortions are allowed as long as they are axisymmetric, we have not encountered problems in applying the dynamical-horizon framework to the horizons found in our simulations.

A fourth method for computing J only applies if an event horizon is found and if its angular velocity has been measured. In a Kerr background, in fact, the generators of the event horizon rotate with a constant angular velocity $\omega \equiv -g_{t\phi}/g_{\phi\phi} = \sqrt{g_{tt}/g_{\phi\phi}}$ and it can be directly related to the angular momentum parameter as $a/M = J/M^2 = \left[A\omega^2\left(1-A\omega^2/4\pi\right)\pi\right]^{1/2}$, where A is the event-horizon proper area (The event horizon is located after the simulation has been completed and the data produced is post-processed using the level-set finder of Diener [7]).

Although the direct comparison of many different methods employed here have provided valuable information on the dynamics of the system, we have found the

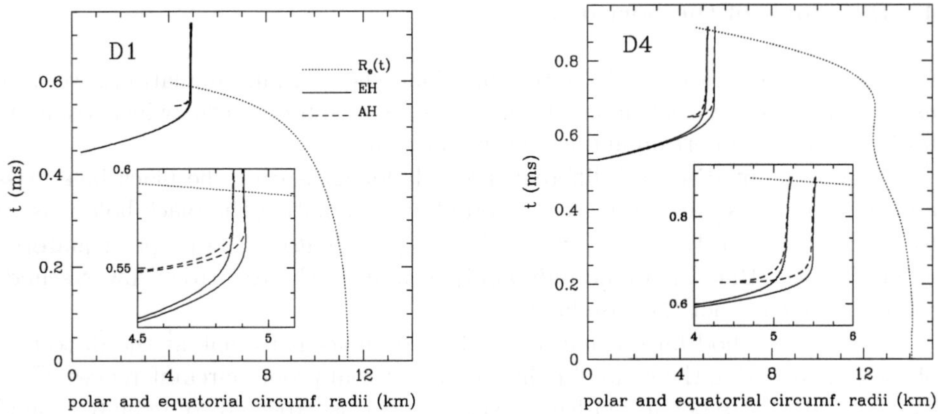

Figure 3. Evolution of the most relevant surfaces during the collapse for the cases D1 and D4.

dynamical-horizon framework to be simple to implement, accurate and not particularly affected by the errors from which equivalent approaches seem to suffer. As a result, we recommend its use as a standard tool in numerical relativity simulations.

5 Reconstructing the global spacetime

All of the results presented and discussed in the previous Sections describe only a small portion of the spacetime which has been solved during the collapse. In addition to this, it is interesting and instructive to collect all of these pieces of information into a *global* description of the spacetime and look for those features which mark the difference between the collapse of slowly and rapidly rotating stellar models. As we discuss below, these features emerge in a very transparent way within a global view of the spacetime.

To construct this view, we use the worldlines of the most representative surfaces during the collapse, namely those of the equatorial stellar surface, of the apparent horizon and of the event horizon. For all of them we need to use properly defined quantities and, in particular, circumferential radii. The results of this spacetime reconstruction are shown in Fig. 3, whose left and right panels refer to the collapse of models D1 and D4, respectively. The different lines indicate the worldlines of the circumferential radius of the stellar surface (dotted line), as well as of the apparent horizon (dashed line) and of the event horizon (solid line). Note that for the horizons we show both the equatorial and the polar circumferential radii, with the latter being always smaller than the former. For the stellar surface, on the other hand, we show the equatorial circumferential radius only. Note that in both panels of Fig. 3 the event horizon grows from an essentially zero size to its asymptotic value. In contrast, the apparent horizon grows from an initially non-zero size and, as it should, is always contained within the event horizon. At late times, the worldlines merge to the precision at which we can compute them. A rapid look at the two panels of Fig. 3 is sufficient to appreciate the different properties in the dynamics of the collapse of slowly and rapidly rotating models.

The worldlines of the stellar equatorial circumferential radius are very different in the two cases. In the slowly rotating model D1, in particular, the star collapses smoothly and the worldline always has a negative slope, thus reaching progressively smaller radii as the evolution proceeds (*cf.* left panel of Fig. 3). By time $t \simeq 0.59$ ms, the stellar equatorial circumferential radius has shrunk below the corresponding value of the event horizon. In the case of the rapidly rotating model D4, on the other hand, this is no longer true and after an initial phase which is similar to the one described for D1, the worldline does not reach smaller radii. Rather, the stellar surface slows its inward motion and, at around $t \sim 0.6$ ms, the stellar equatorial circumferential radius does not vary appreciably. Indeed, the right panel of Fig. 3 shows that at this stage the stellar surface moves to slightly larger radii. This behaviour marks the phase in which a flattened configuration has been produced and the material at the outer edge of the disc experiences a stall. As the collapse proceeds, however, also this material will not be able to sustain its orbital motion and, after $t \sim 0.7$ ms, the worldline moves to smaller radii again. By time $t \simeq 0.9$ ms, the stellar equatorial circumferential radius has shrunk below the corresponding value of the event horizon.

6 Conclusion

We have implemented a hydrodynamical excision technique within our new 3D general-relativistic numerical evolution code that combines state-of-the-art numerical methods for the spacetime evolution (*i.e.* the NOK formulation of the Einstein equations with Gamma-driver shift conditions) with an accurate hydrodynamical evolution employing several high-order HRSC methods. The evolution of the spacetime and of the hydrodynamics is coupled transparently through the method of lines, which allows for the straightforward implementation of various different time-integrators.

As a first astrophysical problem for this novel setup, we have here focused on the collapse of rapidly rotating relativistic stars to Kerr black holes. The stars are assumed to be in uniform rotation and dynamically unstable to axisymmetric perturbations. While the collapse of slowly rotating initial models proceeds with the matter remaining nearly uniformly rotating, the dynamics is shown to be very different in the case of initial models rotating near the mass-shedding limit, for which strong differential rotation develops. Although the stars become highly flattened during collapse, attaining a disc-like shape, the collapse cannot be halted because the specific angular momentum is not sufficient for a stable disc to form. Instead, the matter in the disc spirals towards the black hole and angular momentum is transferred inward to produce a spinning black hole.

Several different approaches have been employed to compute the mass and angular momentum of the newly formed Kerr black hole. Besides more traditional methods involving the measure of the geometrical properties of the apparent *and* event horizons, we have fitted the oscillations of the perturbed Kerr black hole to specific quasi-normal modes obtained by linear perturbation theory. In addition, we have also considered the recently proposed *isolated* and *dynamical* horizon frameworks, finding it to be simple to implement and yielding estimates which are accurate and more robust than those of other methods. This variety of approaches has allowed for the determination of both the mass and angular momentum of the black hole with an accuracy unprecedented for a 3D simulation.

Recent progress has been made in using mesh refinement techniques [10] to move the outer boundary sufficiently far from the source so that important information can be extracted on the gravitational wave emission produced during the collapse. The results of these investigations have been presented in ref. [4].

7 Acknowledgments

It is a pleasure to thank Joachim Frieben, José Mª. Ibáñez and Andrea Nerozzi who have participated to the development and testing of the code. We are also very grateful to Peter Diener, Jonathan Thornburg, Erik Schnetter, Badri Krishnan and Frank Herrmann for their help in the study of the trapped surfaces and to Abhay Ashtekar for clarifying some aspects of the dynamical-horizon framework. We are also grateful to the `Cactus`-code team, for their efforts in producing an efficient infrastructure for numerical relativity. Financial support for this work has been provided by the MIUR and EU Network Programme (Research Training Network contract HPRN-CT-2000-00137). J.A.F. acknowledges financial support from the Spanish Ministerio de Ciencia y Tecnología (grant AYA 2001-3490-C02-01). The computations were performed on the *Albert100* cluster at the University of Parma and the *Peyote* cluster at the Albert Einstein Institute.

References

[1] A. Ashtekar, C. Beetle, S. Fairhurst, Class. Quantum Grav., 17, 253 (2000)

[2] A. Ashtekar, B. Krishnan, Phys. Rev. Lett., 89, 261101 (2002)

[3] L. Baiotti, I. Hawke, P.J. Montero, F. Löffler, L. Rezzolla, N. Stergioulas, J.A. Font, E. Seidel, Phys. Rev. D71, 024035 (2005)

[4] L. Baiotti, I. Hawke, L. Rezzolla, E. Schnetter, Phys. Rev. Lett. *in press* (2005)

[5] S. Brandt, E. Seidel, Phys. Rev. D52, 856 (1995)

[6] S. Brandt, J.A. Font, J. M. Ibáñez, J. Massó, E. Seidel, Comp. Phys. Comm., 124, 169 (2000)

[7] P. Diener, Class. Quantum Grav., 20, 4901 (2003)

[8] O. Dreyer, B. Krishnan, D. Shoemaker, E. Schnetter, Phys. Rev. D67, 024018 (2002)

[9] I. Hawke, F. Löffler, A. Nerozzi, `gr-qc/0501054`

[10] E. Schnetter, S.H. Hawley, I. Hawke, Class. Quantum Grav., 21, 1465 (2004)

[11] N. Stergioulas, J.L. Friedman, Astrophys. J., 444, 306 (1995)

[12] N. Stergioulas, Living Rev. Relativity, 6 (2003)

[13] J. Thornburg, Class. Quantum Grav., 21, 743 (2004)

GRAVITATIONAL WAVE SOURCES

VALERIA FERRARI [a,b]

[a] *Dipartimento di Fisica, Università di Roma "La Sapienza", P.le A. Moro 2, 00185 Roma, Italy*

[b] *INFN, Sezione Roma 1,P.le A. Moro 2, 00185 Roma, Italy*

Abstract

Gravitational waves may reach the Earth in different forms which depend on the nature of the emitting source: they may be short bursts, outcome of a catastrophic event like the gravitational collapse or the coalescence of a binary system; they may be continuous, weak wavetrains emitted by a non axially symmetric, rotating neutron star or by binary systems far from coalescence; or, they may appear as a background created as a polyphonic chorus superposition of different voices: those of populations of astrophysical sources born and evolved in the past, or those coming from matter and energy fluctuations in the first instant of life of the Universe. In this paper we will shortly review some of the most promising astrophysical sources of gravitational waves to be detected by the the resonant or interferometric antennas actually in operation or under construction.

1 Introduction

The detection of gravitational waves (GWs) is one of the most challenging problems of modern astrophysics. The astronomical observations of binary pulsars like PSR 1913+16 [1], or the more recently discovered PSR J0737-3039 [2], provide an indirect proof of their existence: indeed, the orbital period of such systems decreases at a rate which is fully explained if one assumes that the variation of

B.G. Sidharth, F. Honsell and A. de Angelis. (eds.), Frontiers of Fundamental Physics, 83–91.

the orbital energy compensates the energy emitted by the system in GWs, as predicted by General Relativity. However, gravitational waves are very elusive, and have never been observed directly because they are extremely weak, and because they weakly interact with any form of matter-energy. For this reason to detect them directely extremely sophisticated instruments are needed that work at the very edge of modern technology. As an example, few years ago the resonant antenna NAUTILUS in Rome was cooled to a temperature of 100 mK, which is the lowest temperature ever reached for a massive body. At the same time, in studying GW sources theorists have to explore phenomena that occur in regions of spacetime where the known lows of physics have to be applied in some extreme conditions: extreme is indeed the physics of a gravitational collapse and the coalescence of massive bodies, extreme are the densities and pressures reached in the interior of neutron stars, or the state of matter and energy when the universe was born. Thus, in searching for GWs we are certainly operating at the frontiers of fundamental physics. It is instructive to see what is the frequency region that will be explored by the detectors that are in operation, in construction, or planned for the next future. Resonant bar detectors, ALLEGRO, NIOBE, AURIGA, EXPLORER, NAUTILUS, have been operating for years and are now close to the maximum sensitivivity that can be reached by these instruments; they are narrow-band detectors, sensitive in a frequency region of a few Hz centered at about 1 kHz. A new generation of spherical, omnidirectional resonant detectors (MINIGRAIL, GRAIL) is presently under study. For instance MINIGRAIL, which is already in operation, is a sphere of 1400 kg, resonant at 2.9 kHz with a bandwidth of 230 Hz. The ground based interferometers are sensitive to a larger frequency region which extends from a few tens of Hz to a few kHz; TAMA, GEO600, LIGO have already started the first scientific runs, whereas the italian-french experiment VIRGO is in the commissioning phase [3]. Typical sources for these detectors are the gravitational collapse to a neutron star (NS) or to a black hole (BH), stellar pulsations, coalescence of binary systems composed of compact objects like NSs and BHs. At lower frequencies, the sensitivity of ground based detectors is strongly deteriorated by the effect of local disturbances (like wind and atmospheric events, human activities, etc.) and seismic noise that are hard to screen; therefore, space-based experiments are the only possibility to explore this region, which is rich of expected sources. For instance some of the most interesting sources to be detected are binary systems like white dwarfs-white dwarfs (WD-WD) binaries and Low Mass X-Ray Binaries (LMXRB) with short orbital period $T \in [10m, 1hour]$, that are expected to emit gravitational waves at typical frequencies of a few mHz; or, binary systems composed of two black holes with masses ranging from $10^2\ M_\odot$ to $10^6\ M_\odot$ which, during the last phases of coalescence would emit a GW-signal that would span the frequency region $\nu_{GW} \in [10^{-4}, 10^{-1}]$ Hz in a time interval ranging from a tenth of a year to 500 yrs, depending on their mass. LISA (Laser Interferometer Space Antenna) will be the first high sensitivity space-based GW detector, consisting of a constellation of 3 spacecraft in heliocentric orbit, located at the vertices of an equilateral triangle with a $5 \cdot 10^6$ km side. Each spacecraft contains a pair of test-masses in geodesic motion; the test masses are the end-mirror of a single arm interferometer, the other end-mirror being in one of the other two spacecraft. As in all instruments of these kind, gravitational waves will be detected by the strain they induce in the distances among the test-particles. In the following sections we will review some of the most interesting GW sources that emit in the bandwidth of the above mentioned detectors; in particular, we will consider signals that exhibit some well defined features and that last for long enough so that, if detected, will possibly allow the identification of the emitting source.

2 Rotating neutron stars

Neutron stars can radiate their rotational energy essentially in two ways. If the star is triaxial, it has a time varying quadrupole moment and emits gravitational waves at twice the rotation frequency, and with an amplitude which can be parametrized as follows

$$h \sim 4.2 \times 10^{-24} \left(\frac{ms}{P}\right)^2 \left(\frac{r}{10\ kpc}\right)^{-1} \frac{I}{10^{45}\ g\ cm^2} \left(\frac{\epsilon}{10^{-6}}\right)$$

where $I = 10^{45}\ g\ cm^2$, is a typical value for the moment of inertia of a neutron star, P is the rotation period, and ϵ is the oblateness of the star which shows how much it deviates from axial symmetry, here given in units of 10^{-6}. These waves could be detected, for instance, by VIRGO with one year of integration, if the amplitude of the signal were of the order of $h \sim 10^{-26}$; this means that neutron stars outside our Galaxy would practically be out of reach by the interferometers of first generation. In addition, much depends on the value of the oblateness, and several studies have tried to constraint its possible range of variation. For instance in ref. [4] observational data on a number of known pulsars have been used to set an upper limit on ϵ. Assuming that the observed slowing down of their period is entirely due to the emission of GWs, the authors found $\epsilon \in\sim [10^{-2}, 10^{-9}]$; of course this can only be an upper bound, since the rotational energy of pulsars is dissipated also by other mechanisms, like the electromagnetic emission and/or the acceleration of charged particles in the magnetosphere. Further studies on this problem [5] established that if the triaxial shape is due to strains in the crust of the neutron star, the oblateness could be $\epsilon \lesssim 10^{-7} \left(\frac{\sigma}{10^{-2}}\right)$; here σ is the strain needed to break the crust, but it should be said that its value is quite uncertain; for instance $\sigma \in [10^{-2}, 10^{-1}]$ according to [6], and $\sigma \in [10^{-4}, 10^{-3}]$ according to [7]. Thus, in the light of what we presently know we cannot say what is the most likely value of ϵ and gravitational waves will probably tell us how much oblate a neutron star can be.

A time dependent quadrupole moment can also be due to a precession of the star's angular velocity around the symmetry axes. In this case the radiation is emitted at a frequency $\nu_{prec} = \frac{1}{2\pi}(\omega_{rot} + \omega_{prec}) \simeq \nu_{rot}$, but the amplitude of the precessional contribution depends on a further parameter, the "wobble-angle" between the rotation and the symmetry axes which is, however, largely unknown.

3 Binary systems far from coalescence

In 1975, Hulse and Taylor applied the quadrupole formalism [8] to predict the slowing down of the period of the binary pulsar PSR 1913+16 [1]. They found $\frac{dP}{dt} = -2.4 \cdot 10^{-12}$, in excellent agreement with the observed value, $\frac{dP}{dt} = -(2.3 \pm 0.22) \cdot 10^{-12}$, thus providing the first indirect evidence of the existence of gravitational waves. Since LISA will be sensitive to the low frequency region, it is now interesting to ask wether these waves could be seen directly. The quadrupole formalism, used to solve Einstein's equations in the weak-field, low motion limit, shows that when the orbit is circular the radiation is emitted at twice the keplerian orbital frequency. If, as an example, we assume that we are looking at the radiation which emerges in the direction orthogonal to the orbital plane, the wave amplitude, computed in the TT- (Transverse-Traceless) gauge is given by

$$h_{ij}^{TT} = -\frac{4G^2(M_1+M_2)\mu}{l_0 c^4}\frac{1}{d}\begin{pmatrix} \cos 2\omega_K t & \sin 2\omega_K t & 0 \\ \sin 2\omega_K t & -\cos 2\omega_K t & 0 \\ 0 & 0 & 0 \end{pmatrix}, \quad (1)$$

where l_0 is the orbital separation, $\omega_K = \sqrt{\frac{G(M_1+M_2)}{l_0^3}}$ is the orbital frequency, d the distance of the source from Earth and $\mu = \frac{M_1 M_2}{M}$ the reduced mass of the system of total mass $M = M_1 + M_2$. If the orbit is eccentric, waves will be emitted at frequencies multiple of $\nu_k = \omega_k/2\pi$, and the number of equally spaced spectral lines will increase with the eccentricity [9]. For instance, the binary pulsar PSR 1913+16 is composed of two very compact stars with masses $M_1 = 1.4411\ M_\odot$ and $M_2 = 1.3874\ M_\odot$, revolving around their center of mass with an eccentric orbit ($e = 0.617139$), and keplerian frequency $\nu_k = 3.583 \cdot 10^{-5}$ Hz. The system is at a distance $d = 5$ kpc from Earth. In this case the maximum of the GW emission occurs at $\nu_{max} = 1.44 \cdot 10^{-4}\ Hz$, with amplitude $h_{max} \sim 10^{-23}$. A similar calculation for the recently discovered double pulsar PSR J0737-3039, in which case $M_1 = 1.337\ M_\odot$, $M_2 = 1.250\ M_\odot$, the orbital period is $P = 2.4$, h $d = 5 - 600$ pc, $e = 0.088$, shows that the maximum of the GW emission occurs at $\nu_{max} = 2.3 \cdot 10^{-4}\ Hz$, and the amplitude of the corresponding spectral line is $h_{max} \sim 6 - 7 \cdot 10^{-22}$. Thus in both cases the GW emission is in the bandwidth of LISA, but the expected sensitivity curve of this instrument shows that the signal is too low to be detectable. However, other interesting sources of radiation of this kind exist in our Galaxy; they are, for instance, cataclysmic variables that are semi-detached binaries of low mass and very short period, in which the primary star is an accreting degenerate white dwarf, and the secondary is usually a late type-star filling its Roche lobe and transfering matter on the companion; furthermore, there are double-degenerate binary systems, WD-WD or WD-NS binaries; 10 of such systems have been observed so far that have ultra-short periods, smaller than ten minutes. They are typically strong X-ray emitters, and they appear very promising for LISA because the emission frequency in GW is well inside the detector bandwidth. In table I the masses, emission frequencies and characteristic wave amplitude are listed for some of these short period systems, that should serve as calibration sources for LISA.

Table 1. Galactic binaries with component masses M_1 and M_2, gravitational wave frequency ν_{GW}, and gravitational wave amplitude h.

name	$m_1(M_\odot)$	$m_2(M_\odot)$	ν_{GW} (Hz)	$\log h$
WD 0957-666	0.37	0.32	$0.4 \cdot 10^{-3}$	−21.4
WD 1101+364	0.31	0.36	$0.2 \cdot 10^{-3}$	−21.7
WD 1704+481	0.39	0.56	$0.2 \cdot 10^{-3}$	−21.4
KPD 0422+4521	0.51	0.53	$0.3 \cdot 10^{-3}$	−21.2
KPD 1930+2752	0.5	0.97	$0.2 \cdot 10^{-3}$	−21.0
RXJ0806.3+1527	0.4	0.12	$6.2 \cdot 10^{-3}$	−21.4
RXJ1914+245	0.6	0.07	$3.5 \cdot 10^{-3}$	−21.2
KUV05184-0939	0.7	0.092	$3.2 \cdot 10^{-3}$	−22.0
AM CVn	0.5	0.033	$1.9 \cdot 10^{-3}$	−21.7
HP Lib	0.6	0.03	$1.8 \cdot 10^{-3}$	−21.7
CR Boo	0.6	0.02	$1.4 \cdot 10^{-3}$	−22.0
V803 Cen	0.6	0.02	$1.2 \cdot 10^{-3}$	−22.0
U Gem	1.26	0.57	$1.3 \cdot 10^{-4}$	-20.8
IP Peg	1.15	0.67	$1.5 \cdot 10^{-4}$	-20.9
HU Aqr	0.95	0.15	$2.7 \cdot 10^{-4}$	-21.3
VW Hyi	0.63	0.11	$3.1 \cdot 10^{-4}$	-21.3
EX Hya	0.78	0.13	$3.4 \cdot 10^{-4}$	-21.4
WZ Sge	0.45	0.058	$4.1 \cdot 10^{-4}$	-22.1
ST LMi	0.76	0.17	$2.9 \cdot 10^{-4}$	-21.4
SW UMa	0.71	0.10	$4.9 \cdot 10^{-4}$	-21.6
Z Cha	0.84	0.125	$3.1 \cdot 10^{-4}$	-21.5
V 436 Cen	0.7	0.17	$3.7 \cdot 10^{-4}$	-21.6

4 Coalescing binary systems

The target of ground-based interferometers is to detect the signal emitted during the latest phases of inspiraling of a binary system. Assuming that the two coalescing bodies are point masses, by using the quadrupole formalism and including the radiation reaction effects it is possible to show that the loss of gravitational energy induces a circularization of the orbit, and that the radius decreases according to the law $R(t) = R_{in}\left(1 - t/t_{coal}\right)^{1/4}$; where $t_{coal} = \frac{5}{256}\frac{R_{in}^4}{\mu M^2}$ is the time of the final coalescence. The orbital frequency consequently increases, and so does the frequency of the emitted wave, with a time dependence given by $\nu = \frac{1}{\pi}\left[\frac{5}{256}\frac{1}{\mu M^{2/3}}\frac{1}{(t_{coal}-t)}\right]^{3/8}$. Thus, the gravitational signal emitted by a coalescing system resembles the chirp of a singing bird. For instance if the binary is composed of two neutron stars, about 10^8 years after formation the frequency of the emitted signal will enter the bandwidth of the ground based interferometers and in about 15-20 minutes (depending on the star masses) will sweep the

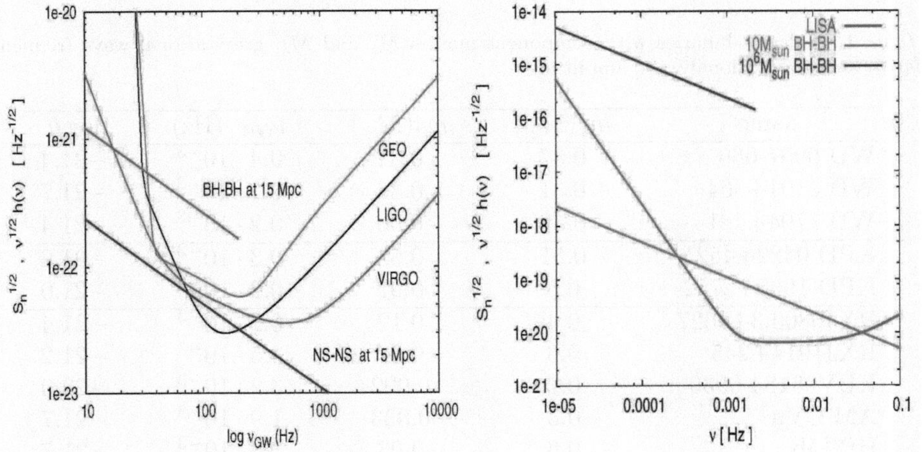

Figure 1. The strain amplitude of the gravitational signal produced by the coalescence of binary systems. In the left panel we plot the noise strain amplitude of the first generation of ground based intrerferometers VIRGO, LIGO and GEO versus frequency, and the strain amplitude of the signal produced by the coalescence of two neutron stars (NS-NS) with mass $M = 1.4\ M_\odot$ each, and two black holes of mass $M = 10\ M_\odot$. In both cases the source is assumed to be at the distance of the Virgo cluster ($d \sim 15$ Mpc). In the right panel the same plot is done for the space-based interfereometr LISA and for two cases of BH-BH coalescence, respectively of $100\ M_\odot$ and $10^6\ M_\odot$. In both cases the source is assumed to be at a distance $d = 3$ Gpc. All signals extend up to the frequency corresponding to the ISCO (see text).

frequency region ranging from $\sim$ 10Hz to $\sim$ 900Hz. The amplitudes of the two polarizations of the wave emitted during this fast inspiralling phase are

$$h_+ = \frac{2(1+\cos^2 i)\mu\,(\pi M\nu)^{2/3}}{r}\cos(2\pi\nu t), \quad h_\times = \pm\frac{4\mu\cos i\,(\pi M\nu)^{2/3}}{r}\sin(2\pi\nu t), \tag{2}$$

where i is the angle of inclination of the orbit to the line of sight, and the emitted energy per unit frequency is given by $\frac{dE}{d\nu} = \frac{\pi^{2/3}}{3}\mu\frac{M^{2/3}}{\nu^{1/3}}$. In figure 1 we plot the strain amplitude (i.e. $\sqrt{\nu}h(\nu)$, where $h(\nu)$ is the Fourier transform of $h(t)$) associated to the GW signal emitted by different types of binary systems, compared to the noise strain amplitude $(S_h)^{1/2}$ of the first generation of ground based (left panel) and space-based (right panel) intereferometric detectors. In all cases the signal extends up to the frequency corresponding to the ISCO (Innermost Stable Circular Orbit $R_{ISCO} \sim \frac{6G(M_1+M_2)}{c^2}$) after which the two bodies merge, and the waveform given in (2) is no longer appropriate to describe the emitted wave. It should be stressed that near the ISCO the signal (2) has to be corrected to take into account tidal interaction effects and corrections due to the orbital motion. It should also be stressed that the signal described by eq. (2) refers to the case of non rotating bodies. If they are rotating, spin-coupling effect may produce signals that are considerably different ([10]).

From figure 1 we see that it is quite unlikely that the first generation of ground-based interferometrs will observe the coalescence of neutron stars, while they would be able to detect the coalescence of sufficiently massive black holes. Conversely, LISA will be a fantastic instrument to detect the coalescence of massive and supermassive black holes out to cosmological distances. It is interesting to mention that given a coalescing source, the larger is the detector bandwidth, the longer will be the time spent by the signal in that bandwidth, and this will increase the detection chances. This is of particular importance for the ground based interferometers: for instance, a signal emitted by two $10M_\odot$ - black holes will stay in the VIRGO bandwidth ($\sim [10Hz, 1kHz]$) about 38 s, whereas it will sweep LIGO's bandwidth ($\sim [40Hz, 1kHz]$) in about 1 s.

The evolution of the system when it approaches the ISCO has to be described by fully relativistic numerical simulations. Although these studies are fastly progressing, they are still far from providing reliable waveforms to confront with the detector sensitivity curves. However we have a clue on what the signal, emitted after a single compact object forms, could be; indeed, the newly born object will violently oscillates in its quasi-normal modes emitting GWs at the corresponding frequencies, which only depend on the mass and angular momentum if it is a black hole [11], or on the equation of state of matter, if it is a neutron star [12]. Unfortunately, we do not know how much energy can be stored in each mode, and consequently the wave amplitude is unknown. A clue on this crucial information will be available when the numerical studies of the merging phase will progress enough to describe the formation and the subsequent initial evolution of the newly born compact object. However, it should be stressed that it will be the detection of such signals which will provide the most interesting insight into the physics of phenomena about which so little is known.

5 Concluding Remarks

The detection of gravitational waves will allow to test the theory of General Relativity in the strong field regime; but beyond that there are many reason why it is of fundamental importance. GWs will allow to test alternative theories of gravity: for example binary pulsars could yield bounds on scalar-tensor theories because they predict dipole gravitational radiation and violation of the strong equivalence principle. The most famous of such theories is the Jordan-Brans-Dicke theory which postulates the existence of a scalar field which couples only to gravity; ω is the coupling constant, such that the larger the value of ω, the weaker the scalar field. The bound from PSR 1913+16 is only $\omega > 100$, because the two stars have nearly equal masses and dipole radiation is suppressed by symmetry. A binary system composed of two compact stars of unequal mass, like a white dwarf and a neutron star, could yield bounds as large as 10^4 [13].

Through the study of radiation reaction effects we will be able to study how the emission of GWs affects the evolution of the emitting sources, as for instance the rotation rate of a non axisymmetric neutron stars, the process of coalescence in a binary system or the evolution of a newly born neutron star.

GWs will provide information on the equation of state of matter at supranuclear densities (unreachable by experiments on Earth) through the frequencies of the quasi normal modes of oscillation of Neutron Stars, which may be excited after the star formation in a gravitational collapse, during a glitch or in the latest phases of binary coalescence [12], [14].

GWs have been produced since the beginning of the Universe, and the existence of relic gravitational waves essentially rely on the validity of General Relativity and on basic principles of quantum field theory. Indeed, quantum fluctuations of the gravitational field are amplified by the accelerated expansion of the universe during the inflationary stage. In the same way as a sudden change in the electromagnetic field produces photons, the rapid change in the space-time curvature produces gravitons, and consequently a stochastic background of gravitational waves. The detection of such background is the only way to learn about the evolution of the very early Universe, up to the limits of the Planck era and the Big Bang. The stochastic background of gravitational waves has also a component of astrophysical origin. Indeed, bursts and continuous signals of gravitational radiation have been emitted by astrophysical sources since the beginning of star formation, and the cumulative effect of a large number of unresolved and uncorrelated sources generates a stochastic background the features of which depend on the specific source. The different contributions depend also on the rate of events of the selected type occurred per unit comoving volume, as a function of the cosmological redshift, i.e. on the star formation rate history (SFRH). For those sources that emit over a timescale small compared to the typical timescales of evolution of the underlying galaxy population, the emission of radiation can be considered as "instantaneous", and directly related to the rate of star formation at that particular value of the cosmological redshift when the star formed. Thus, the detection of such background would provide an insight into the formation and evolution of astrophysical sources in the Universe [15, 16, 17]

Finally, GWs will allow to discover unknown sources unreachable by electromagnetic search, and will enlighten regions of the Universe dominated by strong gravitational fields and high velocities, unaccessible in any other way.

References

[1] R.A. Hulse, J.H. Taylor, *Astrophys. J.* 195, L51, 1975

[2] M. Burgay et al, *Letter to Nature* 426, 531, 2003

[3] For details of the gravitational experiments see the web sites:
www.roma1.infn.it/rog/rogmain.html,
www.auriga.lnl.infn.it,
www.minigrail.nl/index.html,
www.virgo.infn.it, www.ligo.caltech.edu,
www.geo600.uni-hannover.de,
www.tamago.ntk.nao.ac.jp, lisa.jpl.nasa.gov

[4] E. Gourgoulhon, S. Bonazzola, *Proceedings of the International Conference on Gravitational Waves, Sources and Detectors*, World Scientific, Singapore 1996, 51

[5] G. Ushomirsky, C. Cutler, L. Bildsten, *Mon. Not. R. Astron. Soc.* 319 n. 3, 902, 2000

[6] R. Smoluchowski, *Phys. Rev. L.* 24, 923 1970

[7] R. Ruderman, *Ap. J.* 382, 587, 1991

[8] P.C. Peters, J. Mathews, *Phys. Rev.* 131, 435, 1963

[9] V. Ferrari, M. D'Andrea, E. Berti, *Int. J. Mod. Phys.* D9 n. 5, 495, 2000

[10] K. Glampedakis, D. Kennefick, *Phys. Rev. D* 66, 044002, 2002

[11] S.Chandrasekhar *The mathematical theory of black holes.* (Oxford: Claredon Press 1984)

[12] V. Ferrari, G. Miniutti, J.A. Pons *Mon. Not. R. Ast. Soc.* **342** 624, 2003

[13] C. M. Will, in *2001: A Relativistic Spacetime Odyssey*, ed. by I. Ciufolini, D. Dominici, L. Lusanna, World Scientific Pub. , 247, 2003

[14] O. Benhar, V. Ferrari, L. Gualtieri, *Phys. Rev. D* 70 n.12, 124015, 2004

[15] V. Ferrari, S. Matarrese, R. Schneider, *Mon. Not. R. Astron. Soc.* 303, 247, 1999

[16] V. Ferrari, S. Matarrese, R. Schneider, *Mon. Not. R. Astron. Soc.* 303, 258, 1999

[17] R. Schneider, V. Ferrari, S. Matarrese,S.F. Portegies Zwart, *Mon. Not. R. Astron. Soc.* 324, 797, 2001

MODEL ANALYSIS OF GRAVITATIONAL SHELL COLLAPSES

MASAFUMI SERIU [a]

[a] *Department of Physics, Faculty of Engineering, University of Fukui Bunkyo 3-9-1, Fukui 910-8507, Japan*
E-mail: mseriu@edu00.f-edu.fukui-u.ac.jp

Abstract

There are two important conjectures concerning gravitational collapses: *(A) The cosmic censorship conjecture* and *(B) The hoop conjecture.* If both the conjectures are valid, it should follow that any sufficiently elongated object (made of reasonably natural matter) evolves regularly without any singularity formation. It implies that the detailed study of the dynamical evolution of elongated matter provides us with valuable information on the validity of the two conjectures.

Keeping this point in mind, we here investigate a cylindrical shell-contraction with rotational pressure, accompanying the radiation of gravitational waves and massless particles. The model has been introduced previously, and some authors have claimed the possibility of a singularity-forming evolution of the model, which could suggest the invalidity of either (or both) of the above conjectures. Here we analyze this model once again with more rigor and see whether the claimed singular evolution exists.

It is rigorously proved that, *as far as the weak energy condition is satisfied outside the shell, the collapsing shell bounces back at some point irrespective of the initial conditions, and no singularity is ever formed.*

This result of bouncing behavior is compatible with the results for the other known cylindrical shell-contracting models, and confirms the essential importance of the energy condition and the rotational effect in the gravitational collapse.

The two conjectures (A) and (B) in this manner disclose no flaws in all the analytical, sufficiently natural models with cylindrical symmetry known today, suggesting their stable validity in wider class of situations.

B.G. Sidharth, F. Honsell and A. de Angelis. (eds.), Frontiers of Fundamental Physics, 93–102.

1 Introduction

Physics of gravitation is one of the most actively investigated themes in modern theoretical physics. Though gravitation is universal and can be experienced daily, dynamics involved in it is far from simple and gives rise to various complicated phenomena. Gravitation has so many aspects that it may not be possible to provide a unified view of the phenomena. In any case, however, at least it can be said that the nonlinearity always plays the key role in any gravitational phenomenon.

Problems of gravitational contractions are the kind of topics in which the nonlinear nature of gravity is involved in a particularly explicit manner. Indeed, its highly nonlinear nature is still preventing us from clarifying several issues questioned even more than 30 years ago.

Among such unanswered questions, the following two conjectures have been in particular significant and have motivated enormous amount of investigations so far: *(A) The cosmic censorship conjecture* (Penrose 1969 [1]) and *(B) The hoop conjecture* (Thorne 1972 [2]).

The cosmic censorship conjecture [1] is the claim that *no naked singularity is allowed in Nature.* Since standard physical description breaks down at spacetime singularities, it is not a theoretically desirable situation if such singularities are visible to distant observers. Thus this conjecture seems quite reasonable at least conceptually, setting aside technical details. Another way of stating this conjecture is that any matter in *natural*, smooth initial conditions should either evolve regularly or evolve into black holes.

The hoop conjecture [2], on the other hand, claims that *black holes are formed when and only when the inequality $\mathcal{C} \leq 2\pi \cdot \frac{2GM}{c^2}$ holds* (here $\mathcal{C}$ is the largest circumference of the matter, while M is the total mass of the matter). Stated differently, it claims that black holes are never formed unless matter is compressed into a small, compact region with the size comparable to its Schwarzschild radius. The escape velocity v for the matter with mass M and the size R is estimated by the formula $v^2/2 \sim GM/R$ (note $R \sim \mathcal{C}/2\pi$), so that the above inequality describes the situation in which the matter is compressed so compactly that its escape velocity exceeds even the speed of light. Thus, the hoop conjecture also seems quite reasonable.

Now it is very suggestive if the above two conjectures are combined together [3]. Logical consequence of this combination is that *any sufficiently elongated object (made of reasonably natural matter) should evolve regularly without any singularity formation.* As a typical example of the elongated matter, gravitational contractions of cylindrical models have been studied quite intensively [2, 5, 6]. We also investigate one of such cylindrical models in this paper.

At this stage it should be noted that even spherical collapses reveal surprisingly complicated features, and one can even construct spherical models which show naked singularity formations [4], though their relevance in view of naturalness and generality should be questioned. At least it should not be overlooked that some researchers take those solutions very seriously and that investigations are still going on [4]. It is not surprising, then, that cylindrical models reveal even richer features than the spherical models, though the former models have actually been less investigated so far.

In this paper, we focus on a particular cylindrical shell model discussed by Pereira and Wang [7, 8]. This model can be interpreted as describing a shell filled with massless particles rotating in the ϕ-direction; those rotating particles give rise to the rotational pressure p_ϕ of the shell.

Regarding this model, it is claimed that there is a special class of solutions in which the shell develops into line-like singularity, which could possibly be naked [7, 8]. If this claim were true, it would mean that either (or both) of the two conjectures mentioned above should be invalid. Thus this model needs to be investigated in more detail to see whether the claim is true, since the original argument [7, 8] was just a preliminary one and has been awaiting for detailed analysis. Here we analyze the model once again with much more care: We then rigorously prove that the shell always bounces back and singularities are never formed as far as the weak energy condition is satisfied outside the shell [3]. This bouncing behavior is similar to the one for the case of a cylindrical shell of counter-rotating dust particles [6].

The two conjectures [1, 2] in this manner disclose no self-inconsistency in all the analytical models with cylindrical symmetry known today (with sufficiently reasonable matter content and initial conditions), suggesting their validity in wider class of situations.

2 A dynamical shell in a cylindrical spacetime

We start with a metric of a spacetime possessing a cylindrical spatial symmetry,

$$ds^2 = -T(t,r)^2 dt^2 + R(t,r)^2 dr^2 + Z(t,r)^2 dz^2 + \Phi(t,r)^2 d\phi^2 \ , \tag{1}$$

where $(t,\ r,\ z,\ \phi)$ is the standard cylindrical coordinate-system. Let Σ be a *timelike* surface in the spacetime defined by Eq.(1), given by $r = \rho(t)$. It is often convenient to introduce the proper-time τ for an observer sitting on the shell, determined by

$$d\tau^2 = T^2\ (1 - \frac{R^2}{T^2}\ \dot{\rho}^2)\ dt^2 =: X^{-2} dt^2 \ . \tag{2}$$

Here the symbols $X := dt/d\tau$ and $^\circ := \partial_\tau$ will frequently appear below. We note that $(e_\tau, e_z,\ e_\phi) := (\partial_\tau,\ 1/Z.\partial_z,\ 1/\Phi.\partial_\phi)$ forms a set of orthonormal bases on the shell Σ.

Now the so-called extrinsic curvature $K_{\alpha\beta}$ of Σ serves as a key tool for the analysis below. The extrinsic curvature of a surface Σ describes how a 3-dimentional surface Σ is embedded in a 4-dimentional spacetime [9, 10]. Roughly speaking, $K_{\alpha\beta}$ is defined as $K_{\alpha\beta} \sim \partial_\alpha \hat{n}$, where $\hat{n}$ stands for the unit normal vector of Σ, while ∂_α indicates the derivative along Σ. Since $\hat{n}_t$ is proportional to $\dot{\rho}$, then, $K_{tt} \sim \partial_t n$ depends on $\ddot{\rho}$. In this manner, we see that K_{tt} (or $K_{e_\tau e_\tau}$) contains the dynamical information of the shell Σ. More explicitly, the normal unit vector for

Σ is given by

$$\hat{n}_\mu = XTR(-\overset{\circ}{\rho}/X,\ 1,\ 0,\ 0)_{(trz\phi)} \quad ,$$

and it is straightforward to compute the extrinsic curvature of the shell [3];

$$\begin{aligned} K_{e_\tau e_\tau} &= -\frac{R}{XT}\overset{\circ\circ}{\rho} - \frac{R}{2XT}\overset{\circ}{\rho}(\ln R^2)^\circ + \frac{1}{2}(T^2X^2-1)\,\partial_n(\ln R^2) \\ &\quad + \frac{X^2T^2}{2}(X^2T^2-1)\,\partial_n(\ln T^2/R^2) \\ &\quad - \frac{XTR}{2}\overset{\circ}{\rho}\,(X^2T^2-1)\,(\ln T^2/R^2)^\circ \quad , \end{aligned} \tag{3}$$

$$K_{e_z e_z} = \partial_n \ln Z \quad , \quad K_{e_\phi e_\phi} = \partial_n \ln \Phi \quad , \quad \text{others} = 0 \quad . \tag{4}$$

As is already remarked, $K_{e_\tau e_\tau}$ depends on $\rho^{\circ\circ}$ and describes the dynamics of the shell.

According to the standard relativistic junction-condition formulas [10], when the energy-momentum density is concentrated on a 3-surface Σ, it gives rise to the discontinuity in the extrinsic curvature across Σ. This result is quite analogous to the relation of the electric charge induced on the surface of a conducting ball with the discontinuity in the electric field there. In the present case, we get

$$\begin{aligned} &-\frac{1}{\kappa}\left([K_{e_z e_z}] + [K_{e_\phi e_\phi}]\right) = \epsilon \quad , \\ &-\frac{1}{\kappa}\left([K_{e_\tau e_\tau}] - [K_{e_\phi e_\phi}]\right) = p_z \quad , \\ &-\frac{1}{\kappa}\left([K_{e_\tau e_\tau}] - [K_{e_z e_z}]\right) = p_\phi \quad . \end{aligned} \tag{5}$$

Here ϵ, p_z and p_ϕ are interpreted as the energy density of the shell, the pressure in the z-direction and pressure in the ϕ-direction, respectively; $[K]$ indicates discontinuity across Σ (i.e. $K(\text{outside}) - K(\text{inside})$); $\kappa := 8\pi G$ is the Einstein's gravitational constant. As we will see soon, the second equation in Eq.(5) yields the dynamical equation for the shell.

3 The cylindrical shell model with rotational pressure

Let us now consider a particular cylindrical shell model discussed by Pereira and Wang [7, 8]. In this model an interior flat spacetime (described by ds_-) and an exterior cylindrical spacetime (ds_+) are matched together at a timelike shell Σ:

$$ds_-^2 = -dt_-{}^2 + dr_-{}^2 + dz_-{}^2 + r_-{}^2 d\phi_-{}^2 \tag{6}$$

$$ds_+^2 = e^{2\gamma(\xi)}(-dt_+{}^2 + dr_+{}^2) + dz_+{}^2 + r_+{}^2 d\phi_+{}^2 \quad , \tag{7}$$

where γ is a function of $\xi := t_+ - r_+$ only. Let the shell Σ be specified by $r_+ = \rho_+(t_+)$ from the exterior viewpoint[1].

We now note that the C-energy [11] C and the Einstein tensor $\mathbf{G}_{ab}$ outside the shell are given by

$$\begin{aligned} C &= \frac{1}{8\kappa}(1 - e^{-2\gamma}) \ , &(8)\\ \mathbf{G}_{ab} &= \frac{\gamma'}{\alpha^2 r_+} k_a k_b \ , &(9) \end{aligned}$$

where $k_\mu = (\alpha, \ -\alpha, \ 0, \ 0)_{(trz\phi)}$ is a null-vector. Equation (9) suggests that the model may be describing a contracting shell accompanied by outward radiation of massless particles and gravitational waves [7]. Postulating the positivity of the C-energy (see Eq.(8)) and the weak energy condition (see Eq.(9)), then, the function γ should satisfy

$$\gamma \geq 0 \quad , \quad \gamma' \geq 0 \ . \tag{10}$$

The quantity X (see Eq.(2)) becomes [2]

$$X := \frac{dt_+}{d\tau} = \frac{e^{-\gamma_+}}{\sqrt{1 - \dot{\rho}_+^2}}$$

so that $\dot{\rho}_+^2 < 1$. With the metric Eq.(7), it implies that the radial velocity of the shell is always less than the light-velocity, as it should be.

Now the junction conditions Eq.(5) yield

$$\begin{aligned} \kappa\epsilon &= e^{-\gamma_+} \frac{\Delta - 1}{\rho_+ \sqrt{1 - \dot{\rho}_+^2}} &(11)\\ \kappa p_z &= \frac{e^{-\gamma_+}}{\Delta(1 - \dot{\rho}_+^2)^{3/2}} \{ (\Delta - 1)\ddot{\rho}_+ - \gamma'_+ \dot{\rho}_+ (1 - \dot{\rho}_+)[(\Delta + 1)\dot{\rho}_+ + 1] &(12)\\ &\quad -\Delta(\Delta - 1)\frac{1 - \dot{\rho}_+^2}{\rho_+} \} \\ \kappa p_\phi &= \frac{e^{-\gamma_+}}{\Delta(1 - \dot{\rho}_+^2)^{3/2}} \{ (\Delta - 1)\ddot{\rho}_+ - \gamma'_+ \dot{\rho}_+ (1 - \dot{\rho}_+)[(\Delta + 1)\dot{\rho}_+ + 1] \} \ , &(13) \end{aligned}$$

with

$$\Delta := \frac{dt_-}{dt_+} = \left\{ (1 - e^{2\gamma_+})\dot{\rho}_+^2 + e^{2\gamma_+} \right\}^{1/2} \ . \tag{14}$$

It should be noted that $\gamma_+ > 0$ (Eq.(10)) implies $\Delta > 1$ (Eq.(14)), so that it follows $\epsilon > 0$ (Eq.(13)).

[1] The suffix "+" is for the exterior quantities (ds_+, t_+, r_+ and so on), while "-" is for the interior quantities.

[2] Here the suffix + in e.g. ρ_+ or γ_+ implies that the quantity is evaluated on the shell from the viewpoint of the exterior observer.

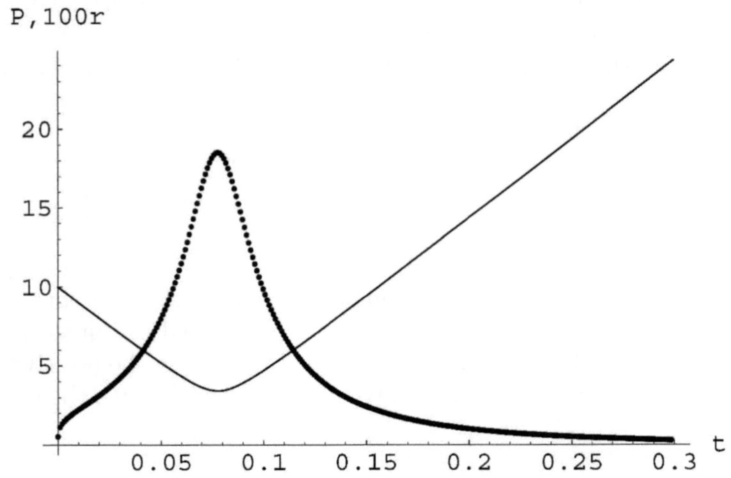

Figure 1. Typical evolutions of the shell-radius $\rho_+(t_+)$ (thinner curve) and the rotational pressure p_ϕ (thicker curve). We have set $\gamma_+(r_+ - t_+) = \frac{1}{10^6}(r_+ - t_+ + 100)^3$. Initial conditions are $\rho_+(0) = 0.1$ and $\dot{\rho}_+(0) = -0.999$. The vertical line indicates $100\rho_+$ and p_ϕ, while the horizontal line indicates t_+.

Let us set

$$P_0 := \kappa p_z \quad \text{(constant)}$$

in Eq.(13), then, we finally get the dynamical equation for the shell,

$$\begin{aligned} \ddot{\rho}_+ &= \Delta \frac{1 - \dot{\rho}_+^2}{\rho_+} + \gamma_+' \frac{\dot{\rho}_+}{\Delta - 1}(1 - \dot{\rho}_+)[(\Delta + 1)\dot{\rho}_+ + 1] \\ &+ \frac{\Delta}{\Delta - 1}(1 - \dot{\rho_+}^2)^{3/2} e^{\gamma_+} P_0. \end{aligned} \tag{15}$$

4 Bouncing of the shell without singularity-formation

As is mentioned in *Introduction*, the problems of gravitational contractions inevitably involve highly nonlinear dynamics. Equation (16) is indeed extremely nonlinear in $\rho_+(t_+)$, considering how $\rho_+(t_+)$ is included in Δ and $\gamma_+ = \gamma_+(t_+ - \rho_+(t_+))$. Before studying the shell dynamics Eq.(16) in detail, it is then helpful to get a rough idea for the dynamics by showing typical numerical results.

Figure 1 shows the evolutions of the shell-radius $\rho_+(t)$ and the rotational pressure p_ϕ for the case $\gamma_+(r_+ - t_+) = \frac{1}{10^6}(r_+ - t_+ + 100)^3$ and $P_0 = 0$ with the initial conditions $\rho_+(0) = 0.1$ and $\dot{\rho}_+(0) = -0.999$ (i.e. very close to the light-velocity). The figure indicates that the rotational pressure prevents the shell from collapsing and that the shell bounces back and its velocity approaches to the light-velocity after the bouncing.

On the other hand, *Figure 2* indicates the behavior of the same model as

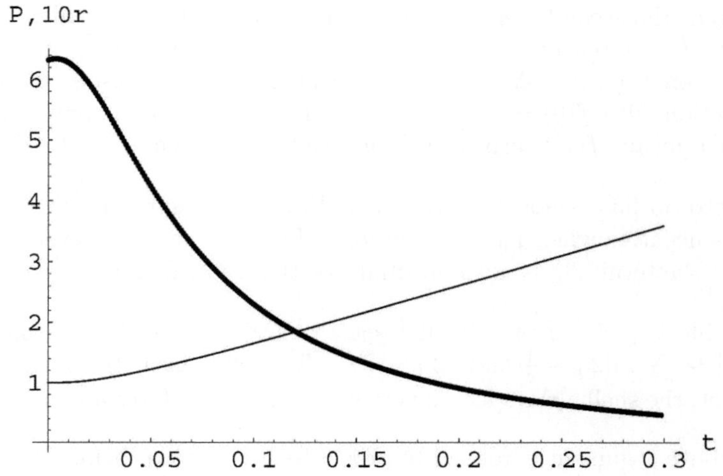

Figure 2. The same model as Figure 1 with initial conditions, $\rho_+(0) = 0.1$ and $\dot{\rho}_+(0) = -0.1$. The vertical line indicates $10\rho_+$ (for the thinner curve) and p_ϕ (for the thicker curve), while the horizontal line indicates t_+.

Figure 1 but with different (milder) initial conditions $\rho_+(0) = 0.1$ and $\dot{\rho}_+(0) = -0.1$. The bouncing behavior is again observed.

Even though we change initial conditions in various ways, the bouncing behavior similar to the above typical numerical results always appears. These results lead us to doubt the claim by the preceding authors [7, 8] that a line-like singularity (possibly naked) could form in this model. Indeed, as we see below, it is rigorously proved that the shell always bounces back and never forms a singularity in this model, irrespective of initial conditions.

Let us set $V := -\dot{\rho}_+$ which takes the value $0 < V < 1$ in the contracting phase. Instead of studying Eq.(16), then, it is more convenient to investigate the following;

$$\ddot{\rho}_+ = \Delta\frac{1-V^2}{\rho_+} + \gamma'_+\frac{V}{\Delta - 1}(1+V)[(\Delta+1)V-1] + \frac{\Delta}{\Delta-1}(1-V^2)^{3/2}e^{\gamma_+}P_0 \quad , \quad (16)$$

where $\Delta = \left[V^2 + e^{2\gamma_+}(1-V^2)\right]^{1/2}$.

Let [[1]], [[2]] and [[3]] be the first, the second and the third term on the right-hand side of Eq.(16), respectively. Then we observe the following:

(i) The first term [[1]] produces a strong repulsive force, which prevents the shell from collapsing. The only possible exception may arise when V approaches to 1 faster than $\rho_+ \to 0$. Only in this case the term [[1]] can be negligible. Let us call, then, the phase-point $(\rho_+ \, , \, V) = (0, 1)$ the *singularity-prone point.* However, near the singularity-prone point, the other two terms ([[2]] + [[3]]) become dominant and they produce a strong repulsive force:

(ii) Indeed, the second term [[2]], which is an attractive force when V is small, turns to a repulsive force as $V \to 1$. This is easily seen by noting that the function $f(V) := [(\Delta + 1)V - 1]$ is a monotonic, increasing and continuous function with $f(0) = -1$ and $f(1) = 1$. Thus $f(V)$ has only zero in (0,1), which means $f(V)$ turns from negative to positive as $V \to 1$.

(iii) The term [[3]] is non-negative when $P_0 \geq 0$ (pressure in z-direction), while it is negative when $P_0 < 0$ (tension). In either case, however, it turns out that the term [[3]] is not important for the whole dynamics.

(iv) Setting $V = 1 - \delta$ ($\delta > 0$), it is easy to see that the three terms behave as $[[1]] \sim \frac{\delta}{\rho_+}$, $[[2]] \sim \frac{1}{\delta}$ and $[[3]] \sim \delta^{1/2}$. Thus even near the singularity-prone point, the shell always feels a very strong repulsive force and never collapses.

The above argument is refined to prove *Theorem* shown below [3].

Definition

We define the *core-region* as a connected region in the phase-space (for the dynamical variable ρ_+) which contains a neighborhood of the singularity-prone point $(\rho_+ \ , \ V\ ) = (0, 1)$ and where $[[2]] + [[3]] > 0$ is satisfied [3].

Now we first show the following *Lemma:*

Lemma

Once the shell enters the core-region at $t = t_*$, it bounces back without reaching zero-radius. Indeed, $\rho_+(t)$ is bounded from below as

$$\rho_+(t) > \rho_+(t_*)\sqrt{1 - \dot{\rho}_+(t_*)^2} \ .$$

Proof:

The claim follows from the fact that

$$\ddot{\rho} = \frac{1 - \dot{\rho}^2}{\rho} \quad \text{, with } \rho(0) = a \ , \ \dot{\rho}(0) = -b \ \ (a > 0, 1 > b > 0)$$

is exactly solved as $\rho(t) = ((t - ab)^2 + +a^2(1 - b^2))^{1/2}$, so that $\rho_+(t) > \rho(t) \geq a\sqrt{1 - b^2} > 0$.

Then we immediately arrive at

Theorem

The shell never reaches $\rho_+ = 0$ irrespective of its initial conditions.

Proof:

If the shell could have ever reached $\rho_+ = 0$, it should have been through the core-region. However, *Lemma* indicates that this never happens.

The above *Theorem* reveals that the singularity-formation "solution" claimed by the other authors is actually not a solution and should be discarded. The reader is referred to Ref. [3] for more detailed arguments. Here it suffices to give a simple example which is analogous to the present case: Let us consider a differential equation,

$$\ddot{y} = 1 \ , \tag{17}$$

the general solution of which is the one indicating a uniform positive acceleration. However, if we artificially rewrite Eq.(17) as

$$\ddot{y} = \beta / \dot{y} \ , \tag{18}$$

with

$$\beta = \dot{y} \ , \tag{19}$$

a formal contracting solution for Eq.(18) emerges as

$$y(t) = y_0 - \frac{2\sqrt{2}}{3}\sqrt{\beta} t^{3/2} \ , \tag{20}$$

provided that β is assumed to be a constant. However, the solution Eq.(20) is in any case inconsistent: Noting that $\beta := \dot{y} = -\sqrt{2\beta t}$, it follows that $\beta = 2t$ so that it contradicts with the starting assumption of β being a constant.

The singularity-formation "solution" is obtained by the same trick as Eqs.(18) and (19). Though much more sophisticated arguments are required than this simple example, it turns out that the singularity-formation "solution" should be discarded in a more or less similar manner [3].

5 Summary

In this paper, we have investigated the dynamics of a cylindrical shell-contraction model with rotational pressure. As far as the weak energy condition is satisfied outside the shell, the collapsing shell always bounces back at some point irrespective of the initial conditions, and escapes from the singularity formation. This result reveals that the singularity-formation solution claimed by the preceding authors is not a relevant solution and that it should be discarded. This bouncing behavior is compatible with the results of other cylindrical shell-collapse models and confirms once again the essential importance of the rotational effect in the gravitational collapse.

The two conjectures ("the cosmic censorship conjecture" and "The hoop conjecture") combined together imply that any sufficiently elongated object (made of reasonably natural matter) should evolve regularly without any singularity formation. The present result is compatible with this statement, and along with other known models with cylindrical symmetry, suggests the stable validity of the two conjectures for wider class of situations.

This work has been supported by the Japan Ministry of Education, Culture, Sports, Science and Technology with the grant #14740162.

References

[1] R. Penrose, Riv. Nuovo Cimento **1** (Numero Special), 252 (1969).

[2] K. S. Thorne, in *Magic Without Magic: John Archibald Wheeler* (J. Klauder (ed.), Freeman, San Francisco, 1972), p.231.

[3] M. Seriu, Phys. Rev. D**69**, #124030 (2004).

[4] For basic accounts on the spherical models of gravitational contractions, see e.g. P.S. Joshi, *Global Aspects in Gravitation and Cosmology* (Clarendon, Oxford, 1993).

[5] S. L. Shapiro and S. A. Teukolsky, Phys. Rev. Lett. **66**, 994 (1991).

[6] T. A. Apostolatos and K. S. Thorne, Physical Review D**46**, 2435 (1992).

[7] P. R. C. T. Pereira and A. Wang, Physical Review D**62**, #124001 (2000).

[8] P. R. C. T. Pereira and A. Wang, Physical Review D**67**, #129902(E) (2003).

[9] For a standard account of the extrinsic curvature, see e.g. R. M. Wald, *General Relativity* (University of Chicago Press, Chicago, 1984), Appendix E.2. Note, however, that we are now concerned with a *timelike* surface and not a *spacelike* surface, and some care is needed for dealing with a timelike surface.

[10] W. Israel, Il Nuovo Cimento **B44**, 1 (1966); **B48**, 463(E) (1967).

[11] K. S. Thorne, Phys. Rev. **138**, B251 (1965).

THE DEUTERON AND THE BIG BANG

MASATAKA MIZUSHIMA [a]

[a] *Department of Physics, University of Colorado, Boulder, Colorado 80309 U.S.A.*
mizushima@colorado.edu

Abstract

Following Einstein's theory of general relativity, the gravitational acceleration of deuteron is shown to be less than that given by Newton's theory. During the primordial nucleosynthesis, deuterons were more important than proposed by the original Big-Bang theory. But these deuterons escaped out of the gravitational fields of the solar system.

1 Gravitational Acceleration of the Deuteron

Newton, in his equation of motion, assumed that the inertial mass is equal to the gravitational mass. Thus everything located at a distance r from a gravitational source with mass M is accelerated toward the source with a gravitational acceleration g,

$$\frac{d^2r}{dt^2} = g. \tag{1}$$

Following Einstein's general relativistic equation of motion [1], using the Schwarzschild metric, we obtain [2] an expression of this gravitational acceleration as

$$g = \left(1 - \left(\frac{dr}{cdt}\right)^2\right) g_N, \tag{2}$$

neglecting a correction factor on the speed of light c. Because Einstein assumed that

$$(ds)^2 = g_{00}(cdt)^2 - g_{\alpha\alpha}(dx^\alpha)^2, \tag{3}$$

B.G. Sidharth, F. Honsell and A. de Angelis. (eds.), Frontiers of Fundamental Physics, 103–106.

that is,

$$\left(\frac{dx^\alpha}{dt}\right)^2 \leq \frac{g_{00}}{g_{\alpha\alpha}}c^2 \sim c^2, \tag{4}$$

where the correction factor, on g_N in eq. (2), is a natural consequence of his theory, because, without this correction factor, the radial speed of a test particle can exceed the speed of light soon, as it falls toward the gravitational source.

Inside a deuteron nucleus, one proton and one neutron are orbiting around each other. However, the ground state of the deuteron is mostly ^3S, which means that the quantum mechanical expectation value of dr/dt is zero. Because of the tensor component of the nuclear force, it is shown that a small contribution of a ^3D-state exists. But the correction factor due to this contribution, in eq. (2), is negligible.

The resonating excited state of the deuteron, d^*, must be an (isospin $=$ 1)-state, which means that its energy is about the same as the ground states of a di-proton and a di-neutron nucleus, both of which are unstable. In the di-proton, the Coulomb repulsive force is too strong, and the binding energy of the $n+n$ system is expected to be less than 0.05 MeV. Even though no di-neutron system is observed, we know a neutron star exists as a pulsar. We thus expect that an excited-state deuteron would exist. Actually, a deuteron coupled with a nonresonating γ ray would be virtually excited. We call such a state a dressed deuteron, d^*. The virtually excited (isospin $=$ 1)-state should be a ^3P-state, in which nucleons are orbiting around each other with an extremely high speed [3].

$$d + \gamma \leftrightarrow d^*. \tag{5}$$

The dressed deuteron is described by a wave function $\psi_P(\mathbf{r})e^{i\omega t} + \psi_S(\mathbf{r})$, where $\psi_P(\mathbf{r})$ and $\psi_S(\mathbf{r})$ are the space parts of the wave functions for the ^3P and ^3S states, respectively, and ω is the excitation energy in $\hbar$. When the intensity of the γ ray is very high the contributions of the ^3S and the ^3P states will be about the same to each other. Thus the expectationvalue of dr/cdt, the correction term of eq. (2), is about $R\omega/c$ where R is the average radius of the nucleon orbits. If the excited ^3P state is at the dissociative level, ω must be 2.73 MeV $= 4.15{\times}10^{21}$ rad/s. Assuming $R = 2 \times 10^{-15}$ m, we obtain the expectationvalue of dr/cdt to be 0.03 when the nucleon orbits are in the radial (vertical) plane.

2 The Big Bang

Gamow, Alpher, and others [4,5,6,7,8] proposed that nuclei were formed from primordial neutrons in a few minutes and were then frozen, as an *unfinished building-up process.* A neutron decays into a proton and an electron, $n \rightarrow p + el$, with a lifetime of 10.28 minutes [9]. But before all neutron decay, the newborn proton interacts with the surviving neutron, forming a deuteron, $p + n \rightarrow d$, in about 3 minutes. Reactions $p + d \rightarrow {}^3He$, $n + {}^3He \rightarrow \alpha$, and $d + d \rightarrow \alpha$ follow very quickly. Thus, the helium/proton ratio is expected to be about 0.3 at this stage of the primordial nucleosynthesis. This ratio is close to what we actually find in the universe [10].

For the abundance of heavy nuclei, the theory implies a relation [8]

$$n(A)\sigma(A) = n(A+1)\sigma(A+1), \tag{6}$$

where σ is the cross section for neutron capture and n is the abundance. The relative abundance of heavy nuclei in meterorites [11,12] can be explained by this simple formula, including the maxima at mass numbers A = 50, 82, and 126.

However, an almost fatal objection to this theory lies in the fact that nuclei with mass numbers 5 and 8 are all short-lived with lifetimes $3 \times 10^{-22}(^5Li)$, $7 \times 10^{-22}(^5He)$, $2 \times 10^{-20}(^8C)$, $0.77(^8B)$, $7 \times 10^{-17}(^8Be)$, $0.845(^8Li)$, and $0.119(^8He)$ in sec, respectively.

Bethe, in his famous theory of the production of solar energy, pointed out [13] that reactions $^3He + \alpha \rightarrow\ ^7Be$ and $^7Be + el \rightarrow\ ^7Li$ are possible. Therefore, the gap at the mass number 5 is no trouble to Gamow's theory. We propose that a large enough number of deuterons or/and dressed deuterons existed during the period of primodial nucleosynthesis so that a reaction

$$d^* +\ ^7Li \rightarrow\ ^9Be, \tag{7}$$

was possible. After ^9Be, synthesis of heavier nuclei follows.

It is known that deuterons exist on Earth (as ocean water) with the abundance ratio, d/p, of 1.5×10^{-4}. However, this abundance ratio must be far below this number in the sun [14]. This isotope ratio in the upper atmosphere of Venus is 0.016 (which is about 100 times the corresponding number in ocean water), although the isotope ratios for Ne, Ar, C, and O are almost the same as those on Earth [15]. On Jupiter, the d/p ratio is 2.8×10^{-5} to 7.5×10^{-5} as CH_3D and is 2.1×10^{-5} as HD, and in meteorites, it is 1.3×10^{-4} to 2.0×10^{-4} [16]. Even on Earth, the d/p ratio varies from 3.2×10^{-5} to 1.85×10^{-4} as HD. The temperature of the sun is so high that some deuterons dissociate. But after the primordial period, the outside region of the solar system cooled down so that some deuterons remained on the planets. The large variations in the isotope ratio, however, suggest that some deuterons have escaped from the gravitational fields of planets (and the sun) by now, because the gravitational acceleration of the dressed deuteron is less than the standard acceleration g_N, as shown above.

The big bang may have taken place at the time of the galactic nuclear explosions that have created galactic spiral arms [17,18]. The Milky Way galaxy has four pairs of spiral arms. Therefore, there were four big bangs in our galaxy during the past. Near the black hole, the gravitational field was very strong, and that situation produced the *yelm*, the primordial compressed neutron gas, from which the big bang process started.

References

[1] A. Einstein, Ann. Phys. **49**, 769 (1916).

[2] G. C. McVittie, *General Relativity and Cosmology*, section 5-1, John Wiley & Sons Inc. New York (1956).

[3] M. Mizushima, Hadronic J. **26**, 5 (2003).

[4] G. Gamow, Phys. Rev. **70**, 572 (1946).

[5] R. A. Alpher, H. Bethe, and G. Gamow, Phys. Rev. **73**, 803 (1948).

[6] G. Gamow, Phys. Rev. **74**, 505 (1948).

[7] R. A. Alpher, R. Herman, and G. Gamow, Phys. Rev. **74**, 1198 (1948).

[8] R. A. Alpher and G. Gamow, Proc. Natl. Acad. Sci. U.S. **61**, 363 (1968).

[9] W. Mampe, P. Ageron, C. Bates, J. M. Pendlebury, A. Steyerl, Phys. Rev. Lett. **63**, 2173 (1989).

[10] C. E. Rolfs and W. S. Rodney, *Cauldrons in the Cosmos*, The Univ. Chicago Press, Chicago (1988).

[11] H. E. Suess and H. C. Urey, Revs. Modern Phys. **28**, 53 (1956).

[12] A. G. W. Cameron (1963) as introduced in *Nucleosynthesis* by W. D. Arnett *et al.*, Gordon and Breach, New York (1965).

[13] H. A. Bethe, Phys. Rev. **55**, 434 (1939).

[14] V. A. Krat and L. M. Pravdjuk, Solar Phys. **61**, 279 (1979).

[15] T. M. Donahue and J. B. Pollack *29. Origin and Evolution of the Atmosphere of Venus* in *Venus* ed. by D. M. Hunten *et al*, Univ. of Arizona Press, Tuscon AZ (1983).

[16] D. N. Scramm and R. V. Wagoner, *What Can Deuterium Tell Us?* Phys. Today, Dec. 41 (1974).

[17] M. Mizushima, *Gravitational Wave and Spiral Galaxy (Gravito-Radiative Force.* p. 179 in *Frontiers of Fundamental Physics 4*, ed. by Sudharth and Altaisky, Kluwer Academic, New York (2001).

[18] M. Mizushima, *Gravito-Radiative Force and the Spiral Galaxy* in *4th International Conference on Dynamic Systems and Applications*, May 21-24, 2003, Atlanta GA.

ASTROPHYSICAL APPLICATIONS OF THE THEORY OF SCALE RELATIVITY

LAURENT NOTTALE [a]

[a] *CNRS, LUTH, Observatoire de Paris-Meudon, F-92195 Meudon Cedex, France*

Abstract

In the framework of the theory of scale relativity, one considers the possibility that space-time be not only curved but also fractal at large scales. Then one can show that the equation of dynamics in such a space (i.e. the geodesics equation), takes the form of a generalized Schrödinger equation. This approach allows one to suggest new solutions to the problems of the formation and evolution of gravitational structures and of the 'missing mass'. Indeed, such a description leads to a natural self-organization process and involves the apparition of a new potential energy which is a manifestation of the fractal geometry (in similarity with the Newtonian potential being a manifestation of curvature). We suggest that this 'dark potential' is the cause of the various effects that have up to now been attributed to a postulated non-baryonic dark matter. We conclude by a brief survey of various observational effects that come in support of such a proposal.

1 Introduction

In its present acceptance, gravitation is understood as the various manifestations of the geometry of space-time at large scales. Up to now, in the framework of Einstein's theory [1], this geometry was considered to be Riemannian, i.e. curved. However, in the new framework of scale relativity, the geometry of space-time is assumed to be characterized not only by curvature, but also by fractality [2] beyond a new relative time-scale and/or space-scale of transition, which is an horizon of predictibility for the classical deterministic description. While the concept of

B.G. Sidharth, F. Honsell and A. de Angelis. (eds.), Frontiers of Fundamental Physics, 107–114.

fractal space-time has been first introduced in connection with the microscopic quantum theory [3, 4, 5, 6, 7], it can indeed be also applied to the macroscopic realm, but with a different interpretation. Now, fractality manifests itself, in the simplest case, in terms of the appearance of a new scalar field. We have suggested that this new field leads to spontaneous self-organization and may also be able to explain [11, 18], without additional non-barionic matter, the various astrophysical effects which have been, up to now, attributed to unseen "dark" matter .

2 Gravitational Schrödinger equation

Let us first briefly recall the basics of the scale-relativistic theoretical approach. Under three general conditions, namely, {(i) infinity of geodesics (which leads to introduce a non-deterministic velocity field), (ii) fractal dimension $D_F = 2$ of each geodesic, on which the elementary displacements are described in terms of the sum $dX = dx + d\xi$ of a classical, differentiable part dx and of a fractal, non-differentiable fluctuation $d\xi$, (iii) two-valuedness of the velocity field, which is a consequence of time irreversibility at the infinitesimal level issued from non-differentiability}, one can construct a complex covariant derivative that reads

$$\frac{\bar{d}}{dt} = \frac{\partial}{\partial t} + \mathcal{V}.\nabla - i\mathcal{D}\Delta \,, \tag{1}$$

where $\mathcal{D}$ is a parameter that characterizes the fractal fluctuation, which is such that $< d\xi^2 >= 2\mathcal{D}dt$, and where the classical part of the velocity field, $\mathcal{V}$ is complex as a consequence of condition (iii) (see [8] for a recent complete demonstration).

Then this covariant derivative, that describes the non-differentiable and fractal geometry of space-time, can be combined with the covariant derivative of general relativity, that describes the curved geometry. We shall briefly consider in what follows only the Newtonian limit. In this case the equation of geodesics keeps the form of Newton's fundamental equation of dynamics in a gravitational field,

$$\frac{\bar{D}\mathcal{V}}{dt} = \frac{\bar{d}\mathcal{V}}{dt} + \nabla\left(\frac{\phi}{m}\right) = 0, \tag{2}$$

where ϕ is the Newtonian potential energy. Introducing the action S, which is now complex, and making the change of variable $\psi = e^{iS/2m\mathcal{D}}$, this equation can be integrated under the form of a generalized Schrödinger equation [3]:

$$\mathcal{D}^2\Delta\psi + i\mathcal{D}\frac{\partial}{\partial t}\psi - \frac{\phi}{2m}\psi = 0. \tag{3}$$

Since the imaginary part of this equation is the equation of continuity (Sec. 3), and basing ourselves on our description of the motion in terms of an infinite family of geodesics, $P = \psi\psi^\dagger$ naturally gives the probability density of the particle position [8]. This result is supported by Hermann's numerical simulations [25].

Even though it takes this Schrödinger-like form, equation (3) is still in essence an equation of gravitation, so that it must come under the equivalence principle [13,

19], i.e., it is independent of the mass of the test-particle. In the Kepler central potential case ($\phi = -GMm/r$), GM provides the natural length-unit of the system under consideration. As a consequence, the parameter $\mathcal{D}$ reads:

$$\mathcal{D} = \frac{GM}{2w}, \tag{4}$$

where w is a constant that has the dimension of a velocity. The ratio $\alpha_g = w/c$ actually plays the role of a macroscopic gravitational coupling constant [19, 17].

3 Formation and evolution of structures

Let us now compare our approach with the standard theory of gravitational structure formation and evolution. By separating the real and imaginary parts of the Schrödinger equation we obtain, after a new change of variables, respectively a generalized Euler-Newton equation and a continuity equation, namely,

$$m\,(\frac{\partial}{\partial t} + V \cdot \nabla)V = -\nabla(\phi + Q), \qquad \frac{\partial P}{\partial t} + \mathrm{div}(PV) = 0, \tag{5}$$

where V is the real part of the complex velocity field $\mathcal{V}$. In the case when the density of probability is proportional to the density of matter, $P \propto \rho$, this system of equations is equivalent to the classical one used in the standard approach of gravitational structure formation, except for the appearance of an extra potential energy term Q that writes:

$$Q = -2m\mathcal{D}^2 \frac{\Delta\sqrt{P}}{\sqrt{P}}. \tag{6}$$

The existence of this potential energy, (which amount to the Bohm potential in standard quantum mechanics) is, in our approach, readily demonstrated and understood: namely, it is the very manifestation of the fractality of space, in similarity with Newton's potential being a manifestation of curvature.

In the case when actual particles achieve the probability density distribution (structure formation), we have $\rho = m_0 P$; then the Poisson equation (i.e., the field equation) becomes $\Delta\phi = 4\pi G m m_0 \psi\psi^\dagger$ and it is therefore strongly interconnected with the Schrödinger equation (which is here a new form for the equation of motion). Such a system of equations is similar to that encountered in the description of superconductivity (Hartree equation). We expect its solutions to provide us with general theoretical predictions for the structures (in position and velocity space) of self-gravitating systems at multiple scales [10, 18]. This expectation is already supported by the observed agreement of several of these solutions with astrophysical observational data [3, 13, 17, 14, 15, 20, 21, 16].

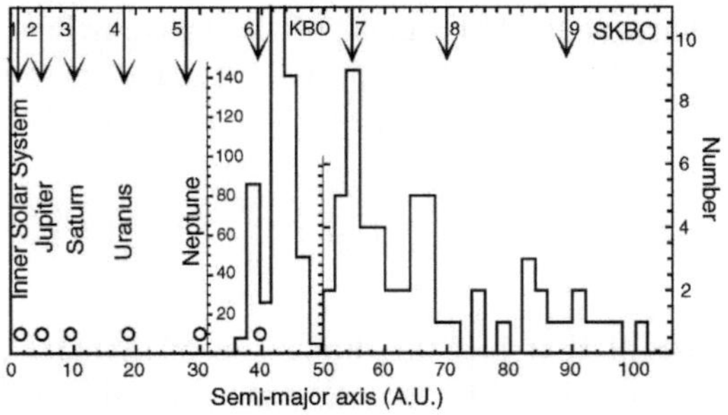

Figure 1. Distribution of the semi-major axis of Kuiper belt objects (KBO) and scattered Kuiper belt objects (SKBO), compared with the theoretical predictions (arrows) of probability density peaks for the outer solar system [18] (see text). The existence of probability density peaks for the Kuiper belt at ≈ 40, 55, 70, 90 AU, etc..., has been theoretically predicted before the discovery of these objects [22], and it is now supported by the observational data.

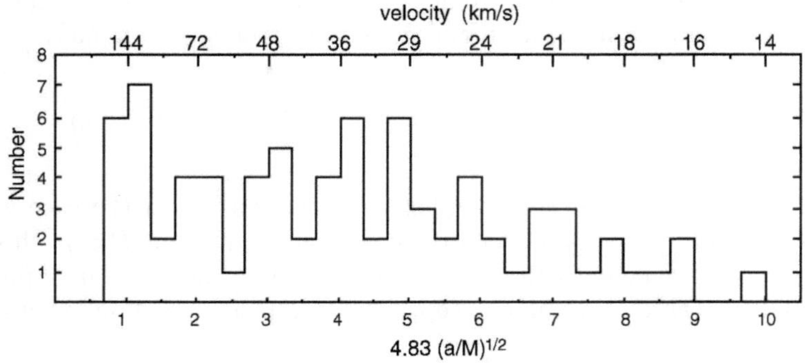

Figure 2. Observed distribution of the semi-major axes of recently discovered exoplanets and inner solar system planets, compared with the theoretical prediction. One predicts the occurence of peaks of probability density for semimajor axes $a_n = GM(n/w_0)^2$, where n is integer, M is the star mass and $w_0 = 144.7 \pm 0.7$ km/s is a gravitational coupling constant (see text). The probability to obtain such an agreement by chance is $P = 4 \times 10^{-5}$.

4 Examples of applications to astrophysics

The theory has been able to predict in a quantitative way several new effects in the domain of gravitational structuring [18]. Moreover, these predictions have been successfully checked in various systems on a large range of scales and in terms of a common gravitational coupling constant (or one of its multiples or submultiples) whose value averaged on these systems was found to be $w_0 = c\,\alpha_g = 144.7 \pm 0.7$ km/s [13]. Indeed, new structures have been theoretically predicted, then checked by the observational data in a statistically significant way, for our solar system, including distances of planets [3, 14] and satellites [16], sungrazer comet peri-

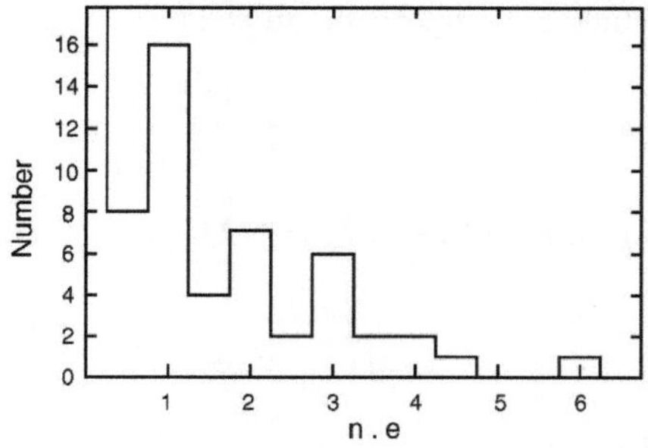

Figure 3. Observed distribution of the eccentricities of exoplanets. The theory predicts that the product of the eccentricity e by the quantity $\tilde{n} = 4.83(a/M)^{1/2}$, where a is the semi-major axis and M the parent star mass, should cluster around integers. The data support this theoretical prediction at a probability level $P = 10^{-4}$ [18].

helions [18], obliquities and inclinations of planets and satellites [20], exoplanets semi-major axes [13, 17] (see Fig. 2) and eccentricities [18] (see Fig. 3), including planets around pulsars, for which a high precision is reached [13, 21], double stars [15], planetary nebula [18], binary galaxies [9] (see Fig. 4), our local group of galaxies [18], clusters of galaxies and large scale structures of the universe [15, 18].

Let us briefly consider the application of the theory to the formation of planetary systems. The standard model of formation of planetary systems can be reconsidered in terms of a fractal description of the motion of planetesimals in the protoplanetary nebula. On length-scales much larger than their mean free path, we have assumed [3] that their highly chaotic motion satisfy the three conditions upon which the derivation of a Schrödinger equation is based (large number of trajectories, fractality and time symmetry breaking).

This description applies to the distribution of planetesimals in the proto-planetary nebula at several embedded levels of hierarchy. Each hierarchical level (k) is characterized by a length-scale defining the parameter $\mathcal{D}_k$ (and therefore the velocity w_k) that appears in the generalized Schrödinger equation describing this sub-system. This hierarchical model has allowed us to recover the mass distribution of planets and small planets in the inner and outer solar systems [14]. It is generally supported by the structure of our own solar system, which is made of several subsystems embedded one in another, namely:

***The Sun**. Through Kepler's third law, the velocity $w = 3 \times 144.7 = 434.1$ km/s is very closely the Keplerian velocity at the Sun radius ($R_\odot = 0.00465$ AU corresponds to $w_\odot = 437.1$ km/s). Moreover, one can also apply our approach to the organization of the sun surface itself. One expect the distribution of the various relevant physical quantities that characterize the solar activity at the Sun surface (sun spot number, magnetic field, etc...) to be described by a wave function whose stationary solutions read $\psi = \psi_0\ e^{iEt/2m\mathcal{D}}$.

The energy E results from the rotational velocity and, to be complete, should also include the turbulent velocity, so that $E = (v_{rot}^2 + v_{turb}^2)/2$. This means that we expect the solar surface activity to be subjected to a fundamental period:

$$\tau = \frac{2\pi m \mathcal{D}}{E} = \frac{4\pi \mathcal{D}}{v_{rot}^2 + v_{turb}^2}, \tag{7}$$

The parameter $\mathcal{D}$ at the Sun radius is $\mathcal{D} = GM_\odot/2w_\odot$, then we obtain:

$$\tau = \frac{2\pi G M_\odot}{w_\odot (v_{rot}^2 + v_{turb}^2)}. \tag{8}$$

The average sideral rotation period of the Sun is 25.38 days, yielding a velocity of 2.01 km/s at equator [23]. The turbulent velocity has been found to be $v_{turb} = 1.4 \pm 0.2$ km/s [24]. Therefore we find numerically

$$\tau = (10.2 \pm 1.0) \text{ yrs}. \tag{9}$$

The observed value of the period of the Solar activity cycle, $\tau_{obs} = 11.0$ yrs, supports this theoretical prediction. We shall in future works test this proposal by a more detailed study of the activity of the Sun and of other stars.

***The intramercurial system**, organized on the constant $w_\odot = 3 \times 144 = 432$ km/s. The existence of an intramercurial subsystem is supported by various stable and transient structures observed in dust, asteroid and comet distributions (see [18]). We have in particular suggested the existence of a new ring of asteroids, the 'Vulcanoid belt', at a preferential distance of about 0.17 AU from the Sun.

***The inner solar system** (earth-like planets), organized with a constant $w_i = 144$ km/s (see Fig. 2).

***The outer solar system** (Jovian planets), organized with a constant $w_o = 144/5 = 29$ km/s (see Fig. 1), as deduced from the fact that the mass peak of the inner solar system lies at the Earth distance ($n = 5$). The recently discovered Kuiper and scattered Kuiper belt objects (Fig. 1) show peaks of probability at $n = 6$ to 9 [18], as predicted before their discovery [22].

We have suggested more than ten years ago [3, 22], before the discovery of exoplanets, that the theoretical predictions from this approach should apply to all planetary systems, not only our own solar system. Meanwhile more than 120 exoplanets have been discovered, and the observational data support this prediction in a highly statistically significant way (see [13, 17, 18] and Figs. 2 and 3).

A full account of this new domain would be too long to be included in the present contribution. We have given here only few typical examples of these effects (see the figures) and we refer the interested reader to the review paper Ref. [18] and references therein for more detail.

5 Possible solution to the "dark matter" problem

In the case (ii) of isolated test particles, the density of matter ρ may be nearly zero while the probability density P does exist, but only as a virtual quantity that determines the potential Q. In this situation, even though there is no or few

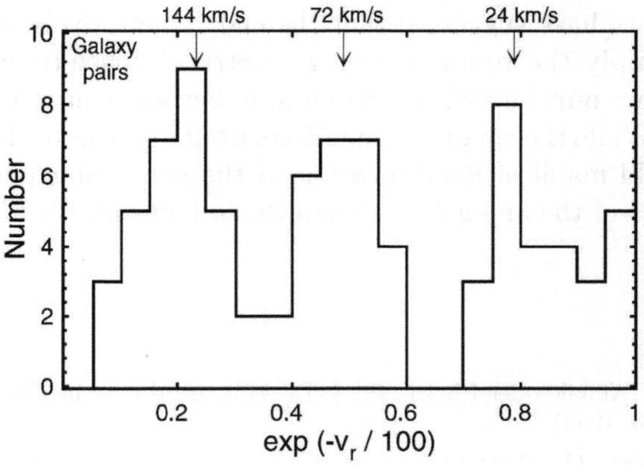

Figure 4. Deprojection of the intervelocity distribution of galaxy pairs (Tricottet and Nottale, reported in [18]) from the Schneider-Salpeter catalog with precision redshifts. The main probability peak is found to lie at 144 km/s (plus secondary peaks at 72=144/2 km/s and 24=144/6 km/s), in agreement with the exoplanet and inner solar system structuring (see Fig. 2).

matter at the point considered (except the test particles that are assumed to have a very low contribution), the effects of the potential Q are real (since it is the result of the structure of the geodesics two-fluid).

We have therefore suggested [11, 18, 12] that this extra scalar field, which is a manifestation of the fractality of space, may be responsible for the various dynamical and lensing effects which are usually attributed to unseen "dark matter". This interpretation is supported by the fact that, for a stationary solution of the gravitational Schrödinger equation, one gets the general energy relation $E = \phi + Q + \frac{1}{2}mV^2$, where E/m can take only quantized values.

This result can be applied, as an example, to the motion of bodies in the outer regions of spiral galaxies. In these regions there is practically no longer any visible matter, so that the Newtonian potential (in the absence of additional dark matter) is Keplerian. While the standard Newtonian theory predicts for the velocity of the halo bodies $v \propto \phi^{1/2}$, i.e. $v \propto r^{-1/2}$, we predict $v \propto |(\phi + Q)/m|^{1/2}$, i.e., $v =$ constant. More specifically, assuming that the gravitational Schrödinger equation is solved for the halo objects in terms of the fundamental level wave function, one finds $Q_{pred} = -(GMm/2r_B)(1 - 2\,r_B/r)$, where $r_B = GM/w_0^2$. This is the result systematically observed in spiral galaxies (i.e., flat rotation curves) which has motivated (among other effects) the assumption of the existence of dark matter.

6 Conclusion

We have recalled in the present contribution how the theory of scale relativity is able to yield equations that describe a natural tendency to make structures, as a very consequence of the fractality of space.

Moreover, we have suggested that the effects tentatively attributed to unseen matter are simply the result of this geometry of space-time. In this proposal, space-time is not only curved but also fractal beyond some given relative time and space-scales. While the curvature manifests itself in terms of the Newton potential, fractality would manifest itself in terms of the new scalar potential Q, and then finally in terms of the anomalous dynamics and lensing effects.

References

[1] A. Einstein, Annalen der Physik **49**, 769 (1916), translated in *The Principle of Relativity* (Dover, 1923, 1952)

[2] B. Mandelbrot, *The Fractal Geometry of Nature* (Freeman, San Francisco, 1982)

[3] L. Nottale, *Fractal Space-Time and Microphysics: Towards a Theory of Scale Relativity* (World Scientific, Singapore, 1993)

[4] L. Nottale and J. Schneider, J. Math. Phys. **25**, 1296 (1984)

[5] G. N. Ord, J. Phys. A: Math. Gen. **16**, 1869 (1983)

[6] L. Nottale, Int. J. Mod. Phys. A **4**, 5047 (1989)

[7] M. S. El Naschie, Chaos, Solitons & Fractals **2**, 211 (1992)

[8] M.N. Célérier & L. Nottale, J. Phys. A: Math. Gen. **37**, 931(2004)

[9] L. Nottale, Chaos, Solitons & Fractals **7**, 877 (1996)

[10] L. Nottale, Astron. Astrophys. 327, 867 (1997)

[11] L. Nottale, in *Frontiers of Fundamental Physics*, Proceedings of Birla Science Center Fourth International Symposium, 11-13 december 2000, edited by B. G. Sidharth and M. V. Altaisky, (Kluwer Academic, 2001) p. 65

[12] L. Nottale, Chaos, Solitons and Fractals 16, 539 (2003)

[13] L. Nottale, Astron. Astrophys. Lett. 315, L9 (1996)

[14] L. Nottale, G. Schumacher and J. Gay, Astron. Astrophys. 322, 1018 (1997)

[15] L. Nottale and G. Schumacher, in *Fractals and beyond: complexities in the sciences*, edited by M. M. Novak (World Scientific, 1998) p. 149

[16] R. Hermann, G. Schumacher, and R. Guyard, Astron. Astrophys. **335**, 281 (1998)

[17] L. Nottale, G. Schumacher, E.T. Lefèvre, Astron. Astrophys. 361, 379 (2000)

[18] D. Da Rocha, L. Nottale, Chaos, Solitons & Fractals 16, 565 (2003)

[19] A.G. Agnese, R. Festa, Phys. Lett. A227, 165 (1997)

[20] L. Nottale, Chaos, Solitons & Fractals **9**, 1035 (1998)

[21] L. Nottale, Chaos, Solitons & Fractals **9**, 1043 (1998)

[22] L. Nottale, in "Chaos and Diffusion in Hamiltonian Systems", ed. D. Benest and C. Froeschlé, (Frontières, 1994), p. 173

[23] J.C. Pecker and E. Schatzman, Astrophysique générale (Masson, Paris, 1959)

[24] K.R. Lang, Astrophysical Formulae (Springer-Verlag 1980)

[25] R. Hermann, J. Phys. A: Math. Gen 30, 3967 (1997)

MAKING MAPS OF THE REES-SCIAMA EFFECT

MÀRIUS J. FULLANA I ALFONSO [a,b,c],
DIEGO P. SÁEZ MILÁN [d]

[a] *Institut de Matemàtica Multidisciplinar Univ. Politècnica de València, 46022 València*
[b] *Departament de Matemàtica Aplicada, Univ. Politècnica de València, 46022 València*
[c] *email adress:mfullana@mat.upv.es*
[d] *Departament d'Astronomia i Astrofísica, Univ. de València, 46100 Burjassot, Spain*

Abstract

The anisotropies of the Cosmic Microwave Background are caused by several physical phenomena. Those anisotropies produced by cosmological inhomogeneities placed far away from (close to) the last scattering surface are called secondary (primary). We focus our attention on one of the secondary anisotropies: the Rees-Sciama (RS) effect due to the time variation of the peculiar gravitational potential of non-linear cosmological inhomogeneities. A full description of a distribution of these structures requires N-body simulations with appropriate resolutions and box sizes. CMB photons cross a periodic universe formed by repeated N-body boxes. Periodicity leads to a magnification of the RS effect. A method to avoid this magnification is described and, then, it is applied to simulate maps of the RS effect which are being statistically analysed. RS is a subdominant effect which appears superimposed to other ones in nature. Each one of these effects must be studied in detail to make possible an eventual separation of components in maps of future CMB experiments as PLANCK.

B.G. Sidharth, F. Honsell and A. de Angelis. (eds.), Frontiers of Fundamental Physics, 115–122.

1 Introduction

The CMB radiation was discovered in 1965 [1]. A few years before, it had been teoretically predicted in the framework of the Big Bang cosmological model (see [2] and references therein). This radiation plays a crucial role in modern cosmology and, consequently, many experiments have been designed to measure its spectrum, polarisation, and anisotropy. These measurements are to be compared with theoretical predictions based on a cosmological background containing some type of cosmological structures. Now, a few words about the nature and importance of the CMB are presented.

In the primordial universe, baryonic matter was strongly coupled to radiation. Protons and electrons were not confined in atoms, in other words, these particles were free in the so-called primeval plasma. In that situation, photons could not travel freely because they frequently interacted with the plasma free electrons and, consequently, the universe was opaque; afterwards, as a result of expansion, the universe cooled and became transparent. In fact, when the temperature was low enough ($\simeq 3500\ K$), protons captured electrons and formed stable hydrogen atoms (recombination). The absence of free electrons led to the decoupling between matter and radiation at redshift $Z \approx 10^3$. The recombination-decoupling process lasted for a time corresponding to a redshift increment $\Delta Z \approx 80$. This short duration allows us to consider a Last Scattering Surface (LSS) separating the opaque ionized Universe from the transparent neutral one. From that period on, CMB photons travelled almost freely. Astrophysical phenomena could have produced a reionization, which would have altered the CMB radiation at low redshifts ($Z \simeq 20$).

Now, some comments about the information supplied by the study of the CMB are made. This radiation appears to be very isotropic (see for example [3]) and, furthermore, it has a black-body spectrum at temperature $T \approx 2.73K$, and a low level of linear polarisation. Small angular temperature fluctuations (anisotropies) have been detected in the CMB (see for instance [4]). In a fully homogeneous universe, the CMB would be absolutely isotropic. The anisotropies are due to the fact that photons coming along distinct directions undergo different frequency shifts, it occurs because these photons cross distinct inhomogeneous regions. Hence, CMB anisotropies give information about the distribution of inhomogeneities (large scale structure) evolving in the universe. On account of these considerations, it is obvious that the observation of the CMB anisotropies allows us to look for the best theory explaining structure formation in the framework of the Big Bang model.

2 The anisotropies of the CMB

The small angular dependence of the CMB temperature can be understood taking into account the following formula for the temperature contrast:

$$\frac{\Delta T}{T}(\vec{n}) \;=\; \left(\frac{\Delta T}{T}(\vec{n})\right)_E \;+\; \vec{n}\cdot(\vec{v}_O - \vec{v}_E) \;-\; \frac{1}{3}(\phi_O - \phi_E) \;-\; 2\int \frac{\partial \phi}{\partial t} dt \qquad (1)$$

where $\vec{n}$ is the unit vector of the line of sight, $\vec{v}$ is the peculiar velocity, and ϕ the peculiar gravitational potential. E stands for emission and O for observation. The terms of the r.h.s. of Eq. (1) have the following interpretation:

- $\left(\frac{\Delta T}{T}\right)_E$ is the initial anisotropy due to temperature fluctuations on the LSS produced by previous evolution.
- $\vec{n} \cdot \vec{v}_E$ is a Doppler effect due to peculiar motions on the LSS.
- $\vec{n} \cdot \vec{v}_O$ is another Doppler effect due to the motion of the observer with respect to the background universe. This component has dipolar structure.
- $-\frac{1}{3}(\phi_O)$ is a monopole due to the peculiar gravitational potential at the position observer.
- $\frac{1}{3}(\phi_E)$ is the Sachs-Wolfe effect due to the peculiar gravitational potential created by the inhomogeneities located close to the LSS.
- $-2\int \frac{\partial \phi}{\partial t} dt$ is the contribution due to time variations of ϕ, this term is called the integrated Sachs-Wolfe (Rees-Sciama) effect when the potential is created by large scale linear structures (non-linear structures).

Hence, the anisotropy given by Eq. (1) is the superimposition of various effects. Among them, the RS effect is a secondary subdominant anisotropy produced by the time variation of the potential ϕ created by non-linear structures at $Z \leq 50$. It is given by the following formula:

$$\frac{\Delta T}{T} \approx -2 \int_{t_{50}}^{t_0} \frac{\partial \phi(\vec{x}, t)}{\partial t} dt \tag{2}$$

The evolution of strongly non-linear structures is a hard problem. Only two methods are known to deal with it: in the first one, only one symmetric structure is described using exact solutions of Einstein equations, and in the second method, a realistic statistical distribution of asymmetric structures is evolved using N-body simulations. Several studies of the anisotropy generated by a unique symmetric cosmological inhomogeneity (Great Attractor, voids) have been done. Two main models have been used for this purpose: the Lemaître-Tolman-Bondi model (see for example [2]) or the Swiss-Cheese model (see for instance [5]). Although these models led to important qualitative ideas about the Rees-Sciama effect with a moderated computational cost, only N-body simulations describing a set of inhomogeneities can lead to a realistic prediction of the RS effect. These predictions are based on a periodic Universe filled by identical N-body boxes where CMB photons move along null geodesics. By this reason ray-tracing through N-body simulations is first used to get maps of the RS effect and, then, these maps are statistically analysed to estimate deviations from gaussianity; see for example [6, 7]; in this last paper, the ray-tracing procedure used here was described in detail. It is based on the existence of a preferred direction minimising the magnification of the RS effect in a periodic universe.

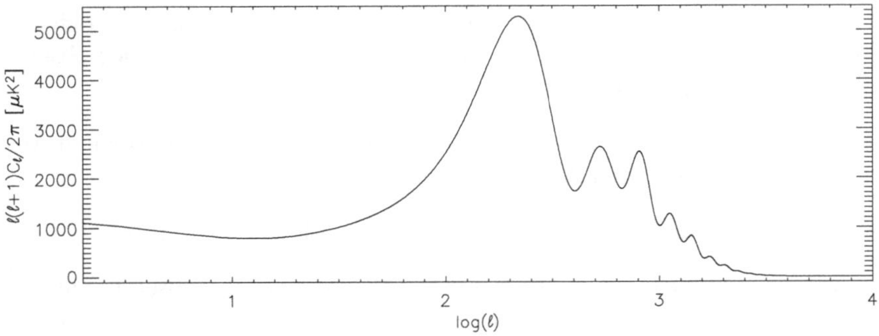

Figure 1. APS of the dominant anisotropies. Model parameters: $\Omega_0 = 1.0$, $\Omega_\Lambda = 0.73$, $\Omega_b = 0.04$, $\Omega_c = 0.23$, $h = 0.71$ and reionization at $Z = 17$.

In this paper, the Angular Power Spectrum (APS) of the CMB is calculated using the code CMBFAST [8] for the model suggested by WMAP observations [9]. This APS is displayed in Fig. 1 and no subdominant anisotropies as RS, lens deformations, Sunyaev-Zel'dovich effect and so on have been considered at all.

3 Ray-tracing methods

In standard ray-tracing procedures, photons move along null geodesics in a periodic universe. Identical N-body boxes are used to tile the space at any given time (periodicity). All the boxes evolve in the same way according to N-body predictions. Equation (2) must be integrated along each null geodesic (direction) in order to estimate the RS effect. This integration must be performed along a set of appropriate directions to get a RS map.

The main problems with this type of ray-tracing is that CMB photons can be affected by the same structures in successive boxes and, consequently, a wrong magnification of the resultingwidth=9.0cm,height=9.0cm,angle=270 effect appears. This magnification depends on the propagation direction and it is particularly great for directions parallel to the box edges. In [10] one of these ray-tracing procedures involving multiple plane projection is described. It is the classical method.

In order to avoid the above problem, the boxes could be moved (roto-traslations) when the photon crosses from a box to the next one. Thus, a given photon would move through separated independent regions in successive boxes. Nevertheless, this method introduces unavoidable discontinuities in the crossing points, which affect the integral of the r.h.s. of Eq. (2).

Another procedure was proposed by [11], it is the tiling ray-tracing method, in which a different N-body simulation describes structure evolution in each box crossed by the CMB photon. In this way, the above periodicity effect is surmounted, but discontinuities at crossing points appear as it occurs for roto-traslations. Furthermore, many N-body simulations –large computational cost– must be performed.

Recently, another method based on a unique N-body simulation has been proposed [7]. It is based on the existence of preferred directions. A photon moving along these directions crosses independent regions of the simulation box through thousands of Megaparsecs and, consequently, no magnification of the RS effect appears; at least, if an appropriate cut-off is performed to avoid scales larger than a certain spatial scale S_{max} satisfying the following conditions: (a) scales smaller than S_{max} are well evolved in the N-body simulation, and (b) the scale S_{max} is greater than the distance, L, between the regions crossed by the CMB photon in two neighbouring boxes. No discontinuities appear in the crossing points because only a simulation is used and no roto-traslations are performed.

Some comments about our simulations are now necessary. A PM (particle-mesh) N-body code designed by V. Quilis and D. Sáez (see [12] and references cited therein) is used. The initial redshift: $z_{in} = 5$ is assumed. The box size is: $L_{box} = 256\ Mpc$. The preferred direction is: $\theta = 77.2°$, $\phi = 12.6°$. The separation, L, between crossed regions in neighbouring boxes is: 52 Mpc. Spatial scale larger than S_{max} are eliminated in the peculiar gravitational potential (not in the N-body simulation). Various scales $S_{max} \leq 60\ Mpc$ have been tried. The photon crosses $\sim$ 30 boxes ($\sim$ 8000 Mpc) before re-entering a box by the zone where it was placed at the initial position (at $z_{in} = 5$). Two types of simulations have been performed, in the first one (hereafter poor resolution simulations) the cell size is of 2 Mpc and 128^3 particles of dark matter are used, whereas in the second type (hereafter medium resolution simulations) the cell size is 1 Mpc and the number of particles is 256^3. Higher resolution simulations are being developed but the computational cost is large. A set of directions ensures a uniform coverage of $\sim 2° \times 2°$ in the sky. A RS map is generated using these directions.

4 Results

Two particular RS simulations (preferred directions plus cut-off) are now displayed.

Fig. 2 corresponds to a low resolution simulation with the following characteristics: (1) spatial resolution 2 Mpc, (2) maximum spatial scale $S_{max} = 60\ Mpc$, (3) number of particles 128^3, (4) $1.82° \times 1.82°$ map, (5) maximum temperature contrast 1.92×10^{-6}, and (6) minimum temperature contrast -1.85×10^{-6}.

Fig. 3 is a particular middle resolution simulation with: (1) spatial resolution 1 Mpc, (2) maximum spatial scale 128 Mpc, (3) number of particles 256^3, (4) $1.82° \times 1.82°$ map, (5) maximum temperature contrast 1.70×10^{-6}, and (6) minimum temperature contrast -1.16×10^{-6}.

The simulations look well, no rare structures appear in them, but what about their APS?

In the left panel of Fig.4, we present the APS of the low resolution simulation of Fig. 2, whereas in the right panel of the same Figure, we show the ratio between this APS and that of Fig. 1. Since this ratio is a few times 10^{-3} in the wide ℓ

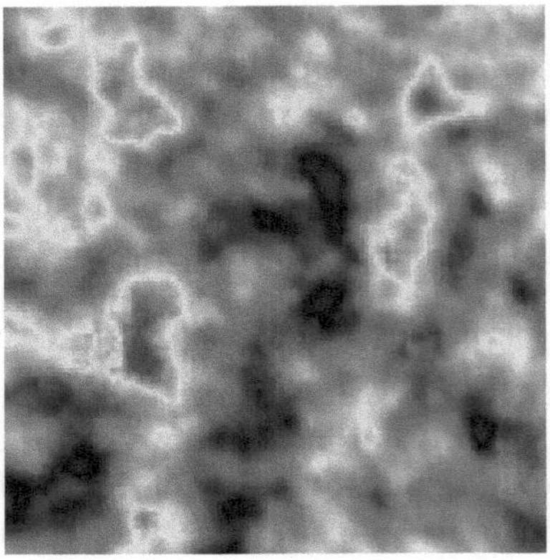

Figure 2. Low resolution RS simulation.

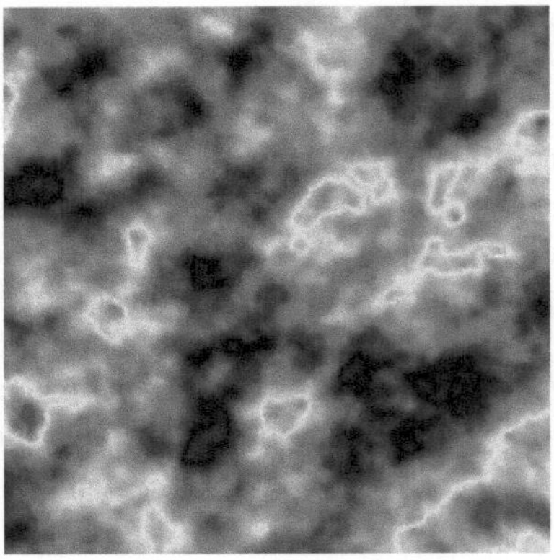

Figure 3. Middle resolution RS simulation.

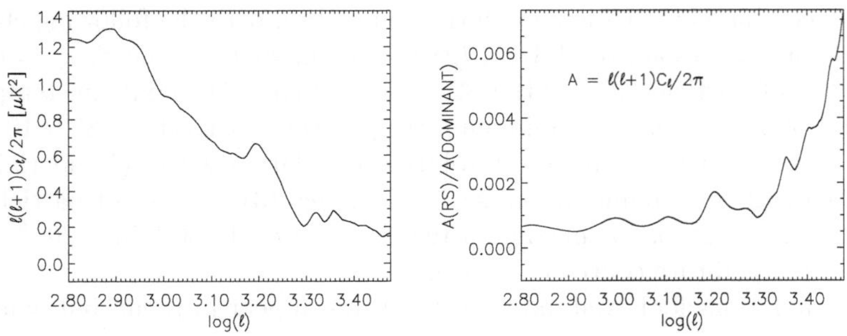

Figure 4. Left panel: APS of Fig. 2 simulation. Right panel: Ratio between APS of left panel and APS of Fig. 1.

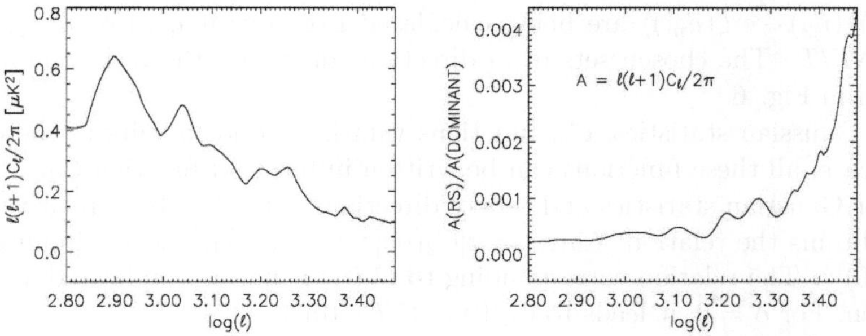

Figure 5. Left panel: APS of Fig. 3 simulation. Right panel: Ratio between APS of left panel and APS of Fig. 1.

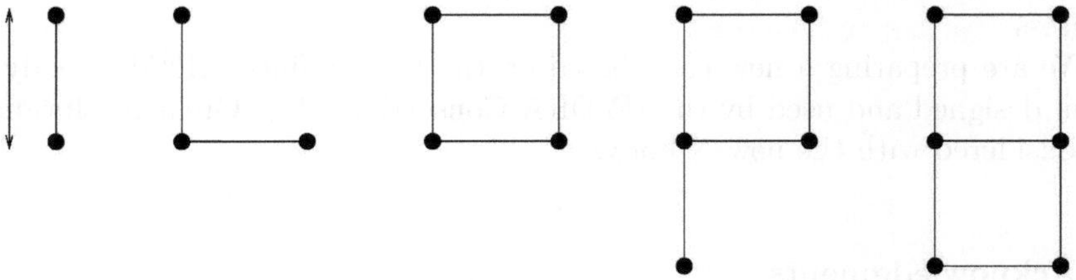

Figure 6. Defining m-direction correlation functions in maps.

interval under consideration, we conclude that the RS effect is much smaller than the dominant anisotropy due to other effect (see Section 2 and Fig. 1).

Fig. 5 has the same structure as Fig. 4, but it corresponds to the medium resolution simulation of Fig. 3. Although the resolution is greater than in the case of Figs. 2 and 4, results are similar. The amplitude of the spectrum (ratio) presented in the left (right) panel of Fig. 5 is smaller than that of the left panel of Fig. 4, but it is only due to the fact that we are comparing two arbitrary simulations and the amplitude (ratio) varies from simulation to simulation for a given resolution. We have verified that the average spectra corresponding to twenty simulations are rather similar for low and medium resolutions.

Simulations with higher resolution are in progress.

5 Conclusions and Perspectives

A ray-tracing procedure based on a preferred direction and an appropriate cut-off has been designed. It has been used to get maps of the RS effect, which appears to be a subdominant contribution to the total anisotropy of the CMB.

The maps are being statistically analysed in order to estimate deviations from Gaussianity due to RS. m-direction correlation functions of the form $C_m =$

$\langle\zeta(\vec{n}_1)\zeta(\vec{n}_2)\cdots\ \zeta(\vec{n}_m)\rangle$ are being calculated from our maps (for $m \leq 6$), where $\zeta = \Delta T/T$. The chosen sets of m directions draw –on the LSS– the figures displayed in Fig. 6.

In Gaussian statistics, C_m functions vanish for odd m values, whereas in the even case, all these functions can be written in terms of function C_2.

For Gaussian statistics and sets of directions with the above relative positions one obtains the relation: $C_4(\alpha) = 2[C_2(\alpha)]^2 + [C_2(\sqrt{\alpha})]^2$, which implies $C_4(0) = 3[C_2(0)]^2$. The relation corresponding to $C_6(\alpha)$ is more complicated and it is not written. For $\alpha = 0$, it leads to $C_6(0) = 15[C_2(0)]^3$.

Non-Gaussianity implies either non vanishing correlations for odd m or violations of the above relations between the even correlations and C_2. We are computing C_m in order to distinguish the RS effect, which is subdominant and non-Gaussian, from the dominant effects, which are Gaussian. This work is in progress.

We are preparing a new code based on the mesh-refined AP^{3}M N-body algorithm designed and used by the HYDRA Consortium [13]. Great resolutions will be considered with the new N-body.

6 Acknowledgments

This work has been partially supported by the Generalitat Valenciana under the projects GRUPOS03/062 and GRUPOS03/170, by the Spanish DGI grant BFM2001-2783, and by the Spanish MCyT (project AYA2003-08739-C02-02 partially funded with FEDER). One of us, M.J. Fullana i Alfonso, wants to dedicate this work to his mother, Eugènia M. Alfonso i Nàcher, who died recently.

References

[1] A.A. Penzias and R.W. Wilson, *ApJ Lett.* **142** (1965) 419

[2] M.J. Fullana, J.V. Arnau and D. Sáez, *Vistas in Astronomy* **41** (1998) 467

[3] J.C. Mather et al., *ApJ* **420** (1994) 439

[4] B.E. Corey and D.T. Wilkinson, *Bull. Am. Astron. Soc.* **8** (1976) 351

[5] M.J. Rees and D.W. Sciama, *Nature* **217** (1968) 511

[6] A.M. Aliaga, V. Quilis, J.V. Arnau, and D. Sáez, *MNRAS* **330** (2002) 625

[7] P. Cerdá-Durán, V. Quilis, and D. Sáez, *Phys. Rev.* **69D** (2004) 043002

[8] U. Seljak and M. Zaldarriaga, *ApJ* **469** (1996) 437

[9] C.L. Bennet et al., *ApJS* **148** (2003) 1

[10] B. Jain, U. Seljak and S. White, *ApJ* **530** (2000) 547

[11] M. White and W. Hu, *ApJ* **537** (2001) 1

[12] V. Quilis, J.M. Ibáñez, and D. Sáez, *ApJ* **502** (1998) 518

[13] H.M.P. Couchman, *ApJ Lett.* **368** (1991) 23

STATIONARY POINTS OF SCALAR FIELDS COUPLED TO GRAVITY

H. KRÖGER [a], G. MELKONYAN [a], F. PARADIS [a],
S.G. RUBIN [b,c]

[a] *Département de Physique, Université Laval, Québec, Québec G1K 7P4, Canada*
[b] *Moscow State Engineering Physics Institute, Kashirskoe sh., 31, Moscow 115409, Russia*
[c] *Center for Cosmoparticle Physics "Cosmio", Moscow, 125147, Russia*

Abstract

We investigate the dynamics of gravity coupled to a scalar field using a non-canonical form of the kinetic term. It is shown that its singular point represents an attractor for classical solutions and the stationary value of the field may occur distant from the minimum of the potential. In this paper properties of universes with such stationary states are considered. We reveal that such state can be responsible for modern dark energy density.

1 Introduction

Scalar fields play an essential role in modern cosmology. A realistic scenario of the origin of our universe is based on the inflationary paradigm and a vast majority of inflationary models use the dynamics of scalar fields. Here we show in a natural way how to produce a class of effective potentials of the scalar field. It is achieved by invoking the simplest form of a potential but non-canonical kinetic terms. The drawback of using scalar fields is the occurence of potentials with unnatural forms. For example, potentials have to be extremely flat to be consistent with the standard inflationary scenario [1].

B.G. Sidharth, F. Honsell and A. de Angelis. (eds.), Frontiers of Fundamental Physics, 123–130.

We consider an action which couples gravity to a scalar field. The latter has a non-trivial kinetic term $K(\varphi) \neq 1$. By supposition, it contains a singular point of the following form

$$K(\varphi) = M^n/(\varphi - \varphi_s)^n \ , n = -1, 1, 2 \ , \qquad (1)$$

and investigate their effect on the scalar field dynamics, see also [6]. Here M is some model parameter. The existence of the singular kinetic term opens a rich variety of possibilities for the construction of cosmological models. The well known Brans - Dicke model [4] is one of the particular case.

It is known that appropriate change of the field variable, leads to the standard form of kinetic term, i.e. $K = \pm 1$ what can be done during inflationary stage. The situation becomes much more complex when the field fluctuates around a singular point. The equation of motion for a uniform field distribution has the form

$$\ddot{\varphi} + 3H\dot{\varphi} - \frac{n}{2(\varphi - \varphi_s)}\dot{\varphi}^2 + V(\varphi_s)'(\varphi - \varphi_s)^n/M^n = 0 \ .$$

in the Friedmann-Robertson-Walker universe, H is the Hubble parameter and expression (1) is taken into account. The field value φ_s is a stationary solution for any smooth potential V and $n > 0$ provided that $\dot{\varphi} = o(\varphi - \varphi_s)$. The cosmological energy density of the vacuum is connected usually with one of its potential minima. Here the situation is different - the vacuum state is connected with the singular point of the kinetic term $K(\varphi)$. To prove this statement, we consider the simplest form of the potential

$$V(\varphi) = V_0 + m^2\varphi^2/2 \ .$$

In the following we will only consider the class of models characterized by the set of parameters m, V_0, M. The stationary state φ_s is chosen in a way such that it fits the cosmological Λ-term (see review [5]),

$$V_0 + m^2\varphi_s^2/2 = V(\varphi_s) = \Lambda \ . \qquad (2)$$

The energy density $\sim \Lambda$ in a modern epoch is small compared to any scale during the inflationary stage, which allows us to neglect it whenever this is possible and obtain the relation

$$\varphi_s \cong \sqrt{2\,|V_0|}/m. \qquad (3)$$

To proceed, an auxiliary variable χ will be taken into account. We suggest the substitution of variables $\varphi \to \chi$ in the form

$$d\chi = \pm\sqrt{K(\varphi)}d\varphi, \quad K(\varphi) > 0 \ , \qquad (4)$$

what leads to the action in terms of the auxiliary field χ

$$S = \int d^4x\sqrt{-g}\left[\frac{R}{16\pi G} + sgn(\chi)\frac{1}{2}\partial_\mu\chi\partial^\mu\chi - U(\chi)\right] \ , \qquad (5)$$

where the potential $U(\chi) \equiv V(\varphi(\chi))$ is a 'partly smooth' function. Its form depends on the form of the initial potential $V(\varphi)$, the form of the kinetic term and the position of the singularities at $\varphi = \varphi_s$. Now let us consider some particular cases of $K(\varphi)$, [7].

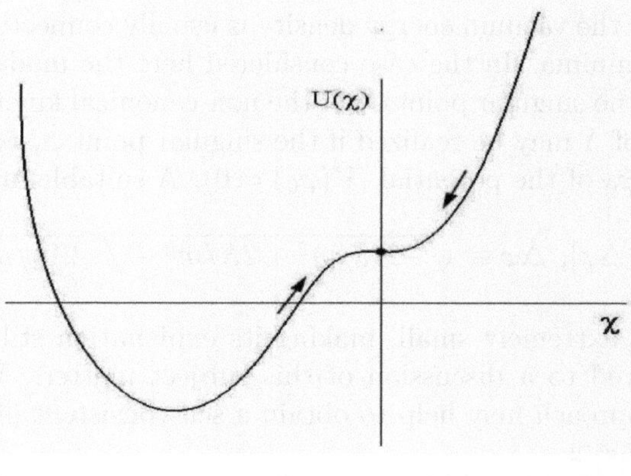

Figure 1. Potential in terms of auxiliary field χ for the case $n = 1$. If $\chi < 0$ the auxiliary field behaves like a phantom field moving classically to the local extremum at $\chi = 0$.

2 Effective potentials

The case $n = 1$

In this case formulas (1,4) give the action (5) with the potential

$$U(\chi) \equiv V(\varphi(\chi)) = V_0 + \frac{1}{2} m^2 (\varphi_s + sgn(\chi) \frac{\chi^2}{4M})^2 \quad \text{for} \quad \varphi_s > 0. \tag{6}$$

Here and below we keep the one - to - one correspondence between the physical variable φ and auxiliary variable χ in the intervals:

$$\begin{aligned} \varphi &< \varphi_s \rightarrow \chi < 0; \\ \varphi &> \varphi_s \rightarrow \chi > 0. \end{aligned}$$

If the auxiliary field starts from $\chi > 0$, it finally approaches the singular point $\chi = 0$ (see Fig.[1]). If the field obeys $\chi < 0$, than the auxiliary field behaves like a phantom field, which climbs up to the top of the potential and hence tends to the singular point as well. Finally, the field settles down in the vicinity of the singular point $\chi = 0 \quad (\varphi = \varphi_s)$. One concludes that this point is the stationary point and the vacuum energy density equals to $V(\varphi_s)$, (see Eq.(2)) rather than to V_0. The value of parameters can be estimated if we interpret the auxiliary field as the inflaton which in addition is responsible for the dark energy. In the course of inflation, a slow roll condition [1] should hold. This happens if the parameters take the values

$$M \sim M_P; \quad |V_0| \sim M_P^4; \quad m \sim 10^{-12} M_P \; . \tag{7}$$

The parameter m is small in order to fit data of large scale temperature fluctuations [8].

The problem of smallness of the vacuum energy density, $\Lambda = 10^{-123} M_P^4$ remains topical in this approach although the situation has changed. As mentioned above, the smallness of the vacuum energy density is usually connected with the smallness of a potential minima. In the case considered here the modern energy density is determined by the singular point φ_s of the non-canonical kinetic term (see Eq.(2)). The smallness of Λ may be realized if the singular point φ_s is placed very close to the zero point φ_0 of the potential ($V(\varphi_0) = 0$). A suitable interval is

$$\varphi_s \in [\varphi_0,\ \varphi_0 + \Delta\varphi],\ \Delta\varphi \equiv \sqrt{-2V_0/m^2 + 2\Lambda/m^2} - \sqrt{-2V_0/m^2} \cong \frac{\Lambda}{m\sqrt{2\,|V_0|}}\ . \tag{8}$$

This interval is extremely small, making its explanation still difficult. The next section is devoted to a discussion of this subject matter. We will show that a probabilistic approach may help to obtain a self-consistent picture.

The case $n = 2$

Now formulas (1, 4) give the action (5) with the potential

$$U(\chi) = \frac{1}{2}m^2\varphi_s^2\left[1 + sgn(\varphi_s)\cdot sgn(\chi)\cdot e^{\chi/M}\right]^2 + V_0\ . \tag{9}$$

In the case $\varphi_s < 0$; $\varphi > \varphi_s$ the potential (9) is highly asymmetric, and the behavior of the inflaton is rather different at $\chi < 0$ from that at $\chi > 0$. If we suppose that the inflation starts with $\chi = \chi_{in} > 0$, the picture is similar to the improved quintessence potential [10]. It is free of problems with the description of the radiation-dominated stage during Big Bang nucleosynthesis which could explain the modern distribution of chemical elements [9]. The chosen parameter values

$$M \sim M_P,\ m \sim M_P,\ |V_0| \sim 10^{-14}M_P^4 \tag{10}$$

permit a suitable inflationary stage and they are in agreement with observations of temperature fluctuations [8].

The case $n = -1$

A nontrivial situation occurs when the kinetic function has not a pole but a root at some point, $K(\varphi) = (\varphi - \varphi_s)/M$. Let the initial field value obey $\varphi = \varphi_{in} > \varphi_s \sim M_P$, which gives rise to the inflation in early universe. Then the potential of the auxiliary field χ becomes

$$U(\chi) = \frac{1}{2}m^2(\varphi_s + sgn(\chi)\cdot\gamma|\chi|^{2/3})^2 + V_0\ . \tag{11}$$

$U(\chi)$ is finite at $\chi = 0$ but its derivative is singular. Classically, the situation looks very strange - the singular point attracts the solution, but forbids it to stay there forever. It looks is similar to quantum mechanics, in particular to the case of an electron in the Coulomb field.

The potential (11) behaves like $\chi^{4/3}$ at large field values. It leads to standard inflation with moderate fine tuning of the parameters. Namely

$$M \sim M_P, \quad m \sim 10^{-6}M_P, \quad V_0 \sim 10^{-12}M_p^4\ . \tag{12}$$

If $\varphi_s > 0$, the field φ will fluctuate around some critical point with energy density (2). This motion never attenuates completely because classical stationary points are absent in this region.

3 Probabilistic approach to the form of action

Here we have investigated several specific forms of effective potentials. Many other potentials and kinetic terms have been discussed in the literature. A substantial number of them does not contradict observational data. In this context the question can be raised and need to be answered: Why is it that particular shape of potential and kinetic term is realized in nature? What are the underlying physical reasons?

Some theoretical hints on the form of the potential have been given in the context of supergravity, which predicts an infinite power series expansion in the scalar field potential [11]. Its minima, if they exist, correspond to stationary states of the field. The potential, due to an infinite number of terms in a power series could correspond to a function with an infinite set of potential minima. This assumption with randomly distributed minima appears to be self-consistent [2]. In the low energy regime it is reasonable to retain only a few terms (lowest powers in the Taylor expansion) of the scalar field [12]. In the vicinity of each of those minima the potential has a particular form. A similar behavior may hold also for the kinetic term. If the scalar field is responsible for the inflation, each local minimum produces an individual universe, different from any other universe. Our own universe is associated with a particular potential minimum, not necessarily located at $\varphi = 0$.

The observed smallness of the value of the Λ-term is explained usually in terms of a more fundamental theory like supergravity or the anthropic principle. Our point of view is that we have to merge these approaches. The more fundamental theory supplies us with an infinite set of minima of the potential. These minima having an individual shape are responsible for the formation of those universes used in the anthropic consideration.

Practically, it could be performed in the framework of the random potential [2, 3] and the kinetic term of the scalar field discussed in sect.(1). A part of such potential and the kinetic term in a finite region of the field φ are represented in Fig.[2]. Fluctuations of the scalar field being generated at high energies in the inflationary stage move classically to stationary points. Those of them who reach stationary points with appropriate energy density could form a universe similar to our Universe. This energy density ($\sim 10^{-123} M_P^4$) is the result of a small value of the concrete potential minimum or a small value of the difference $\varphi_s - \varphi_m$, where φ_m is a zero of the potential ($V(\varphi_m) = 0$). The fraction of such universes is relatively small, but nevertheless is infinite because of an infinite number of stationary states.

How could one decide which of the stationary points is most promising? To get an idea we should recall that the main defect of the inflationary scenario is the smallness of some intrinsic parameter compared to unity. It is the value of selfcoupling $\lambda \sim 10^{-13}$ for the potential $V_4 = \lambda\varphi^4$ or the smallness of the mass of the inflaton field in Planck units, $m/M_P \sim 10^{-6}$ for the potential $V_2 = m^2\varphi^2/2$. Let us consider an infinite set of potential wells corresponding to infinite set of its minima [2] as discussed above. Then we can use the concept of probability to find a potential well with specific properties. To estimate the relative number of specific universes, let us suppose that if there are no observational data on the value of a parameter g, the probability density W for any parameter g is distributed by a random uniform distribution in the range $(0,1)$ in Planck scale. An immediate conclusion is that the probability of a potential $\lambda\varphi^4$ is about 10^{-13} while the probability of a potential $m^2\varphi^2/2$ is about 10^{-6}. It means that the latter is realized 10^6 times more frequently.

In fact the probability is much smaller due to smallness of the cosmological Λ-term. So the probability to find a universe with such small vacuum energy is $P_\Lambda = 10^{-123}$. Recall that the set of potential minima is infinite. It means that the set of universes with an appropriate vacuum energy density is relatively small but still infinite. So the probability to find an appropriate potential V_4 is

$$P(V_4) = 10^{-13} P_\Lambda, \quad V_4 \sim \varphi^4, \tag{13}$$

while the same for the potential V_2 is

$$P(V_2) = 10^{-6} P_\Lambda, \quad V_2 \sim \varphi^2. \tag{14}$$

The lowest stationary state could be a singular point of the kinetic term, rather than a potential minimum. Thus we could expect that singular point(s) φ_s may be found near some minima φ_m of the potential. Now the problem is reformulated as follows: "which part of infinite number of minima contains singular points located closely to them? " This part is very small, but not zero, due to infinite number of the minima. Only this part is important - it represents those vacua where galaxies could be formed due to extremely small value of $\Lambda-$ term [13].

Following the way discussed above we can compare the probability of realization of such potentials. Their common factor is connected with the probability to find the singular point of the kinetic term in a small interval Eq.(8),

$$P_0 = \Delta\varphi_s/M_P \cong \frac{\Lambda}{M_P m\sqrt{2\,|V_0|}} = P_\Lambda \frac{M_P^3}{m\sqrt{2V_0}} \, . \tag{15}$$

For the case $n = 1$ the only additional smallness is dictated by expression (7) and the probability for such universes to occur is

$$P_1 \sim \frac{m}{M} P_0 = P_\Lambda \frac{M_P^3}{M\sqrt{2V_0}} \approx P_\Lambda \, . \tag{16}$$

Universes with the properties described in the case $n = 2$ are distributed with probability

$$P_2 \sim \frac{V_0}{M_P^4} P_0 \simeq P_\Lambda \frac{\sqrt{2V_0}}{mM_P} \sim 10^{-7} P_\Lambda \, , \tag{17}$$

if the inflation starts at the right branch of the potential. Here we assumed $m \sim M_P, V_0 \sim M_P^4$. The last case considered, $n = 2$, has a probability by an order of magnitude larger

$$P_{-1} \sim \frac{m}{M} \frac{V_0}{M_P^4} P_0 \simeq P_\Lambda \frac{\sqrt{2V_0}}{M_P^2} \sim 10^{-6} P_\Lambda \, . \tag{18}$$

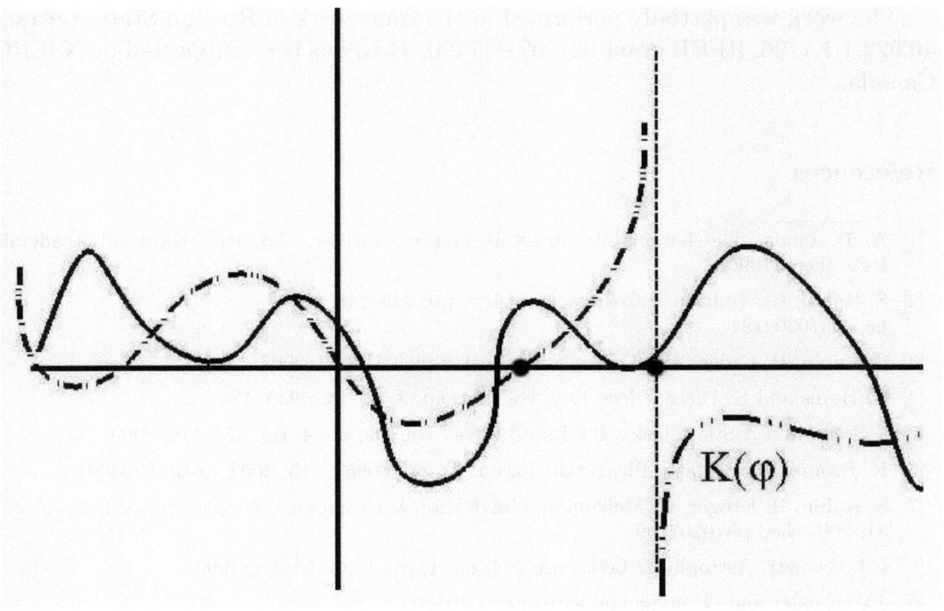

Figure 2. Random potential and kinetic term. Dots denote stationary states of the field φ.

An important conclusion from this consideration is that the model with kinetic term $K \sim (\varphi - \varphi_s)^{-1}$ is much more probable (at least by a factor 10^6) comparing with other models discussed above, including the models with a standard kinetic term and potentials $\sim \varphi^2$ and $\sim \varphi^4$, see expressions (14), (13). It means that our Universe is likely governed by the model with kinetic term $K \sim (\varphi - \varphi_s)^{-1}$.

In conclusion we have discussed several inflationary models having common features like the occurrence of singular points in non-canonical kinetic terms. We have shown that the existence of such points where the kinetic term changes its sign or tends to infinity opens new possibilities for scalar field dynamics. It takes place even for the simplest form of the potential. Depending on a position of the singular point of the kinetic term, specific forms of the potential of the auxiliary field have been obtained. One of the main results is that the stationary value of scalar field could occur at singular points of kinetic term rather than at minima of the potential. We estimated the parameter values for three type of new inflationary models. The probabilities to find universes with specific values of parameters have been estimated. It was shown that the probability is much greater for the model with kinetic term $K \sim (\varphi - \varphi_s)^{-1}$ than for the other models including the most promising model of chaotic inflation with the quadratic potential.

Another interesting result is that if the singular point is a root of the kinetic term, the final state is intrinsically a quantum state.

The work was partially performed in the framework of Russian State contract 40.022.1.1.1106, RFBR grant 02 – 02 – 17490. H.K. has been supported by NSERC Canada.

References

[1] A. D. Linde, The Large-scale Structure of the Universe. London: Harwood Academic Publishers, 1990.

[2] S. Rubin, Gravitation & Cosmology, vol. 9, pp. 243–248, 2003, hep-ph/0309184.

[3] S. Rubin, H. Kröger and G. Melkonian, astro-ph/0310182, 2003.

[4] C. Brans and R. Dicke, Phys. Rev, vol. 124, no. 3, pp. 925–935, 1961.

[5] V.Sahni and A.Starobinsky, Int.J.Mod.Phys., vol. D9, no. 4, pp. 373–444, 2000.

[6] K. Bronnikov, J. Math. Phys., vol. 43, no. 12, pp. 6096–6115, 2002. gr-qc/0204001.

[7] S. Rubin, H. Kröger, G. Melkonian, Gravitation & Cosmology, Supplement, vol. 8, pp. 27–31, 2002; hep-ph/0207109.

[8] C.L.Bennett, Astrophys.J.Lett., vol. 464, no. 1, pt. 2, pp. L1–L4, 1996.

[9] J.P. Kneller and G. Steigman, astro-ph/0210500

[10] A. Albrecht and C. Skordis, , Phys. Rev. Lett., vol. 84, no. 10, pp. 2076–2079, 2000.

[11] H.P.Nilles, Phys.Rept., vol. 110, no. 1-2, pp. 1–162, 1984.

[12] D.H.Lyth and E. D. Stewart, Phys.Rev., vol. D 54, no. 12, pp. 7186-7190, 1996.

[13] S. Weinberg, astro-ph/0005265.

THE DENSITY MATRIX DEFORMATION IN QUANTUM AND STATISTICAL MECHANICS OF THE EARLY UNIVERSE

ALEXANDER E. SHALYT-MARGOLIN [a],
VASSILII I. STRAZHEV [a]

[a] *National Centre of Particles and High Energy Physics, Bogdanovich Str. 153, Minsk 220040, Belarus*

Abstract

In this paper a new approach to investigation of Quantum and Statistical Mechanics of the Early Universe (Planck scale) - density matrix deformation - is proposed. Quantum Mechanics of the Early Universe is chosen as Quantum Mechanics with Fundamental Length. Here Quantum Mechanics with Fundamental Length is obtained as a deformation of well-known Quantum Mechanics. The proposed approach allows to describe dynamics. It is demonstrated that Statistical Mechanics of the Early Universe is a deformation of the conventional Statistical Mechanics. The statistical-mechanics deformation is constructed by analogy to the earlier quantum mechanical results. Some implications of the obtained results are discussed. In particular, the problem of singularity, possible improvement of the definition of statistical entropy and the problem of information loss in black holes are considered. It is noted that the obtained results enable the derivation of the Bekenstein-Hawking's formula for the black hole entropy in a semiclassical approximation in a simple and natural way.

As known, Quantum Mechanics of the Early Universe (Planck scale) differs from the conventional one [1]. The main motivation for this difference is the presence of the General Uncertainty Relations (GUR) [2] appropriated to describe a physical behavior of the Early Universe and unavoidably leading to the fundamental length concept. The derivation of GUR proceeding from the basic principles has been performed relatively recently [3]. By this means the Quantum Mechanics of the

B.G. Sidharth, F. Honsell and A. de Angelis. (eds.), Frontiers of Fundamental Physics, 131–134.

Early Universe should be treated as a deformation of the well-known Quantum Mechanics (QM). The deformation is understood as an extension of a particular theory by inclusion of one or several additional parameters in such a way that the initial theory appears in the limiting transition. The deformation in Quantum Mechanics at Planck scale is realized by the commutator deformation, or more precisely by deformation of the respective Heisenberg algebra [2],[4]. However, this approach suffers from two serious disadvantages:1)the deformation parameter is a dimensional variable κ with a dimension of mass [2] and 2)in the limiting transition to QM this parameter goes to infinity, whereas fluctuations of other values are hardly sensitive to it. In this paper it is proposed to study the quantum mechanics of the Early Universe proceeding from the density matrix deformation approach [5]-[7]. First starting from the existence of a minimum length $l_{min} \sim L_{pl}$, we give strong validation for the dependence of the density matrix on the additional dimensionless parameter $\alpha = l_{min}^2/x^2$ (where x is the scale of measurement) varying over the interval $0 < \alpha \leq 1/4$. An exact definition and the principal properties of the characteristic deformed density matrix $\rho(\alpha)$ referred to as the density pro-matrix may be found in [5]. Of particular importance is the fact that in the low-energy limit

$$\lim_{\alpha\rightarrow 0} \rho(\alpha) = \rho$$

transforms to the conventional density matrix in Quantum Mechanics. Besides, the following condition must be fulfilled:

$$Sp[\rho(\alpha)] - Sp^2[\rho(\alpha)] \approx \alpha. \tag{1}$$

As a matter of fact, the deformation parameter α should assume the value $0 < \alpha \leq 1$. As seen from (1), however, $Sp[\rho(\alpha)]$ is well defined only for $0 < \alpha \leq 1/4$. That is if $x = il_{min}$ and $i \geq 2$, then there is no problem at all. At the point of $x = l_{min}$ there is a singularity related to the complex values following from $Sp[\rho(\alpha)]$, i.e. to the impossibility of obtaining a diagonalized density pro-matrix at this point over the field of real numbers. In [5, 7] it has been noted that this singularity is directly associated with the description problem of space-time singularities, at any rate in case of Schwarzschild black holes.

We consider possible solutions for (1). For instance, one of the solutions of (1), at least to the first order in α, is

$$\rho^*(\alpha) = \sum_i \alpha_i exp(-\alpha)|i><i|, \tag{2}$$

where all $\alpha_i > 0$ are independent of α and their sum is equal to 1. In this way

$$Sp[\rho^*(\alpha)] = exp(-\alpha)$$

We can easily verify that

$$Sp[\rho^*(\alpha)] - Sp^2[\rho^*(\alpha)] = \alpha + O(\alpha^2)$$

Note that in the momentum representation $\alpha \sim p^2/p_{pl}^2$, where p_{pl} is the Planck momentum. When present in the matrix elements, $exp(-\alpha)$ can damp the contribution of great momenta in a perturbation theory.

The applications of the obtained results may be as follows [5, 7]:
1) An additional component in the deformed Liouville's equation arises in two cases: at transfer out of inflation to low energies; on absorption of the matter by a black hole and its motion to singularity [8].
Note that the appropriate additional component in the right-hand side of Schrodinger equation may be obtained in similar way [5, 7].
2)The approach used allows for introduction of the following matrix value that is a complete analog for the statistical entropy in case under consideration and is meaning the entropy density over a minimum unit area [5, 7]:

$$S_{\alpha_2}^{\alpha_1} = -Sp[\rho(\alpha_1)\ln(\rho(\alpha_2))] = -<\ln(\rho(\alpha_2))>_{\alpha_1}, \quad (3)$$

where $0 < \alpha_1, \alpha_2 \leq 1/4$.
$S_{\alpha_2}^{\alpha_1}$ has a clear physical meaning: the entropy density is computed on the scale associated with the deformation parameter α_2 by the observer who is at a scale corresponding to the deformation parameter α_1. Note that with this approach the diagonal element $S_\alpha = S_\alpha^\alpha$ of the described matrix $S_{\alpha_2}^{\alpha_1}$ is the density of entropy measured by the observer, who is at the same scale as the measured object associated with the deformation parameter α. In [5, 7] such a construction and exponential ansatz (2) were used in derivation of a semiclassical Bekenstein-Hawking formula [9] for the Black Hole entropy from the basic principles.
3) Besides, value $S_{\alpha_2}^{\alpha_1}$ may be used in analysis of the Information Paradox Problem [8]. In [5, 7] it has been demonstrated that for any observer in the vicinity of a black hole the information loss close to the Big Bang initial singularity and close to the final singularity of a black hole is comparable. But then in both cases the entropy density should be the same:

$$S(in) = S_{\frac{1}{4}}^{\alpha_1} = S(out)$$

and in fact there is no any information loss.
Proceeding to the Statistical Mechanics [10], we further assume that an internal energy of any ensemble U could not be in excess of $E_{max} \sim E_p$ which appears from the generalized uncertainty relations for the "energy - time" pair and hence temperature T could not be in excess of $T_{max} = E_{max}/k_B \sim T_p$. On this basis the density matrix deformation in Statistical Mechanics at Planck scale is constructed in the [6, 7]. In this way ρ_{stat} at very high $T \gg 0$ becomes dependent on the parameter $\tau = T^2/T_{max}^2$, i.e. in the most general case $\rho_{stat} = \rho_{stat}(\tau)$,$Sp[\rho_{stat}(\tau)] < 1$ and for $\tau \ll 1$ we have $\rho_{stat}(\tau) \approx \rho_{stat}$ in the well-known Statistical Mechanics [10].The following condition must be fulfilled in the same way as in case of Quantum Mechanics of of the Early Universe:

$$Sp[\rho_{stat}(\tau)] - Sp^2[\rho_{stat}(\tau)] \approx \tau. \quad (4)$$

Again similar to QMFL, as a possible solution for (4) we have an exponential ansatz [6, 7] with the help of which the deformation of a canonical Gibbs distribution at Planck scale (up to factor $1/Q$) takes an elegant and completed form:

$$\omega_{\mathbf{n}}(\tau) = \mathbf{exp}(-\tau)\omega_{\mathbf{n}} = \mathbf{exp}(-\frac{\mathbf{T^2}}{\mathbf{T^2_{max}}} - \beta \mathbf{E_n}) \tag{5}$$

References

[1] L. Garay, *Int.J.Mod.Phys.A10(1995)145*,
D.V. Ahluwalia, *Mod.Phys.Lett.A17(2002)1135.*

[2] M. Maggiore, *Phys.Lett.B319(1993)83*,
A. Kempf,G. Mangano,R.B. Mann, *Phys.Rev.D52(1995)1108.*

[3] R.J. Adler,D.I. Santiago,*Mod.Phys.Lett.A14(1999)1371.*

[4] M. Maggiore,*Phys.Rev.D49(1994)5182*

[5] A.E. Shalyt-Margolin,J. Suarez, *Intern.Journ.of Mod.Phys.D.12(2003)1265*,
A.E. Shalyt-Margolin, *Mod.Phys.Lett.A19(2004)391*,
A.E. Shalyt-Margolin, *Intern.Journ. of Mod.Phys.D.13 (2004)853.*

[6] A.E. Shalyt-Margolin,A.Ya. Tregubovich,*Mod. Phys. Lett. A19(2004)71.*

[7] A.E. Shalyt-Margolin, *The Density Matrix Deformation in Physics of the Early Universe and Some of its Implications*, will be published in the book "Progress in General Relativity and Quantum Cosmology Research" (Nova Science Publishers, Inc., Hauppauge, NY)

[8] S. Hawking,*Phys.Rev.D14(1976)2460*,
A. Strominger, *Les Houches Lectures on Black Holes*, [hep-th/9501071].

[9] J.D. Bekenstein,*Phys.Rev.D7(1973)2333*,
S. Hawking, *Phys.Rev.D13(1976)191.*

[10] R.P. Feynman,*Statistical Mechanics,A Set of Lectures,California, Institute of Technology* W.A.Benjamin,Massachusets 1972.

Part II

Foundations of Physics

DARK ENERGY, CHAOTIC FIELDS, AND FUNDAMENTAL CONSTANTS

CHRISTIAN BECK

School of Mathematical Sciences, Queen Mary, University of London, Mile End Road, London E1 4NS, UK

Abstract

To explain the currently observed accelerated expansion of the universe, a large number of different theoretical models are presently being discussed. In one way or another, all of these contain 'new physics', though at different levels. The big question is how to select out of infinitely many possible models the right one. We here discuss a possibility that has so far been somewhat neglected, namely that the new physics underlying dark energy arises out of a gravitationally active amendment of the electroweak and strong sector of the standard model. This amendment basically consists of a rapidly fluctuating gravitationally active dynamics of vacuum fluctuations with a cutoff of the order of the neutrino mass scale. We consider a concrete model for this based on second-quantized self-interacting scalar fields, which evolve in a chaotic way. It is shown that expectations with respect to the chaotic dynamics yield statements on the observed numerical values of the electroweak coupling constants with amazing precision, thus providing evidence for the physical relevance of this model.

1 Introduction

To understand the fact that the universe is currently in a phase of accelerated expansion [1, 2, 3, 4], an enormous amount of theoretical and experimental work is currently being performed. The favoured explanation for the acceleration is the existence of dark energy, though other possibilities might exist as well. An amazing number of models has been developed in the mean time, and basically

B.G. Sidharth, F. Honsell and A. de Angelis. (eds.), Frontiers of Fundamental Physics, 137–146.

every week one finds in the preprint archives some new idea concerning the nature of dark energy. The most popular models are currently quintessence models of various kinds [5], phantom fields [6], Born-Infeld quantum condensates [7], the Chaplygin gas [8], fields with nonstandard kinetic terms [9], to name just a few. All of these approaches contain 'new physics' in one way or another, though at different levels. However, it should also be clear that the number of possible dark energy models that are based on new physics is infinite, and in that sense there is much more to study than the above, currently most popular, models.

When trying to select the most suitable theoretical model for dark energy out of many possibilities, the most straightforward idea would be to compare the various predictions of different theoretical models with the experimental observations (supernovae, cosmic microwave background, large scale structure, etc.) to single out the most relevant model. However, there is a huge degeneracy in the sense that completely different theoretical models cannot be distinguished with the currently available observational data, they often make the same or indistinguishable predictions (e.g. for the equation of state as a function of redshift) or no prediction at all. This situation of degeneracy is not expected to improve significantly in the near future, though the precision of the observational data will increase.

We are thus lead to single out good dark energy models by other, more theoretical criteria. Let us here suggest the following checklist applicable to *any* dark energy model:

1. Does the model explain why the current value of dark energy density is so small (the cosmological constant problem) ?

2. Does the model explain why the current value of dark energy density is of the same order as the matter density (the cosmological coincidence problem) ?

3. Does the model have a reasonable quantum field theoretical background, or does it represent a reasonable extension of quantum field theory?

4. Does the model fit into Einstein's gravity, or does it represent a reasonable extension of it?

5. Is the model compatible with observations?

6. Does the new physics contained in the model explain some further phenomena that are so far unexplained, for example why the fundamental constants of nature (coupling constants, mass ratios, mixing angles) take on the values we observe and not some other values?

7. Does the model give some sense to dark energy (for example, by explaining it as a relic of inflation), rather than letting it look like an unnecessary curiosity of the universe?

8. Besides dealing with dark energy, does the model also explain why there is dark matter?

9. Is the theory aesthetic and accessible to many physicists, or is it just so complicated that hardly anybody understands it?

10. Are there any laboratory experiments that can verify or disprove this theory?

Every theoretical model builder may rank his or her favourite dark energy model on a scale from 0–10, depending on how much of the above criteria are satisfied. The perfect theory, from which we are still far away, reaches a mark of 10. Current models, perhaps, reach something in the region 1–5, at best.

As mentioned before, with the assumption that new physics is relevant there is an enormous number of possible models. In the following I will restrict myself to a model introduced in [10], which scores relatively high on points 3,6,7,10. With some additional assumptions described in [10], it also scores high on points 1,2 and 4. The basic idea is that there is a rather 'sterile' amendment of the standard model of electroweak and strong interactions which just consists of a scalar dynamics of vacuum fluctuations with a finite cutoff. The expectation of the underlying potentials produces the currently observed dark energy. Amazingly, the above model seems to distinguish the observed values of the electroweak coupling parameters as local minima in the dark energy landscape. It is well known that quintessence fields can produce a very slow time variation of fundamental constants of nature, e.g. of the fine structure constant [11]. Here we go a step further and show that not only a possible variation of the fine structure constant but also its currently observed equilibrium value can be understood by a suitable scalar field dynamics underlying dark energy.

2 Amending the standard model by gravitationally active vacuum fluctuations

Let us consider the standard model of electroweak and strong interactions. It is a second-quantized field theory and it allows for vacuum fluctuations. The vacuum energy density associated with a particle of mass m and spin j is given by

$$\rho_{vac} = \frac{1}{2}(-1)^{2j}(2j+1)\int \frac{d^3k}{(2\pi)^3}\sqrt{\mathbf{k}^2+m^2} \tag{1}$$

in units where $\hbar = c = 1$. Here $\mathbf{k}$ represents the momentum and the energy is given by $E = \sqrt{\mathbf{k}^2+m^2}$. Unfortunately, the above integral is divergent. One has to introduce a suitable upper cutoff. Choosing as an upper cutoff the Planck mass m_{Pl}, one gets an enormous amount of vacuum energy density of the order m_{Pl}^4, larger than the currently observed dark energy density by a factor of $\sim 10^{120}$. This is the famous cosmological constant problem. To circumvent it, the common view is that the absolute value of the above vacuum energy (e.g. in QED) is not observable, it is 'renormalized away', which in a sense means that one adds an infinite constant to get rid of the vacuum energy. This works as long as one does not consider gravity.

However, ultimately we have to unify the standard model with gravity. Also note that almost all particles in the standard model do have mass, so they know what gravity is. It looks a bit like a 'dirty trick' to say that gravity is decoupled from the standard model if the particles have mass. So let us for the moment assume that the vacuum energy in the standard model is cancelled by some kind of symmetry, for example some kind of supersymmetry. This still doesn't explain why we do observe some tiny positive amount of vacuum energy density in the universe, corresponding to the currently measured dark energy density, which is more of the order m_ν^4 rather than m_{Pl}^4, where m_ν is a typical neutrino mass scale. Hence let us assume that there is something more to the standard model: In addition to the ordinary standard model fields (whose vacuum energies are cancelled by some symmetry) there could be other fields that just show up in a rather sterile way in form of vacuum fluctuations, with a rather small cutoff scale of the order of the neutrino mass. The vacuum energy of these fields is not cancelled by symmetry, there is a symmetry breaking towards positive vacuum energy, at least at the current stage of the universe.

Of course, the above assumption represents new physics, but any decent dark energy model seems to require new physics, in one way or another. The advantages of the above idea are straightforward:

- Since we associate dark energy with a broken symmetry in some sector of this extended standard model, it is not too surprising that the relevant scale of the dark energy density is of the order of some typical particle mass to the power 4 in this model, in this case a neutrino.

- There is increasing experimental evidence [12] for the existence of sterile neutrinos in addition to the known three ordinary neutrino flavours, so apparently there is something more to the standard model than we know. Sterile neutrinos may have something to do with the above gravitationally active sector of the standard model.

- Since our dark energy model deals with vacuum fluctuations that are part of the electroweak sector, there is a chance to measure the effects of these fluctuations in laboratory experiments on the earth, such as in Josephson junction experiments, which do probe the spectrum of vacuum fluctuations near the neutrino mass scale due to a nonlinear mixing effect in the junction [13].

In the following, we want to consider a concrete model for vacuum fluctuations with a small cutoff, as introduced in [10]. For quantum field theories with a cutoff, a particular quantization scheme is very convenient to choose, namely the stochastic quantization scheme introduced by Parisi and Wu [14]. This scheme is based on a stochastic differential equation, which naturally embeds various kinds of cutoffs and is by far simpler to deal with than the canonical quantization procedure. For that reason, our gravitationally active amendment of the standard model will be formulated in terms of stochastic quantization. Amazingly, the model will turn out to distinguish the numerical values of the electroweak coupling constants as corresponding to local minima in the dark energy landscape. This can be seen as an indication that one is on the the right track with these kinds of models.

3 Chaotic model of vacuum fluctuations

Let us consider a homogeneous self-interacting scalar field φ with potential $V(\varphi)$ that forms the basis for our gravitationally active amendment of the standard model. Our amendment should have rather 'sterile' properties, so it is in good approximation sufficient to look at the scalar field equations of this sector on its own, rather than coupling them to the ordinary standard model field equations. We also need the amended sector to consist mainly of vacuum fluctuations with a suitable cutoff, rather than containing stable observable particles, with the possible exception of sterile neutrinos and/or dark matter. In fact, the only connection to the ordinary standard model is that the virtual particles underlying the vacuum fluctuations could potentially interact with the same electroweak and strong coupling constants as in the ordinary standard model.

We quantize the scalar field underlying the steril sector using the Parisi-Wu approach of stochastic quantization. The 2nd quantized equation of motion is

$$\frac{\partial}{\partial s}\varphi = \ddot{\varphi} + 3H\dot{\varphi} + V'(\varphi) + L(s,t), \tag{2}$$

where H is the Hubble parameter, t is physical time, s is fictitious time (just a formal coordinate to do quantization) and $L(s,t)$ is Gaussian white noise, δ-correlated both in s and t. The fictitious time s is just introduced as a formal tool for stochastic quantization, it has dimensions GeV^{-2}. Quantum mechanical expectations can be calculated as expectations of the above stochastic process for $s \to \infty$. The simplest way to introduce a cutoff is by making t and s discrete (as in any numerical simulation). Hence we write

$$s = n\tau \tag{3}$$

$$t = i\delta, \tag{4}$$

where n and i are integers and τ is a fictitious time lattice constant, δ is a physical time lattice constant. Note that the uncertainty relation $\Delta E \Delta t = O(\hbar)$ always implies an effective lattice constant Δt for a given finite energy ΔE. We also introduce a dimensionless field variable Φ_n^i depending on i and n by writing $\varphi_n^i = \Phi_n^i p_{max}$, where p_{max} is some (so far) arbitrary energy scale. The discretized scalar field dynamics (2) can be written as the following discrete dynamical system [15, 3, 17]

$$\Phi_{n+1}^i = (1-\alpha)T(\Phi_n^i) + \frac{3}{2}H\delta\alpha(\Phi_n^i - \Phi_n^{i-1}) + \frac{\alpha}{2}(\Phi_n^{i+1} + \Phi_n^{i-1}) + \tau \cdot noise, \tag{5}$$

where the local map T is given by

$$T(\Phi) = \Phi + \frac{\tau}{p_{max}(1-\alpha)} V'(p_{max}\Phi) \tag{6}$$

and α is defined by

$$\alpha := \frac{2\tau}{\delta^2}. \tag{7}$$

For old universes, one can neglect the term proportional to H, obtaining

$$\Phi_{n+1}^i = (1-\alpha)T(\Phi_n^i) + \frac{\alpha}{2}(\Phi_n^{i+1} + \Phi_n^{i-1}) + \tau \cdot noise. \tag{8}$$

We now want to construct a field φ_n^i that is different from ordinary fields: Rather than evolving smoothly it should exhibit strongly fluctuating behaviour, so that we may be able to interpret it in terms of vacuum fluctuations. As a distinguished example of a φ^4-theory generating such behaviour, let us consider the map

$$\Phi_{n+1} = T_{-3}(\Phi_n) = -4\Phi_n^3 + 3\Phi_n \tag{9}$$

on the interval $\Phi \in [-1, 1]$. T_{-3} is the negative third-order Tchebyscheff map, a standard example of a map exhibiting strongly chaotic behaviour. It is conjugated to a Bernoulli shift, and is distinguished as generating the strongest possible chaotic behaviour possible for a smooth low-dimensional deterministic dynamical system [18]. The corresponding potential is given by

$$V_{-3}(\varphi) = \frac{1-\alpha}{\tau}\left\{\varphi^2 - \frac{1}{p_{max}^2}\varphi^4\right\} + const, \tag{10}$$

or, in terms of the dimensionless field Φ,

$$V_{-3}(\varphi) = \frac{1-\alpha}{\tau} p_{max}^2 (\Phi^2 - \Phi^4) + const. \tag{11}$$

The important point is that starting from this potential we obtain by second quantization a field φ that rapidly fluctuates in fictitious time on some finite interval, choosing initially $\varphi_0 \in [-p_{max}, p_{max}]$. Since these chaotic fluctuations are bounded, there is a natural cutoff.

The idea is now that the expectation of the potential of this and similar chaotic fields (plus possibly kinetic terms) underlie the measured dark energy density in the universe. Expectations $\langle \cdots \rangle$ can be easily numerically determined by iterating the dynamics (8) for random initial conditions. One has

$$\langle V_{-3}(\varphi)\rangle = \frac{1-\alpha}{\tau} p_{max}^2 (\langle \Phi^2 \rangle - \langle \Phi^4 \rangle) + const, \tag{12}$$

which for $\alpha = 0$ can be analytically evaluated [10] to give

$$\langle V_{-3}(\varphi)\rangle = \frac{1}{8}\frac{p_{max}^2}{\tau} + const. \tag{13}$$

Alternatively, we may consider the positive Tchebyscheff map $T_3(\Phi) = 4\Phi^3 - 3\Phi$. It is easy to show that this generates vacuum energy of opposite sign. Symmetry considerations between T_{-3} and T_3 suggest to take the additive constant *const* as

$$const = +\frac{1-\alpha}{\tau} p_{max}^2 \frac{1}{2} \langle \Phi^2 \rangle. \tag{14}$$

In this case one obtains the fully symmetric equation

$$\langle V_{\pm 3}(\varphi)\rangle = \pm\frac{1-\alpha}{\tau}p_{max}^2\left\{-\frac{3}{2}\langle\Phi^2\rangle+\langle\Phi^4\rangle\right\}, \tag{15}$$

which for $\alpha \to 0$ reduces to

$$\langle V_{\pm 3}(\varphi)\rangle = \pm\frac{p_{max}^2}{\tau}\left(-\frac{3}{8}\right). \tag{16}$$

To reproduce the currently measured dark energy, we only need to fix the ratio of the parameters τ and p_{max} as

$$\frac{3}{8}\frac{p_{max}^2}{\tau} = \rho_\Lambda \sim m_\nu^4. \tag{17}$$

This is the simplest model of steril vacuum fluctuations one can think of, a 2nd quantized field theory underlying the cosmological constant Λ. It is easy to show [10] that for $\alpha = 0$ the equation of state of this field is $w = -1$. For small α, it is close to $w = -1$. More complicated models, with $w \neq -1$, as well as symmetry breaking breaking between T_{+3} and T_{-3} can be worked out in detail [10]. These can produce tracking behaviour of dark energy during the evolution of the universe, and mimic some of the properties of quintessence fields.

4 Electroweak couplings as local minima in the dark energy landscape

Let us now give a heuristic argument why the coupling constant α in the above chaotic field equations could have the physical meaning of a gauge coupling. Consider two charges of opposite sign, say, a virtual electron-positron pair, which exists for a short time interval due to a vacuum fluctuation. If the charges are at distance r, the Coulomb potential between them is given by

$$V_C(r) = \alpha\frac{1}{r} \tag{18}$$

(in units where $\hbar = c = 1$), where α is the fine structure constant. Now for any vacuum fluctuation the inverse distance $1/r$ is certainly a fluctuating random variable. Motivated by our interpretation of vacuum fluctuations of the field φ_n^i we may choose

$$\varphi_n^i - \varphi_n^{i-1} = \frac{1}{r}, \tag{19}$$

which has the right dimension and is allowing for both attracting and repelling forces. The above choice basically means that the field difference $\varphi_n^i - \varphi_n^{i-1}$ determines the inverse interaction distance to neighbours in this chaotically evolving discrete model of vacuum fluctuations. Eq. (19) just represents the uncertainty relation $\Delta p \Delta r = O(\hbar)$, interpreting $\varphi_n^i - \varphi_n^{i-1}$ as a momentum uncertainty. Combining eq. (18) and (19), the fluctuating Coulomb potential can thus be written as

$$V_C(\varphi_n^{i-1}, \varphi_n^i) = \alpha p_{max}(\Phi_n^i - \Phi_n^{i-1}). \tag{20}$$

Summing the two contributions of the pair $(i, i-1)$ and $(i, i+1)$ we just get the linear interaction terms of the nearest neighbours in the discrete dynamics (8) (for more details on this model, see [16], chapter 5). The remarkable thing is that in this interpretation α has now the physical meaning of a gauge coupling. Of course, a similar consideration applies to all kinds of coupling constants (electroweak and strong) in the standard model, not only the fine structure constant. Our central hypothesis is thus that the chaotic fields, though being different from the ordinary standard model fields, interact with *the same* coupling constants as in the ordinary standard model.

There may be various degrees of freedom of the chaotic fields underlying dark energy. For example, from a dynamical systems point of view, it makes sense to generalize the chaotic field dynamics (8) to

$$\Phi_{n+1}^i = (1-\alpha)T(\Phi_n^i) + \sigma\frac{\alpha}{2}(T^b(\Phi_n^{i-1}) + T^b(\Phi_n^{i+1})) \tag{21}$$

(the noise term can usually be neglected for chaotic maps). The case $\sigma = +1$ is called 'diffusive coupling', the case $\sigma = -1$ 'anti-diffusive coupling'. Chaotic fields with $b = 1$ are called to be of 'type A' ($T^1(\Phi) =: T(\Phi)$), chaotic fields with $b = 0$ to be of 'type B' ($T^0(\Phi) =: \Phi$). There are two different types of vacuum energies for the chaotic fields, namely the self energy $V(\alpha) := \frac{p_{max}^2}{\tau}\left(\frac{3}{2}\langle\Phi^2\rangle - \langle\Phi^4\rangle\right)$ and the interaction energy $W(\alpha) := \frac{p_{max}^2}{2\tau}\langle\Phi_n^i\Phi_n^{i+1}\rangle$ (see [17] for more details).

Fig. 1 shows that the self energy $V(\alpha)$ of the chaotic fields indeed distinguishes electroweak coupling constants known from the standard model. Everybody can easily reproduce this plot, by simply iterating the dynamics (8) for random initial conditions $\Phi_0^i \in [-1, 1]$ for a long time on a large lattice and averaging the variable $1.5(\Phi_n^i)^2 - (\Phi_n^i)^4$. We observe that $V(\alpha)$ has local minima at $a_1 = 0.000246(2)$, $a_2 = 0.00102(1)$, $a_3 = 0.00220(1)$ (a_1 and a_3 are actually small local minima on top of the hill).

On the other hand, in the standard model of electroweak interactions the weak coupling constant is given by

$$\alpha_{weak} = \alpha_{el}\frac{(T_3 - Q\sin^2\theta_W)^2}{\sin^2\theta_W\cos^2\theta_W} \tag{22}$$

Here Q is the electric charge of the particle ($Q = -1$ for electrons, $Q = 2/3$ for u-like quarks, $Q = -1/3$ for d-like quarks), and T_3 is the third component of the isospin ($T_3 = 0$ for right-handed particles, $T_3 = -\frac{1}{2}$ for e_L and d_L, $T_3 = +\frac{1}{2}$ for ν_L and u_L). Consider right-handed fermions f_R. With $\sin^2\theta_W = \bar{s}_l^2 = 0.2318$ (as experimentally measured) and the running electric coupling $\alpha_{el}(E)$ taken at energy scale $E = 3m_f$ we obtain from eq. (22) the numerical values $\alpha_{weak}^{d_R}(3m_d) = 0.000246$, $\alpha_{weak}^{c_R}(3m_c) = 0.001013$, $\alpha_{weak}^{e_R}(3m_e) = 0.00220$. There is an amazing numerical coincidence between the local minima a_1, a_2, a_3 of $V(\alpha)$ and the experimentally measured weak coupling constants of $f_R = u_R, c_R, e_R$, respectively. The factor 3 of the energy scale can be related to the index of the Tchebyscheff polynomial [16].

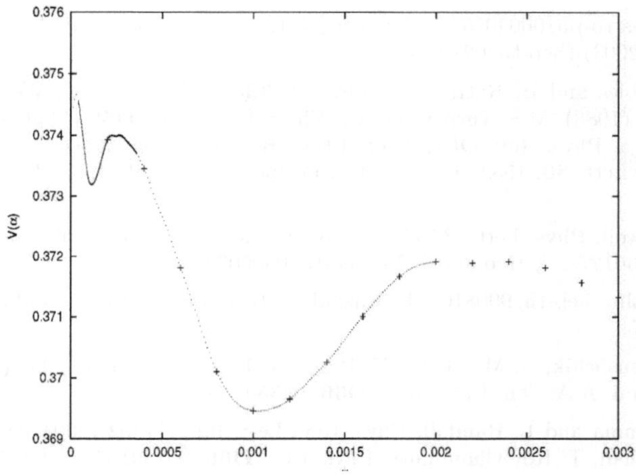

Figure 1. Self energy $V(\alpha)$ of the type-A chaotic field in the low-coupling region. There are local minima at couplings a_i that coincide with the weak coupling constants of right-handed fermions in the standard model.

Now regard the fine structure constant α_{el} and the Weinberg angle $\sin^2\theta_W$ as a priori free parameters. Suppose these parameters would change to slightly different values. Then immediately this would produce larger vacuum energy $V(\alpha)$ in our sterile amendment of the standard model, since we get out of the local minima. The system is expected to be driven back to the local minima, and the fundamental parameters are fixed and stabilized in this way.

Further coincidences of this type have been observed for various other observables associated with the chaotic fields, allowing for a fixing of further fundamental constants such as mass ratios and strong couplings at bosonic mass scales. See [3, 17] for details. All these numerically observed coincidences are not explainable as a random coincidence. Rather, they suggest to interpret the coupling constant α of our second-quantized chaotic fields φ as a running gauge coupling. The chaotic fields are most naturally associated with an additional, sterile sector of the standard model, which just consists of vacuum fluctuations of a scalar field with a cutoff. This sector generates dark energy, and its sense is to fix and stabilize the fundamental constants of nature.

References

[1] A.G. Riess et al., Astron. J. **116**, 1009 (1998), S. Perlmutter et al., Astrophys. J. **517**, 565 (1999), J.L. Tonry et al., astro-ph/0305008

[2] N.W. Halverson et al., Astrophys. J. **568**, 38 (2002), C.B. Netterfield et al., Astrophys. J. **571**, 604 (2002),

[3] D.N. Spergel et al., Astrophys. J. Suppl. **148**, 175 (2003) (astro-ph/0302209), C.L. Bennett et al., Astrophys. J. Suppl. **148**, 1 (2003) (astro-ph/0302207)

[4] S. Caroll, astro-ph/0004075, S. Weinberg, astro-ph/0005265, T. Padmanabhan, Phys. Rep. **380**, 235 (2003) (hep-th/0212290)

[5] P.J.E. Peebles and B. Ratra, Astrophys. J. **325**, L17 (1988), C. Wetterich, Nucl. Phys. **B312**, 668 (1988), M.S. Turner and M. White, Phys. Rev. **D56**, 4439 (1997), J.A. Frieman and I. Waga, Phys. Rev. **D57**, 4642 (1998), R.R. Caldwell, R. Dave, and P.J. Steinhardt, Phys. Rev. Lett. **80**, 1582 (1998), P.J.E. Peebles and A. Vilenkin, Phys. Rev. **D59**, 063505 (1999)

[6] R.R. Caldwell, Phys. Lett. **B545**, 23 (2002), S.M. Caroll, M. Hoffmann, and M. Trodden, astro-ph/0301273, J. Hao and X Li, hep-th/0306033

[7] A.A. Tseytlin, hep-th/9908105, E. Elizalde, J.E. Lidsey, S. Nojiri, and S.D. Odintsov, hep-th/0307177

[8] A.Y. Kamenshchik, U. Moschella, V. Pasquier, Phys. Lett. **B511**, 265 (2001), M.C. Bento, O. Bertolami, A.A. Sen, Phys. Rev. **D66**, 043507 (2002)

[9] S. Mukohyama and L. Randall, Phys. Rev. Lett. **92**, 211302 (2004) (hep-th/0306108), T. Padmanabhan, T. Roy Choudhury, Phys. Rev. **D66**, 081301 (hep-th/0205055)

[10] C. Beck, Phys. Rev. **D69**, 123515 (2004) (astro-ph/0310479)

[11] C. Wetterich, Nucl. Phys. **D302**, 645 (1988), H.B. Sandvik, J.D. Barrow and J. Magueijo, Phys. Rev. Lett. **88**, 031302 (2002) (astro-ph/0107512), C. Wetterich, Phys. Lett. **B561**, 10 (2003) (hep-ph/0301261), M. Doran, astro-ph/0411606, N.J. Nunes and J.E. Lidsey, Phys. Rev. **D69**, 123511 (2004) (astro-ph/0310882)

[12] See, e.g., neutrino reviews at pdg.lbl.gov

[13] C. Beck and M.C. Mackey, Phys. Lett. **B605**, 295 (2005) (astro-ph/0406505)

[14] G. Parisi, Y. Wu, Sci Sin. **24**, 483 (1981), P.H. Damgaard, H. Hüffel (eds.), *Stochastic Quantization*, World Scientific, Singapore (1988)

[15] K. Kaneko, Progr. Theor. Phys. **72**, 480 (1984), R. Kapral, Phys. Rev. **A31**, 3868 (1985), C. Beck, Phys. Lett. **A248**, 386 (1998)

[16] C. Beck, *Spatio-temporal Chaos and Vacuum Fluctuations of Quantized Fields*, World Scientific, Singapore (2002) (50-page summary at hep-th/0207081)

[17] C. Beck, Physica **D171**, 72 (2002) (hep-th/0105152)

[18] C. Beck and F. Schlögl, *Thermodynamics of Chaotic Systems*, Cambridge University Press, Cambridge (1993), C. Beck, Nonlinearity **4**, 1131 (1991), A. Hilgers, C. Beck, Physica **D156**, 1 (2001)

HOW FUNDAMENTAL IS GRAVITATION?

B.G. SIDHARTH

International Institute for Applicable Mathematics,
and Information Sciences,Hyderabad(India) and Udine(Italy),
B.M. Birla Science Centre,
Adarsh Nagar, Hyderabad - 500 063 (India)

Abstract

We consider a model in which the Universe has an underpinning of oscillators in the Quantum Vacuum (or dark energy) at the Planck scale and deduce a number of otherwise inexplicable Large Number relations which have been considered to be empirical accidents. The analysis shows that the gravitational energy is the residual energy of the Planck oscillators constituting the Universe at large on the one hand, and elementary particles on the other. This explains a mysterious puzzle first pointed out by Weinberg several years ago, in a formula relating the pion mass to the Hubble Constant, a puzzle which has remained unexplained ever since.

1 Introduction

More than five thousand years ago, the Rig Veda repeatedly raised the question: "How is it that *though unbound* the sun does not fall down?"

This was a question that puzzled thinking man over the millennia. Indian scholars right up to Bhaskaracharya who lived about a thousand years ago believed in some attractive force which was responsible for keeping the celestial bodies from falling down.

The same problem was addressed by Greek thinkers about two thousand five hundred years ago. They devised transparent material spheres to which each of the celestial objects were attached - the material spheres prevented them from falling down.

B.G. Sidharth, F. Honsell and A. de Angelis. (eds.), Frontiers of Fundamental Physics, 147–154.

Unfortunately it was this answer to the age old question, which held up further scientific progress during the time of Kepler, for even Copernicus accepted the transparent material spheres.

Kepler had a powerful tool in the form of the accurate observations of Tycho Brahe. He also had the advantage of the Indian numeral system, which via the Arabs reached Europe just a few centuries earlier. These lead him to his famous laws of elliptical orbits with definite periods correlated to distances from the Sun.

This was the beginning of modern science.

The important point was that the Greek answer to the problem of why heavenly objects do not fall down was now demolished. Kepler himself speculated about some type of a magnetic force between the Sun and the Planets, rather on the lines of speculations in India.

It was Newton who provided the next chapter in the story.

To quote Hawking [1] "The Philosophiae Naturalis Principia Mathematica by Isaac Newton, first published in Latin in 1687, is probably the most important single work ever published in the physical sciences. Its significance is equalled in the biological sciences only by *The Origin of Species by Charles Darwin.* The original impulse which caused Newton to write the *Principia* was a question from Edmund Halley as to whether the elliptical orbits of the planets could be accounted for on the hypothesis of an inverse square force directed towards the Sun. This was something that Newton had worked out some years earlier but had not published, like most of his work on mathematics and physics. However, Halley's challenge, and the desire to refute the suggestions of others such as Hooke and Descartes, spurred Newton to try to write a proper account of this result."

Newton using Galileo's ideas of Mechanics, thus stumbled upon the universal Law of Gravitation.

This held sway for nearly two hundred and twenty five years, before Einstein came out with his own theory of gravitation. This was no force in the sense that Newton's and preceding scholars had envisaged it to be. Rather it was due to the curvature of spacetime itself. Einstein's bizarre ideas have had some experimental verification while there are some other experimental consequences, such as gravitational waves, which need to be confirmed.

After Einstein's formulation of gravitation a problem that has challenged and defied solution has been that of providing a unified description of gravitation along with other fundamental interactions. One of the earliest attempts was that of Hermann Weyl - the gauge geometry [2] , which though elegant was rejected on the grounds that in the final analysis, it was not really a unification of gravitation with electromagnetism.

Modern approaches to this problem have lead to the abandonment of a smooth spacetime manifold. Instead, the Planck scale is now taken to be a minimum fundamental scale.

We had already argued from different points of view to arrive at the otherwise empirically known equations [3, 4, 5]

$$R = \sqrt{N} l_P = \sqrt{N} l$$

$$l = \sqrt{n} l_P \tag{1}$$

where l_P, l and R are the Planck length, the pion Compton wavelength and the radius of the universe and $N, \bar{N}$ and n are certain large numbers. Some of these are well known empirically for example $\bar{N} \sim 10^{80}$ being the number of elementary particles, which typically are taken to be pions in the literature, in the universe.

One way of arriving at the above relations is by considering a series of N Planck mass oscillators which are created out of the Quantum Vacuum. In this case (Cf. also ref.[6]) we have

$$r = \sqrt{N a^2} \tag{2}$$

In (2) a is the distance between the oscillators and r is the extent. Equations (1) follow from equation (2).

There is another way of arriving at equations (1) (Cf.ref.[5]). For this, we observe that the position operator for the Klein-Gordan equation is given by [7],

$$\vec{X}_{op} = \vec{x}_{op} - \frac{\imath \hbar c^2}{2} \frac{\vec{p}}{E^2}$$

Whence we get

$$\hat{X}^2_{op} \equiv \frac{2m^3c^4}{\hbar^2} X^2_{op} = \frac{2m^3c^6}{\hbar^2} x^2 + \frac{p^2}{2m} \tag{3}$$

It can be seen that purely mathematically (3) for $\hat{X}^2_{op}$ defines the Harmonic oscillator equation, this time with quantized, what may be called space levels. It turns out that these levels are all multiples of $(\frac{\hbar}{mc})^2$. This Compton length is the Planck length for a Planck mass particle. Accordingly we have for any system of extension r,

$$r^2 \sim N l^2$$

which gives back equation (1). It is also known that the Planck length is also the Schwarzschild radius of a Planck mass, that is we have

$$l_P = G m_P / c^2 \tag{4}$$

Using equations (1) and (4), we will now deduce a few new and valid and a number of otherwise empirically known relations involving the various microphysical parameters and large scale parameters. Some of these relations are deducible from the others. Many of these relations featured (empirically) in Dirac's Large Number Cosmology. We follow Dirac and Melnikov in considering $l, m, \hbar, l_P, m_P$ and e as microphysical parameters [8, 9]. Large scale parameters include the radius and the mass of the universe, the number of elementary particles in the universe and so on.

In the process we will also examine the nature of gravitation. It must also be observed that the Large Number relations below are to be considered in the Dirac sense, wherein for example the difference between the electron and pion (or proton) masses is irrelevant [10].

2 Interrelationships

We will use the following well known equation which has been obtained through several routes:

$$\frac{GM}{c^2} = R \tag{5}$$

For example in an uniformly expanding flat Friedman spacetime, we have [10]

$$\dot{R}^2 = \frac{8\pi G\rho R^2}{3}$$

If we substitute $\dot{R} = c$ at the radius of the universe in the above we recover (5).

We now observe that from the first two relations of (1), using the Compton wavelength expression we get

$$m = m_P/\sqrt{n} \tag{6}$$

Using also the second relation in (1) we can easily deduce

$$N = \bar{N}n \tag{7}$$

Using (1) and (5) we have

$$M = \sqrt{N}m_P \tag{8}$$

Interestingly (8) can be obtained directly, without recourse to (5), from the energy of the Planck oscillators (Cf.ref.[4]). Combining (8) and (6) we get

$$M = \left(\sqrt{Nn}\right) m \tag{9}$$

Further if we use in the last of equation (1) the fact that l_P is the Schwarzchild radius that is equation (4), we get,

$$G = \frac{lc^2}{nm} \tag{10}$$

We now observe that if we consider the gravitational energy of the N Planck masses (which do not have any other interactions) we get,

$$\text{Gravitational Energy} = \frac{GNm_P^2}{R}$$

If this is equated to the inertial energy in the universe, Mc^2, as can be easily verified we get back (5). In other words the inertial energy content of the universe equals the gravitational energy of all the N Planck oscillators.

Similarly if we equate the gravitational energy of the n Planck oscillators constituting the pion we get

$$\frac{Gm_P^2 n}{R} = mc^2 \tag{11}$$

Using in (11) equation (4) we get

$$\frac{l_P m_P n}{R} = m$$

Whence it follows on using (7), (6) and (1),

$$n^{3/2} = \sqrt{N}, \, n = \sqrt{\bar{N}} \tag{12}$$

Substituting the value for n from (12) into (10) we will get

$$G = \frac{lc^2}{\sqrt{\bar{N}}m} \tag{13}$$

If we use (12) in (9) we will get

$$M = \bar{N}m \tag{14}$$

Alternatively we could use (14) which expresses the fact that the mass of the universe is given by the mass of the $\bar{N}$ elementary particles in it and deduce equations (11), (12) and (13). Indeed a rationale for this is the fact that the Universe at large is electrically neutral and so it is the gravitational force which predominates, and this very weak in comparison. Using the expressions for the Planck length as a Compton wavelength and equating it to (4) we can easily deduce

$$Gm^2 = \frac{e^2}{n} = \frac{e^2}{\sqrt{\bar{N}}} \tag{15}$$

wherein we have also used $\hbar c \sim e^2$ and (6). Equation (15) is another empirically well known equation which was used by Dirac in his Cosmology. Interestingly, as we have deduced (15), rather than use it empirically, this points to a unified description of electromagnetism and gravitation.

Interestingly also rewriting (13) as

$$G = \frac{l^2 c^2}{Rm}$$

wherein we have used (1) and further using the fact that $H = c/R$, where H is the Hubble constant we can deduce

$$m \approx \left(\frac{H\hbar^2}{Gc}\right)^{\frac{1}{3}} \tag{16}$$

Equation (16) is the so called mysterious Weinberg formula, known empirically [10]. As Weinberg put it, "...it should be noted that the particular combination of $\hbar, H, G$, and c appearing (in the formula) is very much closer to a typical elementary particle mass than other random combinations of these quantities; for instance, from $\hbar, G$, and c alone one can form a single quantity $(\hbar c/G)^{1/2}$ with the dimensions of a mass, but this has the value $1.22 \times 10^{22} MeV/c^2$, more than a typical particle mass by about 20 orders of magnitude!

"In considering the possible interpretations (of the formula), one should be careful to distinguish it from other numerical "coincidences"... In contrast, (the formula) relates a single cosmological parameter, H, to the fundamental constants $\hbar, G, c$ and m, and is so far unexplained."

We will come back to this point but remark that (13) brings out gravitation in a different light– somewhat on the lines of Sakharov. In fact it shows up gravitation as the excess or residual energy in the universe.

Finally it may be observed that (13) can also be rewritten as

$$\bar{N} = \left(\frac{c^2 l}{mG}\right)^2 \sim 10^{80} \tag{17}$$

and so also (10) can be rewritten as

$$n = \left(\frac{lc^2}{Gm}\right) \sim 10^{40}$$

It now immediately follows that

$$N \sim 10^{120}$$

Looking at it this way, given G and the microphysical parameters we can deduce the numbers $N, \bar{N}$ and n.

3 Discussion

Thus the many so called large number coincidences and the mysterious Weinberg formula can be deduced on the basis of a Planck scale underpinning for the elementary particles and the whole universe. This was done from a completely different point of view, namely using fuzzy spacetime and fluctuations in a 1997 model that successfully predicted a dark energy driven accelerating universe with a small cosmological constant [5, 11].

However the above treatment brings out the role of the Planck scale particles in the Quantum Vacuum. It resembles, as remarked earlier the Sakharov-Zeldovich metric elasticity of space approach [12]. Essentially Sakharov argues that the renormalization process in Quantum Field Theory which removes the Zero Point energies is altered in General Relativity due to the curvature of spacetime, that is the renormalization or subtraction no longer gives zero but rather there is a residual energy similar to the modification in the molecular bonding energy due to deformation of the solids. We see this in a little more detail following Wheeler [13]. The contribution to the Lagrangian of the Zero Point energies can be given in a power series as follows

$$L(r) = A\hbar \int k^3 dk + B\hbar^{(4)} r \int k dk$$

$$+\hbar[C({}^{(4)}r)^2 + Dr^{\alpha\beta}r_{\alpha\beta}]\int k^{-1}dk$$

$$+(\text{higher-order terms}). \tag{18}$$

where A, B, C etc. are of the order of unity and r denotes the curvature. By renormalization the first term in (18) is eliminated. According to Sakharov, the second term is the action principle term, with the exception of some multiplicated factors. (The higher terms in (18) lead to corrections in Einstein's equations). Finally Sakharov gets

$$G = \frac{c^3}{16\pi B\hbar \int kdk} \tag{19}$$

Sakharov then takes a Planck scale cut off for the divergent integral in the denominator of (19). This immediately yields

$$G \approx \frac{c^3 l_P^2}{\hbar} \tag{20}$$

In fact using relations like (1), (6) and (12), it is easy to verify that (20) gives us back (10) (and (13)).

According to Sakharov (and (20)), the value of G is governed by the Physics of Fields and Particles and is a measure of the metrical elasticity at small spacetime intervals. It is a microphysical constant.

However in our interpretation of (13) (which is essentially the same as Sakharov's equation (20)), G appears as the expression of a residual energy over the entire universe: The entire universe has an underpinning of the N Planck oscillators and is made up of $\bar{N}$ elementary particles, which again each have an underpinning of n Planck oscillators. It must be reiterated that (20) obtained from Sakharov's analysis shows up G as a microphysical parameter because it is expressed in their terms. This is also the case in Dirac's cosmology. This is also true of (10) because n relates to the micro particles exclusively.

However when we use the relation (12), which gives n in terms of $\bar{N}$, that is links up the microphysical domain to the large scale domain, then we get (13). With Sakharov's equation (20), the mysterious nature of the Weinberg formula remains. But once we use (13), we are effectively using the large scale character of G – it is not a microphysical parameter. This is brought out by (17), which is another form of (13). If G were a microphysical parameter, then the number of elementary particles in the universe would depend solely on the microphysical parameters and would not be a large scale parameter. The important point is that G relates to elementary particles and the whole universe [14]. That is why (13) or equivalently the Weinberg formula (16) relate supposedly microphysical parameters to a cosmological parameter. Once the character of G as brought out by (13) is recognized, the mystery disappears.

Finally it may be remarked that attempts to unite gravitation with other interactions have been unsuccessful for several decades. However, it is possible to get a description of gravitation in an extended gauge field formulation using noncommutative geometry (to take account of the fact that the graviton is a spin 2 particle)

[15, 16]. It is important to recognize what Witten has said, "The existence of gravity clashes with our description of the rest of physics by quantum fields" [17]. This sentiment was echoed much earlier by Pauli who even went so far as to say that we should not try to unite what God had intended to be separate.

References

[1] S.W. Hawking, and Werner Israel, "300 Years of Gravitation", Cambridge University Press, Cambridge, 1987.

[2] H. Weyl, "Space-Time Matter", Denver Publications Inc., New York, 1962, 282ff.

[3] B.G. Sidharth, Found.Phys.Letters., 15 (6), 2002, 577-583.

[4] B.G. Sidharth, Found.Phys.Letters., 17 (5), 2004, 503-506.

[5] B.G. Sidharth, Int.J.Mod.PhysA., 13 (15), 1998, p.2599ff.

[6] Y. Jack Ng and H. Van Dam, Mod.Phys.Lett.A., 9 (4), 1994, p.335-340.

[7] S.S. Schweber, "An Introduction to Relativistic Quantum Field Theory", Harper and Row, New York, 1961.

[8] J.V. Narlikar, "Introduction to Cosmology", Cambridge University Press, Cambridge, 1993, p.57.

[9] V.N. Melnikov, Int.J.Th.Phys., 33 (7), 1994, 1569-1579.

[10] S. Weinberg, "Gravitation and Cosmology", John Wiley & Sons, New York, 1972, pp.619ff.

[11] B.G. Sidharth, "Chaotic Universe: From the Planck to the Hubble Scale", Nova Science Publishers, Inc., New York, 2001.

[12] A.D. Sakharov, Soviet Physics - Doklady, Vol.12, No.11, 1968, p.1040-1041.

[13] C.W. Misner, K.S. Thorne and J.A. Wheeler, "Gravitation", W.H. Freeman, San Francisco, 1973, p.819ff.

[14] B.G. Sidharth, "A Note on the Characterization of Gravitation", To appear in Foundation of Physics Letters.

[15] B.G. Sidharth, "Gravitation from a Gauge like Formulation", to appear in Annales de la Fondation Louis De Broglie.

[16] B.G. Sidharth, "Universe of Fluctuations", Springer (under publication).

[17] W. Witten, Physics Today, April 1996, pp.24-30.

SCALE-DEPENDENT STOCHASTIC QUANTIZATION

MIKHAIL ALTAISKY [a,b]

[a] *Joint Institute for Nuclear Research, Dubna, 141980, Russia*
[b] *Space Research Institute, Moscow, 117997, Russia*

Abstract

Based on the wavelet-defined multiscale random noise proposed in [2], a multiscale version of the stochastic quantization procedure is considered. A new type of the commutation relations emerging from the multiscale decomposition of the operator-valued fields is derived.

1 Introduction

A highly original method of stochastic quantization of gauge fields proposed by G.Parisi and Y.Wu [1] have been attracting attention for more than 20 years. Let $S_E[\phi]$ be the action Euclidean field theory in $\mathbb{R}^d$. Then, instead of direct calculation of the Green functions from the generation functional of the field theory, it is possible to introduce a *fictitious time* variable τ, make the quantum fields into stochastic fields $\phi(x) \to \phi(x,\tau), x \in \mathbb{R}^d, \tau \in \mathbb{R}$ and evaluate the moments $\langle \phi(x_1,\tau_1)\ldots\phi(x_m,\tau_m)\rangle$ by averaging over a random process $\phi(x,\tau,\cdot)$ governed by the Langevin equation

$$\frac{\partial\phi(x,\tau)}{\partial\tau} + \frac{\delta S}{\delta\phi(x,\tau)} = \eta(x,\tau). \tag{1}$$

The Gaussian random force is δ-correlated in both the $\mathbb{R}^d$ coordinate and the fictitious time:

$$\langle \eta(x,\tau)\eta(x',\tau')\rangle = 2D_0\delta(x-x')\delta(\tau-\tau'), \quad \langle\eta(x,\tau)\rangle = 0. \tag{2}$$

B.G. Sidharth, F. Honsell and A. de Angelis. (eds.), Frontiers of Fundamental Physics, 155–161.

The physical Green functions are obtained by taking the steady state limit

$$G(x_1,\ldots,x_m) = \lim_{\tau\to\infty} \langle \phi(x_1,\tau)\ldots\phi(x_m,\tau)\rangle.$$

Following [3] we extend the method of stochastic quantization by introducing the scale-dependent random processes $W(a,b,\cdot)$, where $b \in \mathbb{R}^d$ is a spatial coordinate, and a is the spatial resolution. For a square-integrable function $f(x,\cdot)$ the wavelet coefficients are

$$W(a,b,\cdot) = \int |a|^{-d} \overline{\psi\left(\frac{x-b}{a}\right)} f(x,\cdot) d^d x. \tag{3}$$

Hereafter they will be referred to as the *scale components* of f with respect to the basic wavelet ψ. The reconstruction of a function from its scale components is given by the inverse wavelet transform

$$f(x,\cdot) = \frac{2}{C_\psi}\int_0^\infty \frac{da}{a^{d+1}} \int d^d b \psi\left(\frac{x-b}{a}\right) W(a,b,\cdot), \quad C_\psi = \int \frac{|\tilde\psi(k)|^2}{S_d |k|^d} d^d k, \tag{4}$$

with S_d being the are of the unit sphere in d dimensions, is the normalization for the isotropic wavelets. Performing the wavelet transform (in spatial coordinate) of the fields and the random force in the Langevin equation, we get the possibility to substitute the white noise (2) by a scale-dependet random force

$$\langle \tilde\eta(a_1,k_1,\tau_1)\tilde\eta(a_2,k_2,\tau_2)\rangle = C_\psi (2\pi)^d \delta^d(k_1+k_2)\delta(\tau_1-\tau_2) a_1 \delta(a_1-a_2) D(a_1,k_1) \tag{5}$$

In case the spectral density of the random force is a constant $D(a_1,k_1) = D_0$, the inverse wavelet transform

$$\phi(x) = \frac{2}{C_\psi}\int_0^\infty \frac{da}{a}\int \frac{d^d k}{(2\pi)^d}\frac{d\omega}{2\pi} \exp(\imath(kx-\omega\tau))\tilde\psi(ak)\tilde\phi(a,k,\omega), \tag{6}$$

drives the process (5) into the white noise (2).

In case of arbitrary functions $\phi(a,x,\cdot)$ we have more possibilities. In particular, we can define a narrow band forcing that acts at a single scale

$$D(a,k) = a_0 \delta(a-a_0) D_0. \tag{7}$$

The contribution of the scales with the wave vectors apart from the the typical scale a_0^{-1} is suppressed by rapidly vanishing wings of the compactly supported wavelet $\tilde\psi(k)$.

Here we present two examples of the divergence free stochastic perturbation expansion: (i) the scalar field theory ϕ^3, (ii) the non-Abelian gauge field theory.

2 Scalar field theory

Let us turn to the stochastic quantization of the ϕ^3 theory with the scale-dependent noise [3]. The Euclidean action of the ϕ^3 theory is

$$S_E[\phi(x)] = \int d^d x \left[\frac{1}{2}(\partial\phi)^2 + \frac{m^2}{2}\phi^2 + \frac{\lambda}{3!}\phi^3\right]. \tag{8}$$

The corresponding Langevin equation is written as

$$\frac{\partial\phi(x,\tau)}{\partial\tau} + \left(-\Delta\phi + m^2\phi + \frac{\lambda}{2!}\phi^2\right) = \eta(x,\tau). \tag{9}$$

Substituting the scale components in representation (6) we get the integral equation for the stochastic fields

$$(-\imath\omega + k^2 + m^2)\phi(a,k,\omega) = \eta(a,k,\omega) - \tfrac{\lambda}{2}\overline{\tilde{\psi}(ak)}\left(\tfrac{2}{C_\psi}\right)^2 \int \tfrac{d^dk_1}{(2\pi)^d}\tfrac{d\omega_1}{2\pi}\tfrac{da_1}{a_1{}^{d+1}}\tfrac{da_2}{a_2{}^{d+1}}$$
$$\tilde{\psi}(a_1k_1)\tilde{\psi}(a_2(k-k_1))\phi(a_1,k_1,\omega_1)\phi(a_2,k-k_1,\omega-\omega_1). \tag{10}$$

Starting from the zero-th order approximation $\phi_0 = G_0\eta$ with the bare Green function $G_0(k,\omega) = 1/(-\imath\omega + k^2 + m^2)$ and iterating the integral equation (10), we get the one-loop correction to the stochastic Green function

$$G(k,\omega) = G_0(k,\omega) + \lambda^2 G_0^2(k,\omega)\int \frac{d^dq}{(2\pi)^d}\frac{d\Omega}{2\pi} 2\Delta(q)|G_0(q,\Omega)|^2 G_0(k-q,\omega-\Omega) + \ldots, \tag{11}$$

where $\Delta(k)$ is the scale averaged effective force correlator

$$\Delta(k) \equiv \frac{2}{C_\psi}\int_0^\infty \frac{da}{a}|\hat{\psi}(ak)|^2 D(a,k). \tag{12}$$

In the same way the other stochastic momenta can be evaluated. Thus the common stochastic diagram technique is reproduced with the scale-dependent random force (5) instead of the standard one (2). The 1PI diagrams corresponding to the stochastic Green function decomposition (11) are shown in Fig. 1.

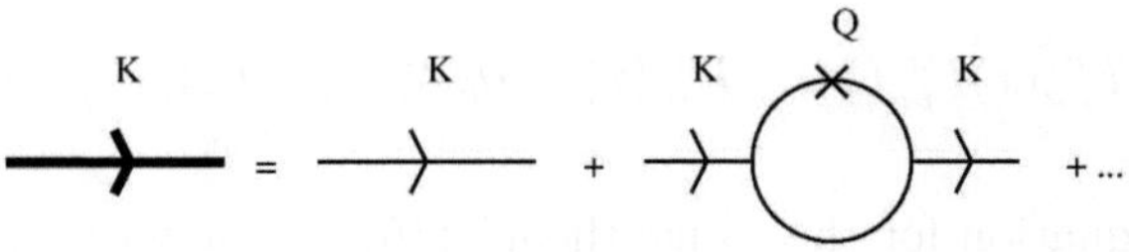

Figure 1. Diagram expansion of the stochastic Green function in ϕ^3-model

It can be easily seen that for a single-band forcing (7) and a suitably chosen wavelet the loop divergences are suppressed. For instance, the use of the Mexican hat wavelet

$$\hat{\psi}(k) = (2\pi)^{d/2}(-\imath k)^2 \exp(-k^2/2), \quad C_\psi = (2\pi)^d \tag{13}$$

for the single band random force (7) gives the effective force correlator

$$\Delta(q) = (a_0q)^4 e^{-(a_0q)^2} D_0. \tag{14}$$

Table 1. Stochastic diagrams for the non-Abelian gauge fields. Redrawn from [4]

Diagram	Notation	Formula
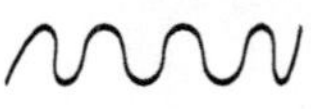	$G^{ab}_{\mu\nu}(k,\tau-\tau')$	$\delta^{ab}\theta(\tau-\tau')\left[\left(\delta_{\mu\nu}-\frac{k_\mu k_\nu}{k^2}\right)e^{-k^2(\tau-\tau')}+\frac{k_\mu k_\nu}{k^2}\right]$
	$D^{ab}_{\mu\nu}(k,\tau-\tau')$	$\delta^{ab}\Big[\left(\delta_{\mu\nu}-\frac{k_\mu k_\nu}{k^2}\right)\left(e^{-k^2\lvert\tau-\tau'\rvert}-e^{-k^2(\tau+\tau')}\right)$ $+2min(\tau,\tau')\frac{k_\mu k_\nu}{k^2}+\frac{k_\mu k_\nu}{k^2}\Big]$
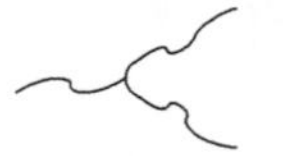	$\frac{g}{2}V^{abc}_{\mu\kappa\lambda}(k_1,k_2,k_3)$	$\frac{\imath g}{2}f^{abc}\big[(k_1-k_2)_\lambda\delta_{\mu\kappa}+(k_2-k_3)_\mu\delta_{\kappa\lambda}$ $+(k_3-k_1)_\kappa\delta_{\mu\lambda}\big]$
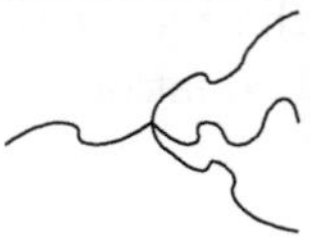	$\frac{g^2}{6}W^{abcd}_{\mu\nu\kappa\lambda}$	$-\frac{g^2}{6}\big[f^{xab}f^{xcd}(\delta_{\mu\kappa}\delta_{\nu\lambda}-\delta_{\mu\lambda}\delta_{\nu\kappa})$ $+f^{xac}f^{xbd}(\delta_{\mu\nu}\delta_{\kappa\lambda}-\delta_{\mu\lambda}\delta_{\nu\kappa})$ $+f^{xad}f^{xbc}(\delta_{\mu\nu}\delta_{\kappa\lambda}-\delta_{\mu\kappa}\delta_{\nu\lambda})\big]$

The loop integrals taken with this effective force correlator (14) can be easily seen to be free of ultra-violet divergences

$$
\begin{aligned}
G_2(k,\omega) &= G_0^2(k,\omega)\int\frac{d^dq}{(2\pi)^d}2\Delta(q)\int_{-\infty}^{\infty}\frac{d\Omega}{2\pi}\frac{1}{\Omega^2+(q^2+m^2)^2} \\
&\times \frac{1}{-\imath(\omega-\Omega)+(k-q)^2+m^2}
\end{aligned}
\tag{15}
$$

3 Non-Abelian gauge theory

The Euclidean action of a non-Abelian field is given by

$$
S[A]=\frac{1}{4}\int d^dxF^a_{\mu\nu}(x)F^a_{\mu\nu}(x),\quad F^a_{\mu\nu}(x)=\partial_\mu A^a_\nu(x)-\partial_\nu A^a_\mu(x)+gf^{abc}A^b_\mu(x)A^c_\nu(x).
\tag{16}
$$

The Langevin equation for the gauge theory (16) can be written as

$$
\frac{\partial A^a_\mu(x,\tau)}{\partial\tau}+\left(-\delta_{\mu\nu}\partial^2+\partial_\mu\partial_\nu\right)A^a_\nu(x,\tau)=\eta^a_\mu(x,\tau)+U^a_\mu(x,\tau),
\tag{17}
$$

where $\eta^a_\mu(x,\tau)$ is the random force and $U^a_\mu(x,\tau)$ is the nonlinear interaction term

$$
U[A]=\frac{g}{2}V^0(A,A)+\frac{g^2}{6}W^0(A,A,A).
$$

The stochastic diagram technique for the gauge field Langevin equation (17) is summarized in the Table 1. The two terms standing in the free field Green function correspond to the transversal and the longtitudal mode propagation:

$$
G^{ab}_{\mu\nu}(k)=\frac{T_{\mu\nu}(k)\delta_{ab}}{-\imath\omega+k^2}+\frac{L_{\mu\nu}(k)\delta_{ab}}{-\imath\omega},\qquad T_{\mu\nu}(k)=\delta_{\mu\nu}-\frac{k_\mu k_\nu}{k^2},L_{\mu\nu}(k)=\frac{k_\mu k_\nu}{k^2}.
$$

(Here we are concerned with divergences and do not touch any gauge fixing.)

Similarly the scalar field theory, we can use the scale-dependent forcing (18) in the Langevin equation (17). Since there is no dynamic evolution for the longtitudal modes in the Langevin equation (17), it is natural to use the transversal scale-dependent random force

$$\begin{aligned}\langle\eta^a_\mu(a_1,k_1,\tau_1)\eta^b_\nu(a_2,k_2,\tau_2)\rangle &= (2\pi)^d\delta^d(k_1+k_2)\delta(\tau_1-\tau_2)T_{\mu\nu}(k_1)\\ &\times\ C_\psi a_1\delta(a_1-a_2)D(a_1,k_1). \qquad (18)\end{aligned}$$

Let us consider a gluon loop with two cubic vertices. Summing up over the gauge group indices $\left(\frac{\imath}{2}g\right)^2 f^{abc}\delta_{bd}f^{der}\delta_{cr} = \frac{g^2}{4}\delta_{ae}C_2$, with $C_2 = N$ for SU_N groups, we can wright the gluon loop as a sum of two diagrams – those with the transversal and the longtitudal stochastic Green functions

$$G^{ab}_{2\mu\nu}(k,\omega) = g^2\delta_{ab}C_2|G_0(k,\omega)|^2\sum_{I=T,L}\int\frac{d\Omega}{2\pi}\frac{d^dq}{(2\pi)^d}N^I(k,\omega,q,\Omega)l^I_{\mu\nu}(k,q)2\Delta(q) \qquad (19)$$

where

$$N(k,q) = \left|\frac{1}{-\imath\Omega+q^2}\right|^2\begin{pmatrix}\frac{1}{-\imath(\omega-\Omega)+(k-q)^2}\\ \frac{1}{-\imath(\omega-\Omega)}\end{pmatrix}$$

$$l_{\mu\nu}(k,q) = V_{\mu\kappa\lambda}(k,k-q,q)T_{\lambda\gamma}(q)V_{\sigma\nu\gamma}(k-q,k,-q)\begin{pmatrix}T_{\kappa\sigma}(k-q)\\ L_{\kappa\sigma}(k-q)\end{pmatrix}$$

As it can be observed after explicit evaluation of the tensor structures $l^T_{\mu\nu}$ and $l^L_{\mu\nu}$, and integration over $d\Omega$, the wavelet factor in the effective force correlator $\Delta(q)$ will suppress the divergences for a narrow-band forcing (7). The power factor k^n of the basic wavelet ψ, that provides $\tilde\psi(0) = 0$, also makes the IR behavior softer. In this respect the wavelet regularization is different from the continuous regularization $\int d^dyR_\Lambda(\partial^2)\eta(y,\tau)$, see e.g. [6], that makes UV behavior softer by the factor $e^{-\frac{k^2}{\Lambda^2}}$, but do not affect the IR behavior.

4 Commutation relation

The stochastic quantization with a forcing localized at a given scale a_0 is in some way similar to the lattice regularization with the mesh size of order a_0. However there is a question what is the physical sense of the scale components, and what are the implications for canonical quantization of these fields? The answer to the first question stems from the definition of wavelet transform: the scale component $\phi(a,x)$ is a projection of the state vector ϕ to a certain multiresolution space [7], where ψ is a basis, i.e., the basic wavelet stands for the apperture of the microscope by which we perceive the system ϕ. To clarify the second question one can use the wavelet decomposition

$$\phi(x) = \frac{2}{C_\psi} \int_0^\infty \frac{da}{a} \int_{k_0>0} \frac{d^d k}{(2\pi)^d} \tilde{\psi}(ak) \left[e^{\imath kx} u(a,k) + (-1)^d e^{-\imath kx} u(a,-k) \right],$$

$$u^+(a,k) \equiv u(a,k)|_{k_0>0}, \quad u^-(a,k) \equiv u(a,-k)|_{k_0>0}, \qquad (20)$$

where the positive and the negative energy components (20) are summed up into the known plane wave components

$$u^\pm(k) = \frac{2}{C_\psi} \int_0^\infty \frac{da}{a} \tilde{\psi}(ak) u^\pm(a,k).$$

The canonical quantization of a scalar massless field, the implies the commutation relations

$$[u^+(k_1), u^-(k_2)] = (2\pi)^d \delta(k_1 - k_2), \qquad (21)$$

that can be maintained if we set [5]

$$[u^+(a_1,k_1), u^-(a_2,k_2)] = (2\pi)^d \delta(k_1 - k_2) \frac{C_\psi}{2} a_1 \delta(a_1 - a_2). \qquad (22)$$

For a massive field, with the given energy of the free particle $\omega_k = \sqrt{k^2 + m^2}$, the commutation relations for creation and annihilation operators

$$[b(a_1,k_1), b^+(a_2,k_2)] = (2\pi)^{d-1} \omega_k \delta^{d-1}(k_1 - k_2) C_\psi a_1 \delta(a_1 - a_2). \qquad (23)$$

To keep the Lorentz invariance at all scales the basic wavelet ψ can depend only on Lorentz scalars, such as $k_\mu k^\mu = m^2$. Being compactly supported in both x and k spaces the wavelet filter $\tilde{\psi}(ak) \equiv \mu(a^2 m^2)$ suppresses the contribution of the scale components which are far from the typical scale $a_m = m^{-1}$.

It should be emphasized that the commutation relations for scale components (22,23) are not unique: there may be constructed some other commutation relations is wavelet space that maintain the same canonical commutation relations in wavenumber space.

As it concerns the causality and operator ordering, the introduction of the scale argument in operator-valued functions implies the operators should be ordered in both the time and the scale. Extending the causality in this way it was suggested [5] to arrange the operator products by decreasing scale from right to left; so that the rightmost operator should correspond to the largest outermost object

$$T(A(\Delta_x,x)B(\Delta_y,y)) = \begin{cases} A(\Delta_x,x)B(\Delta_y,y) & y_0 < x_0 \\ \pm B(\Delta_y,y)A(\Delta_x,x) & x_0 < y_0 \\ A(\Delta_x,x)B(\Delta_y,y) & \Delta_y > \Delta_x, y_0 = x_0 \\ \pm B(\Delta_y,y)A(\Delta_x,x) & \Delta_y < \Delta_x, x_0 = y_0. \end{cases} \qquad (24)$$

References

[1] G.Parisi and Y.-S.Wu. *Scientica Sinica*, v.24, p.483 (1981)

[2] M.V.Altaisky. *Doklady Physics*, v.48, p. 478 (2003)

[3] M.V.Altaisky. in *Group 24: Physical and mathematical aspects of symmetries* pp.893-897, IoP, (2002)

[4] M. Namiki. *Stochastic quantization*, Springer (1992)

[5] M.V.Altaisky. *Renormalization group and geometry.* Proc. Int. Conf. FFP5, Hyderabad (2003)

[6] M.B.Halpern. *Progr. Theor. Phys. Suppl.* v.111, p.163 (1993)

[7] S. Mallat, *A theory for multiresolution signal decomposition: wavelet transform.* Preprint GRASP Lab. Dept. of Computer an Information Science, Univ. of Pensilvania. (1986)

PRINCIPIA GEOMETRICA PHYSICAE

JAIME KELLER

Department of Physics and Theoretical Chemistry,
Facultad de Química and Facultad de Estudios Superiores Cuautitlán,
Universidad Nacional Autónoma de México 04510
México D.F., apartado 70-528, MÉXICO

Abstract

We show that quadratic spaces are a fundamental clue to understand the structure of theoretical physics. Classical Physics is embedded within a 5-D flat quadratic space, 3 Space-, one Time- and one Action- like basis manifolds (Lorentz signature 1,4) faithfully providing a Relativistic Theory (START) describing Newtonian, Maxwell, geometrical optics and General Relativity as particular linear and quadratic forms of this (flat) START space. The 5-D space has a quadratic form which maps into the real quadratic form of a (hyperbolic-complex) 4-D space-time $dS^2 = dS \bullet dS^\dagger$. Otherwise the understanding of a many electron quantum mechanical (QM) system, with all its QM intricacies, is presented considering the QM "density" as a quadratic form and the QM "wave function" as its corresponding linear form.

Space-Time-Action, General Relativity, Geometric Optics.
83D05, 83Exx, 11E88, 15A66(67).
E-mail: keller@servidor.unam.mx and keller@cms.tuwien.ac.at

1 Physics and Quadratic Spaces

Physics is understood as the science describing nature as a whole in a useful way. For the Scientific Method (SM) the usefulness requires that two observers will find

B.G. Sidharth, F. Honsell and A. de Angelis. (eds.), Frontiers of Fundamental Physics, 163–173.

in their experiments similar phenomena and describe them with similar logical structures. Theoretical Physics provides a mathematical, SM acceptable, framework for this purpose. In this paper we show that a flat 5-D Lorentz Geometry is a useful formulation to consider frames of reference. Additionally we show the relation between quadratic spaces and Quantum Mechanics.

- The central purpose of the here formulated START Theory is twofold: first to have a description of nature, the way we perceive it with our senses and experiments, which could be useful in accordance with the Scientific Method, that is which could be a sound basis for physics. Second START is aimed to be a valid general mathematical theory for all the fundamental physical objects and the frame of reference we use for their description. The physical objects are in general aspects of matter, this concept enlarged to englobe also what previously was known as radiation or particles.

- The quadratic form of the Minkowski space $s^2 = (ct)^2 - (x^2 + y^2 + z^2)$ describes kinematics (motion) in **Space–Time**. To faithfully include matter and interaction (dynamics) we propose a generalization of this quadratic form $S^2 = s^2 - w^2$, where $w = \kappa_{(0)}\overline{a}$ is an equivalent distance expressing the physical **Action** corresponding to a time interval t during which a distance $l = \sqrt{(x^2 + y^2 + z^2)}$ has been covered. The definitions being such that energy–momentum corresponds to (Planck constant times) the space–time derivatives of w. We show that the resulting space $s^2 \to S^2 = g_{uv}x^u x^v$ can also be pictured as a curved space–time. The corresponding linear form (in Clifford algebra a 5-dimensional vector) is $S = \boldsymbol{e}_u x^u$, where the first four components correspond to the (Lorentz-)Minkowski vector $s = e_\mu x^\mu$. The usefulness, and in fact the motivation, of this geometry is illustrated through the analysis of light propagation in a medium and in a gravitational field. We adopt the Poincaré's principle of **Relativity** for the formulation of the **Theory** [2] (set of initials: **START**), in all cases the base space for the description is flat. In our presentation in the present paper optics was taken as an example to guide the reader.

- The 5-D START geometry is presented as a flat simple connected Lorentzian manifold time-like oriented in which all light rays correspond to null lines. Massive objects are described as bundles of null trajectories in START. Observers to 4-D time-like trajectories.

- START provides: first a unified presentation of the main structures of theoretical physics, second a common mathematical language for seemingly different branches as geometric optics, optics in a medium, or general relativity.

1.1 THE QUADRATIC FORM IN PHYSICS

We claim that quadratic spaces are the fundamental clue to understand the structure of theoretical physics. We show several examples: optics in a medium, general relativity and, on a different level, the understanding of a many electron quantum mechanical (QM) system, with all its QM intricacies, considering the QM "density" as a quadratic form and the QM "wave function" as its corresponding linear form.

Our frames of reference formulation (see [1][2][3][4][5]) claims that there is a useful generalization of the quadratic form which historically started with the Pythagorean formulation:

Quadratic form	Dim/diff. Op	Group
$l^2 = x^2 + y^2 + z^2 \qquad (\triangle t)$	3-D ∇, ∇^2	Statics Galileo
$s^2 = (ct)^2 - (x^2 + y^2 + z^2)$	4-D $D, \Box^2$	Kinetics Poincaré
$S^2 = (ct)^2 - \left(x^2 + y^2 + z^2\right) - w^2$	5-D K, K^2	Dynamics
$w = \kappa_{(0)}\overline{a}, \qquad \kappa_{(0)} = \frac{d_{(0)}}{h} = \frac{c}{E_{(0)}}$	$(\overline{a})^2 = \sum_\mu a_\mu^2$	START

where l, (x, y, z), c, h, w, a, $E_{(0)}$ are distance, distance components, vacuum speed of light, Planck's constant, distance equivalent to action, action of a system and characteristic energy of the system respectively.

The development of the theory follows this scheme in such a form to obtain a comprehensive theory:

$$SPACE + TIME + ACTION$$
$$+$$
$$FUNDAMENTAL\ PRINCIPLES$$
$$and\ POSTULATES \text{ (Carriers)}$$
$$\Downarrow$$
$$START\ GEOMETRY \text{ were } \|dS\|^2 = dS \bullet dS^\dagger \text{ and } carriers \text{ were } \rho = \Psi \bullet \Psi^\dagger$$
$$\swarrow \qquad \swarrow \qquad \searrow$$
$$General\ Relativity \quad Quantum\ Mechanics \quad Standard\ Model$$

In our formulation [2] of the Standard Model all symmetries are Space-Time-Action symmetries (no need for "internal" symmetries). In the START space we introduce physics through the **START Relativity Principle**

"All trajectories are null for all observers"

"The vacuum Speed of Light is c for all observers"

The set of principles we have introduced are [2]: **START Relativity** (5-D Poincaré group and 5-D Lorentz transformations); **Existence** (physical objects are represented by energy densities); **Least Action** (null, optimal possible, trajectories in START); **Quantized Exchange of Action** (defines systems or subsystems as those among a quanta of action can be exchanged) and, **Choice of Descriptions** (allows all useful physical models to be employed).

1.2 MULTIVECTOR REPRESENTATION

The base space $\mathbb{R}^5$ corresponds to the real variables set $\{ct, x, y, z, \kappa_0\alpha\} \leftrightarrow \{x^u; u = 0, 1, 2, 3, 4\}$ that is: time, 3-D space and action (in units of distance introducing the universal speed of light in vacuum c and the system under observation dependent $\kappa_0 = \lambda_{Compton}^{system\, with\, energy\, mc^2}/h = 1/m_0c)$. For physics time is usually an independent evolution coordinate and action (matter and interaction) is distributed in space, then we consider the functions $x(t), y(t), z(t)$ and $w(t,x,y,z) = \kappa_0\alpha(t,x,y,z)$. The four (space-time) partial derivatives of $w(t,x,y,z)$ gives to dS^2 an structure similar to a, modified, special relativity ds^2 world line element. The nested vectors

$$dS = \sum_\mu dx^u e_u\, ; \; u = 0,1,2,3,4 \qquad 5-D$$
$$ds = \sum_\mu dx^\mu e_\mu\, ; \; \mu = 0,1,2,3 \qquad 4-D$$
$$d\mathbf{x} = \sum_i dx^i \mathbf{e}_i\, ; \; i = 1,2,3\, ; \; \mathbf{e}_i = e_0 e_i \quad 3-D$$

are members of a Clifford algebra generated by the definition of a quadratic form

$$dS^2 \equiv (dS)^2 = \left(\sum_\mu dx^u e_u\right)^2 = \sum_{\mu\nu} g_{uv}^{START} dx^u dx^v\,,$$
$$g_{uv}^{START} = diag\,(1,-1,-1,-1,-1)\,,\; e_u e_v = -e_v e_u$$
$$e = e_0 e_1 e_2 e_3 e_4 = -e^\dagger \qquad e_u e = e e_u$$

2 5-D Formulation of OPTICS

Geometric optics in a medium is used as a guiding concept to enlarge the representation space of physical phenomena to 5-dimensions. Optics was used in the XX century to create a logical 4-dimensional geometric representation of the frame of reference to describe events in nature, the basic consideration was that of "free space" light rays. We show that considering the concept of light propagation in the medium, where the speed of light is lower, a geometry in 5-dimensions appears as a natural frame of reference. This 5-D space, a quadratic space, is constructed from a set of 5 variables: one time-like variable, three space-like variables and one action-like variable. The Lorentz Geometry fundamental quadratic form of this space is $dS^2 = ds^2 - (\kappa_0 da)^2$, here ds^2 is the Poincaré–Einstein-Minkowski 4-D space–time quadratic form. Below we show that General Relativity is also faithfully represented in this quadratic space.

2.1 FROM SPACE–TIME GEOMETRIC OPTICS TO START

The trajectory of a light ray in a medium can be represented in two equivalent forms. Consider $dl = v_e dt = (c/\eta)dt$, for **light propagation in a medium** (refraction index η) and use the quadratic form $dl^2 = dx^2 + dy^2 + dz^2$ for the elementary trajectory in space $dl^2 = (c/\eta)^2 dt^2$, or $(cdt)^2 = \eta^2 dl^2 = (1 + 4\pi\chi)dl^2$, to define

$$dS^2_{light-\eta} = (cdt)^2 - dl^2 - 4\pi\chi dl^2 = (cdt)^2 - dl^2 - \frac{4\pi\chi}{1+4\pi\chi}c^2dt^2 = 0,$$

$$\boxed{dS^2_{light-\eta} = \left(1 - \tfrac{\eta^2-1}{\eta^2}\right) c^2dt^2 - \left(dx^2 + dy^2 + dz^2\right)}$$

The relations above allow a new interpretation of the light propagation in a medium as a **propagation with the vacuum speed of light but in a 5-dimensional quadratic manifold.**

In the 5-D form the fifth term $\left[\left(\eta^2 - 1\right)/\eta^2\right] c^2dt^2$ represents a distance equivalent to the interaction of the medium, with refraction index η, upon the light ray. We now propose the following 5-dimensional construction, including the **action** variable, in the space–time-action quadratic form

$$dS^2 = ds^2 - (\kappa_0 da)^2 = (cdt)^2 - (dx^2 + dy^2 + dz^2) - (\kappa_0 da)^2 \tag{1}$$

where in the case of light the constant $\kappa_0 = c/h\nu$ the inverse of the momentum associated (in vacuum) with the light (one carrier's, one photon) energy $h\nu$. If we associate then the last term above with the last term in the description of the light ray propagation

$$(\kappa_0 da)^2 = \frac{\eta^2-1}{\eta^2}c^2dt^2 = \frac{c^2}{(h\nu)^2}\Delta\left(\mathcal{E}^2\right)dt^2,$$

we can define the equivalent interaction potential as (here $\mathcal{E}$ represents light→media interaction energy)

$$\frac{\eta^2-1}{\eta^2} = \frac{\Delta\left(\mathcal{E}^2\right)}{(h\nu)^2},$$

therefore the **action** $\mathcal{E}dt$ equivalent to the interaction between light and the medium. In this discussion it is fundamental that these processes take place when the carrier of light is **in interaction with a medium**, the medium being described by an index of refraction η.

2.2 THE ACTION DIFFERENTIAL TERM

The quadratic form which is more relevant for Physics considers that observable objects are extended in space and then an action density α in space-time is required. Then, defining $m(\mathbf{x}, t)c^2 = \varepsilon_{total}(\mathbf{x}, t)$, and the (Clifford algebra valued) inverse of the space-time volume $e_0e_1e_2e_3/\triangle x\triangle y\triangle z\triangle t$, and the space-time Laplacian operator $\square = \sum_\mu e^\mu\partial_\mu$ such that along $b = \sum_\mu b^\mu e_\mu$ the directional change operator is $db\cdot\square = \sum_\mu db^\mu\partial_\mu$ (we apply four times for $b = cte_0, xe_1, ye_2, ze_3$), we can obtain the sum of the **directed** changes of the density of w:

$$\mathbf{a}(\mathbf{x},t)e_4 = \kappa_0\alpha(\mathbf{x},t)e_4 = \kappa_0\frac{m(\mathbf{x},t)c^2\Delta t}{\Delta x\Delta y\Delta z\Delta t}e = \frac{1}{m_0 c}\frac{m(\mathbf{x},t)c^2\Delta t}{\Delta x\Delta y\Delta z\Delta t}e = \frac{(m(\mathbf{x},t)/m_0)c\Delta t}{\Delta x\Delta y\Delta z\Delta t}e$$
$$\mathbf{a}(\mathbf{x},t)e_4 = \frac{(m(\mathbf{x},t)/m_0)c\Delta t}{\Delta x\Delta y\Delta z\Delta t}e = \frac{w(\mathbf{x},t)}{\Delta x\Delta y\Delta z\Delta t}e = \mathfrak{w}(\mathbf{x},t)e$$
$$e d\mathfrak{w} = \sum_\mu\left[(\partial_\mu\mathfrak{w}(\mathbf{x},t))\,dx^\mu\right]e_\mu e$$
$$(dS)^2 = (dS)(dS)^\dagger = \left(1-(\kappa_0 p_0)^2\right)(cdt)^2 +$$
$$-\left(\left(1-(\kappa_0 p_1)^2\right)(dx)^2 + \left(1-(\kappa_0 p_2)^2\right)(dy)^2 + \left(1-(\kappa_0 p_3)^2\right)(dz)^2\right)$$

here $\mathfrak{p}_\mu = \partial_\mu\alpha(\mathbf{x},t)$ is a momentum density. Notice that $\mathfrak{w}(\mathbf{x},t)$ is the distance equivalent to a **reduced action** density, this makes the approach universal for all systems.

3 The photon as a general relativity test particle

General Relativity is considered a comprehensive theory, the best known solutions are developed for the so called matter-free space and a test particle. We show that (1) corresponds to a description of the action distribution which agrees with the conceptual development of General Relativity (GR), this last theory itself being based on the physical postulate that all observers have the right to consider their measurements equally valid..

3.1 THE BASIC START SOLUTION

There are two fundamental (energy-)carrier structures: the massless (as the photon) and the massive fields with basic relation

$$\mathcal{E}^2 = (\mathcal{E}_0 + \Delta\mathcal{E})^2,\ \mathcal{E}^2 - \mathcal{E}_0^2 = (pc)^2, \tag{2}$$

where $\Delta\mathcal{E}$ is any gauge-free energy contribution and $\mathcal{E}_0 = m_0c^2 \Longrightarrow h\nu$ (for a photon).

The concept of test particle (at position $\{r,\theta,\phi\}$) in general relativity is compatible with the Newtonian limit for the interaction gravitational energy

$$\Delta\mathcal{E}(r) = -m_0\frac{GM}{r}, \tag{3}$$

where M is the total mass of 'the external system' (confined within a radius r_s) which we are exploring with the test particle. START uses the action square difference, writing $\mathcal{E} = \mathcal{E}_0 + \Delta\mathcal{E}$ for large (classical limit) values of $r > r_s$

$$\mathcal{E}^2 - \mathcal{E}_0^2 = \mathcal{E}_0^2 + 2\mathcal{E}_0\Delta\mathcal{E} + (\Delta\mathcal{E})^2 - \mathcal{E}_0^2 = (pc)^2 \tag{4}$$
$$= 2\mathcal{E}_0\Delta\mathcal{E} + (\Delta\mathcal{E})^2 \to -2m_0c^2m_0\frac{GM}{r} + \left(m_0\frac{GM}{r}\right)^2,$$

this corresponds both to the energy (difference square) and radial momentum terms in $(d\mathcal{A})^2 - (d\mathcal{A}')^2$ if $(d\mathcal{A}')^2 = (m_0c^2dt)^2$, and according to the GR basic description principles, substituting (the negative of (4)) in (1) using $\kappa_0 = 1/m_0c$ and space–time spherically symmetric coordinates t, r, θ, ϕ we obtain

$$(d\mathbf{S})^2 = \left(1 - 2\frac{GM}{c^2r} + \left(\frac{GM}{c^2r}\right)^2\right)\left(c^2\,(dt)^2 - \left\{(dr)^2 + r^2\left[(d\theta)^2 + \sin^2\theta\,(d\phi)^2\right]\right\}\right),$$

with the same physical consequences for the so called "tests of GR" as those of the Schwarzschild [1916] metric in the limit of $r \gg GM/c^2$. The speed of light in this metric remains to be c.

3.2 GENERAL RELATIVITY VIEWED AS A LORENTZ TRANSFORMATION

The gravitational field induces a coordinate transformation where it can be required that the vacuum speed of light is c for all observers. It is shown that this transformation has a formal structure similar to, and includes, the Lorentz transformation. The new (Lorentz-Keller) transformation simplifies General Relativity (GR), giving to this theory a Special Relativity structure. It also removes some conceptual difficult points of GR and simplifies the computation of solutions to the problem of finding the corresponding quadratic form.

In the Lorentz transformation, written as a $ds^2 = (cdt')^2 - (dx')^2$ form

$$ds^2 = \left((cdt)^2[1 - 2\frac{vdx}{ccdt} + \left(\frac{vdx}{ccdt}\right)^2] - (dx)^2[1 - 2\frac{vcdt}{cdx} + \left(\frac{vcdt}{cdx}\right)^2]\right)\frac{c^2}{c^2 - v^2}$$

consider the case where $dx = dl = cdt$, the case for light propagation, to write

$$(cdt)^2[1 - 2\left(\frac{v}{c}\right) + \left(\frac{v}{c}\right)^2] - (dl)^2[1 - 2\left(\frac{v}{c}\right) + \left(\frac{v}{c}\right)^2] = 0$$

a transformation symmetric in t and x, with a form similar to a solution of the Einstein's equations. Notice that v is an independent variable $0 \leq v \leq c$. Our examples above have the same form and the addition of the action term dw^2 has imposed that dS^2 should be null: $dS^2 = ds^2 - dw^2 = 0$.

4 Action Carriers and Quantum Mechanics

We proceed in the following logical sequence: formally define a carrier as a distribution field, decompose this field into the product of an action per unit density and a density ρ, impose an observability condition to this density, as a result an auxiliary amplitude function Ψ is required, show that Ψ is a vector (multivector) to be defined as a locally dependent sum of basis vectors (multivectors).

4.1 DEFINITION OF CARRIER FIELDS

Within our fundamental formulation we will have to define properties of the fields we call "carriers" (see Keller and Weinberger [4]).

A carrier-domain $\mathcal{B}$ is a connected open set whose elements can be put into bijective correspondence with the points of a region (domain in some instances) $\mathbf{B}$ of a Euclidean point space $\mathbb{E}$. Use $\mathcal{X}$ to denotes a representative *element* of $\mathcal{B}$ and $\mathbf{x}$ the position relative to an origin $\mathbf{0}$ of the point $\mathbf{x}$ occupied by $\mathcal{X}$ in $\mathbf{B}$.

At each $\mathcal{X}$ a scalar quantity is given, called carrier density $\varrho(\mathcal{X})$, such that if $\mathbf{x} = \phi(\mathcal{X}, t)$ then $\varrho(\mathcal{X}) \to \rho(\mathbf{x}, t)$ defines a scalar field called local carrier density.

A carrier will have physical significance through its set of properties. The value of the density of a carrier field can be defined through a set of fundamental scalar constants (main examples: mass, electric charges, weak charge, strong charge, spin) such that the integral of the product of these constants and the density gives the experimentally attributed value of a property for that carrier. The density might become an indirect observable through the repeated measurement of those properties, but it is not an observable in itself. We will use an example. A carrier field identified with an electron will have a density $\rho(\mathbf{x}, t)$ and if the property is Q we will obtain the definition

$Q = \int_{\mathrm{V}} q(\mathbf{x}, t) d\mathbf{x} = \int_{\mathrm{V}} Q\rho(\mathbf{x}, t) d\mathbf{x}$

for all t in the system's volume V, which defines that Q is a constant property (in space and time) for that field (otherwise the variable quantity $q(\mathbf{x}, t)$ can be called "the density of Q ") The set of properties $\{Q\}$ characterizes a carrier field and in turn establishes the conditions for a density field to correspond to an acceptable carrier.

4.2 COMPOSITE, DECOMPOSABLE, AVERAGE AND AVERAGE DESCRIPTION OF CARRIERS

There are several forms of analyzing the density. Each one allows a physical interpretation. For example:

- A ***composite*** carrier is defined as one for which the density

$$\rho_C(\mathbf{x}, t) = \sum_c A^c \rho_c(\mathbf{x}, t), \tag{5}$$

  with the definition of each of the $\rho_c(\mathbf{x}, t)$ being meaningful as a description of a carrier itself. In particular we can choose $\int_V \rho_c(\mathbf{x}, t) d\mathbf{x} = \mathbf{1}$ and $A^c = N_c$.

- Similarly a ***non-decomposable, (non-decomposable) elementary, average,*** etc. carrier can be defined.

4.3 THE DENSITY

To properly describe systems in nature the conditions to be obeyed by the analytical function carrier density $\rho_c(\mathbf{x}, t)$ are:

D1.- $\rho_c(\mathbf{x},t)$ is a real quantity $\rho_c(\mathbf{x},t) \subset \mathbb{R}$.

D2.- The density $0 \leq \rho_c(\mathbf{x},t) < \infty$ in order to represent a finite amount of action.

D3.- The derivatives of the density $-\infty < \partial_\mu \rho_c(\mathbf{x},t) < +\infty$ in order to represent a finite amount of energy–momentum.

Theorem (Keller and Weinberger 2002) If $\Psi(\mathbf{x},t)$ is an analytical quadratic integrable complex or multivector function, conditions D1, D2 and D3 are fulfilled identically if $\rho_c(\mathbf{x},t) = |\Psi_c(\mathbf{x},t)|^2$. Here $|f|^2$ means the real quadratic form of any more general function f, even if f itself is not necessarily a real function and we define: if $|f|^2 = f^+ f$ then $\partial_\mu |f|^2 = (\partial_\mu f^+)f + f^+(\partial_\mu f)$.

Condition D1 is fulfilled by the definition $\rho_c(\mathbf{x},t) = |\Psi_c(\mathbf{x},t)|^2$, D2 by the requirement of quadratic integrability, D3 by the definition $\partial_\mu |f|^2 = (\partial_\mu f^+)f + f^+(\partial_\mu f)$ and the analytical properties of $\Psi(\mathbf{x},t)$. It is seen that the conditions D1, D2, D3 and $\int_V \rho_c(\mathbf{x},t)d\mathbf{x} = N_c$ correspond to the $\Psi(\mathbf{x},t)$ being quadratic integrable Hilbert functions. Thus, even if we selected the density (current in the general formulation) as the basic function, the auxiliary amplitude function Ψ is unavoidably necessary.

4.4 THE LINEAR FORM CHARACTER OF THE MANY ELECTRON AUXILIARY AMPLITUDE FUNCTION

Besides all the conditions mentioned above that the auxiliary amplitude function should fulfill, in the case of a many carriers system there are some additional conditions. Those conditions arise from two fundamental principles: equivalent carriers in the same system should correspond to the same density and, second, the statistics which the carriers should obey among themselves are to be included in the auxiliary amplitude function.

We consider the case of electrons, fermions, then the statistics are the Fermi-Dirac statistics and require:

- The density for the N equivalent carriers system is to be constructed as a sum of M independent alternative contributions $\rho(\mathbf{x},t) = \sum_{i=1}^{M \geq N} \rho_i(\mathbf{x},t,s_i)$.
- There should be at least one linearly independent function (pseudo-carrier amplitude) contributing to the density for each of the N equivalent carriers in the system

$$\begin{aligned} \psi_c(\mathbf{x},t) &= \frac{1}{\sqrt{N}} \sum_{i=1}^{M \geq N} a_c^i \varphi_i(\mathbf{x},t,s_i) \qquad a_c^i = b_c \alpha_i \\ a_c^i a_c^{i'} &= -a_c^{i'} a_c^i \qquad \left\|a_c^i\right\|^2 \leq 1\,;\ \sum\nolimits_{i=1}^{M} \left\|a_c^i\right\|^2 = N \end{aligned} \tag{6}$$

- The total amplitude function should be a sum of single (pseudo-)carrier amplitude functions $\psi_c(x,t,s)$, such that the exchange among two carrier's of the space–time–spin (x,t,s) descriptions, in fact the exchange of the indexes

c and c', should correspond, to avoid two descriptions being identical, to a change in sign of the many carrier amplitude function Ψ

$$\Psi = \sum_c \omega^c \psi_c(\mathbf{x}, t, s) \text{ and } \overline{\Psi} = \sum_{c'} \varpi^{c'} \psi^\dagger_{c'}(\mathbf{x}, t, s) \tag{7}$$

This defines Ψ as a vector, linear form, expressed in the basis $\{\omega^c\}$.

- The $\{\Psi, \overline{\Psi}\}$ are defined and the products ordered to obtain

$$\int \|\Psi\|^2 \, d\mathbf{x} = \int \overline{\Psi}\Psi d\mathbf{x} = N \tag{8}$$

The conditions which the $\varphi_i(x, t, s)$ and the ω^c should obey are simply those which correspond to the two fundamental principles above:

$$\int \varphi_{i'}(\mathbf{x}, t, s_{i'})\varphi_i(\mathbf{x}, t, s_i)d\mathbf{x} = \delta_{i'i} \,; \qquad \omega^c\omega^{c'} = -\omega^{c'}\omega^c \,; \; also \; \varpi^c\omega^{c'} = \delta^{cc'} \tag{9}$$

The 1st condition is double: first the orthonormality among the φ_i functions (requiring them to be eigenfunctions of the same differential equation operator) to fulfill the condition of making linear independent combinations ψ_c and second the Grassmann character of the a^i_c coefficients, to make the local density per carrier correspond tho the sum of the squares $\left|a^i_c\varphi_i(\mathbf{x}, t, s_i)\right|^2$, and the 2nd condition, equivalent to the Pauli principle, defines the ω^c as Grassmann variables and the ϖ^c as their Grassmann conjugates. Full presentation see [4].

Acknowledgments: The technical assistance of Mrs. Irma Vigil de Aragón is gratefully acknowledged. This project was additionally supported by the UNAM program PAPIIT number IN113102-3.

References

[1] J. KELLER, The Foundations of Density Functional Theory and Wave Quantum Mechanics, *Rev. Soc. Quim. Mex.* **44**(1) 2000, 22–28

[2] J. KELLER, *The Theory of the Electron* ***A Theory of Matter from START,*** From the series: Fundamental Theories of Physics, Kluwer Academic Publishers, Dordrecht, Vol. 115 (2001), e-book http://www.wkap.nl/prod/b/0-7923-6819-3

[3] J. KELLER, General Relativity from START, in: *Clifford Analysis and Related Topics, K. Gürlebeck X. Ji, W. Sprössig Eds., Adv. in Applied Clifford Algebras* **11**(S2) (2001), 183-204. Also J. KELLER, START: 4-D to 5-D generalization of the (Minkowski-)Lorentz Geometry, *Proceedings of the Meeting Lorentzian Geometry-Murcia 2003* in the series: *Publicaciones de la Real Sociedad Matemática Española*, vol. 6 (2004).

[4] J. KELLER, General Relativity as a Symmetry of a Unified Space-Time-Action Geometrical Space, *Proceedings of Institute of Mathematics of NAS of Ukraine,* **43**(2) also J. KELLER, A Derivation of Quantum Mechanics from START, *idem* **50**, Editors A.G. Nikitin, V.M. Boyko and R.O. Popovych (and I.A.Yehorchenko), Kyiv, Institute of Mathematics, (2002). 557-568 and (2004).811-820, also J. KELLER AND WEINBERGER P., A Formal Definition of Carriers , *Advances in Applied Clifford Algebras*, 2002, **12**(1), 39-62. http://www.imath.kiev.ua/activities.php

[5] J. KELLER, Unification of Electrodynamics and Gravity from START, *Annales de la Fond. Louis de Broglie*, **27**(3) (2002) 359.
http://www.ensmp.fr/aflb/AFLB-272/aflb272p359.htm

DO REAL NUMBERS OBSCURE REAL PHYSICS?

G. N. ORD, [a,b]

[a] *Perimeter Institute of Theoretical Physics, Waterloo, Ont. Canada*

[b] Permanent address: *Mathematics Department, Ryerson University, Toronto Ont. Canada*

Abstract

The real number line has served classical physics well as a model continuum for both space and time. Quantum mechanics preserves this continuum picture of space and time while extending the range of functions on spacetime to complex numbers. The resulting theory is precise, elegant, experimentally verified and universally accepted. However, since the existence of a 'realistic' basis for quantum mechanics remains an open question, there is as yet no consensus on a satisfactory interpretation of the theory.

The step in the transition from classical to quantum physics that characteristically impedes attempts to construct realistic models is the special extension of functions from $\mathbb{R}$ to $\mathbb{C}$. In quantum mechanics this step is formal and algebraic in nature. This talk discusses an alternative in which the geometry of a spacetime path automatically builds the extension from $\mathbb{Z}$ to $\mathbb{C}$, with all spacetime functions simply counting contributions of a single path. The model produces the correct continuum propagator in the continuum limit, but the realistic origin of the propagator is only apparent when the smallest scale in the system (Planck scale) remains finite. This type of model is consequently missed by current formulations of quantum mechanics because the continuum limit is assumed at the outset. This suggests that it may be fruitful to look to discrete formulations of quantum mechanics in the quest for an ontological version of the theory.

B.G. Sidharth, F. Honsell and A. de Angelis. (eds.), Frontiers of Fundamental Physics, 175–182.

1 Introduction

In Classical physics, the correspondence between the physical world we describe, and the mathematics we use, is usually very close. When we write an equation such as $F = m\frac{d^2x}{dt^2}$, we are tracking the position of an object x as a function of time t, and here the differential equation seems a very natural language to use. The extra assumption brought in by the language is that the trajectory of our physical object is smooth in space and time. In terms of classical physics this seems harmless enough. However there is reason to question the smoothness of space and time in quantum physics. In quantum mechanics, both momentum and energy vary with an inverse length, be it space or time, so the concept of localization over small distances or times is problematic. The language we use, for example Schrödinger's equation, maintains the classical paradigm of a smooth background continuum by changing from a description of point-like objects to wave propagation. This has worked spectacularly well and quantum mechanics provides a theory with unequalled empirical accuracy and generality. The price we pay is that we no longer know exactly what it is that the mathematics is describing. Whereas we can easily believe that the $x(t)$ in Newton's law above describes the centre of mass of an object. It is not easy to say what is described by a solution of Schrödinger's equation.

When we look at how we quantize a classical system, we can see in the prescription where the break with classical ontology occurs. For example, for the one dimensional free particle the energy is related to the momentum by $E = \frac{p^2}{2m}$. This system is quantized by replacing E and p by operators: $E \to i\hbar\frac{\partial}{\partial t}$ and $p \to -i\hbar\frac{\partial}{\partial x}$ to give the wave equation $i\hbar\frac{\partial\psi}{\partial t} = -\frac{\hbar^2}{2m}\frac{\partial^2\psi}{\partial t^2}$. Here the main harbinger of quantum mechanics is the explicit presence of $i = \sqrt{-1}$ in the partial differential equation. Without the i, the above equation is just a diffusion equation and may be derived directly from a discrete random walk model.

The presence of i in Schroedinger's equation and the resulting extension of the number system from $\mathbb{R}$ to $\mathbb{C}$ is responsible for wave-particle duality and all the interference effects associated with quantum mechanics. What is unclear is what physical process is behind the algebra. Note that whatever the process itself, it is unlikely to be directly related to the conception of a continuum of phase. That is, if we relate the i in Schrödingers equation to a real phase ϕ via $i = e^{i\phi}$, the number $\phi = \pi/2$ would have to be extremely accurately maintained by nature in order for solutions of the Schroedinger equation to survive domination by diffusion. Whereas we can think of the 'diffusion constant' $\frac{\hbar}{2m}$ as a real number because our solutions to Schroedinger's equation would be stable under a variation of the constant to say one part in 10^{20}, we cannot say the same for ϕ. If it is varied by one part in 10^{20} the solution of the resulting equation changes qualitatively. It would seem that, in spite of the solutions of Schrödinger's equation having continuous phase, the origin of phase in nature is more likely to be discrete than continuous. This means that if we are to find a physical origin for the complex algebra of quantum mechanics, it makes sense to look for it in a discrete space *prior* to taking a continuum limit.

This talk gives an example of the above scenario. We construct a realistic sub-quantum dynamic in 1+1 dimensions through a completely deterministic discrete process. We do this by essentially drawing a picture on spacetime with a single path using a severely constrained geometry. The constrained geometry leads to the required algebra in a fairly transparent way. A free particle propagator may be constructed so that, in the continuum limit, it is appropriate for a Dirac particle. However the continuum limit itself essentially erases the evidence for the underlying dynamical process.

The technical details of the model are not difficult but will appear elsewhere. [1] What I want to do is to sketch the ideas involved. First of all we might ask:

Q. Is i really necessary? There are many famous statements stressing the necessity of complex numbers in QM but in the end, surely i is just a notational convenience?

A. Yes, but because the empirical accuracy of QM is so convincing, the *algebra* of quantum mechanics is something that is clearly recognized by nature. What I would like to argue is that in terms of number systems, the origin of Complex numbers in QM is based on integer counting and geometry. The idea is that, for example, in the Euler formula:

$$\begin{aligned} e^{imt} &= \cos(mt) + i\sin(mt) \\ &= \sum_{k=0}^{\infty}(-1)^k\frac{(mt)^{2k}}{(2k)!} + i\sum_{k=0}^{\infty}(-1)^k\frac{(mt)^{2k+1}}{(2k+1)!} \end{aligned} \tag{1}$$

Nature does not construct the alternating signs in the series expansion by having an algebraic analog of i specifically, but through geometry she does have a way of implementing the alternating signs in the series. That is, if Nature can add and subtract *integers*, she can build the even and odd components of e^{imt} through geometry, thus imitating a continuum of phase.

Q. If the alternating signs in the above series expansion have a geometric significance, where does the addition and subtraction come from.

A. We can at least imagine a scenario that would appear to observers moving forward through time that Nature is able to add and subtract. (This is a well-known example due to Feynman.) Consider Fig. 1 where we sketch a spacetime trajectory that progresses from past to future but contains a single closed loop. An observer moving forward in time sees a single particle before time t_A. At time t_A two new particles appear and move away from each other. At time t_B they both reverse direction and at time t_C they both disappear. Notice that continuous closed loops in spacetime, if they occur in Nature, have the appearance of particle pair creation and annihilation. We would say that the pair of particles created consist of a particle and its anti-particle. The creation and annihilation then give us a physical analog of addition and subtraction of integers. In the figure, if we want to count the number of continuous trajectories in spacetime, we would associate a +1 with forward moving path segments and a -1 for reversed time segments. The total count for the figure, for any t, is then just +1, even though there appears to be more than one particle in the system between t_A and t_C. That

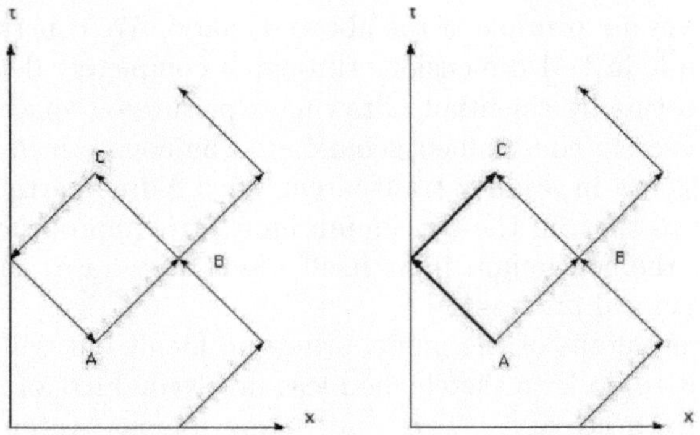

Figure 1. On the left is a continuous spacetime trajectory that has a single closed loop. When viewed by an observer moving forward in time (right sketch) the corner in the path at A looks like a pair creation event with a particle moving off to the right and an anti-particle moving off to the left. Particle and anti-particle annihilate at C. Anti-particles provide a geometric mechanism for subtraction and can be thought of as a discrete precursor of phase.

is, the one continuous trajectory in the system is represented by the integer 1 for $t < t_A$ and by $1 + 1 - 1$ for $t_A < t < t_C$. Notice that the natural subtraction of integers signalled by this simple spacetime loop is *not* present in spacetime paths that are constrained to have no loops. Thus, for example, the Brownian paths of diffusion have no loops and are incapable of imitating subtraction.

Simply put, we can use this integer counting of particles and antiparticles to simulate continuous phase and ultimately the 'quantum mechanical' extension from $\mathbb{R}$ to $\mathbb{C}$. All we need is this natural addition–subtraction mechanism combined with a simple spacetime geometry. To see how this works consider Fig. 2. Here a

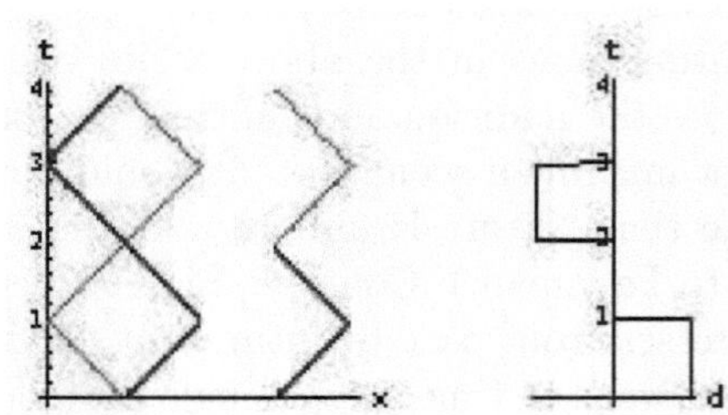

Figure 2. An entwined path with two loops appears on the left. The particle travels from $t = 0$ to $t = 4$ on the zig-zag path that starts off moving to the right(dark line in figure). At $t = 4$ the particle zig-zags back to the origin. The time reversed portions of the trajectory contribute to a counting process with opposite sign to forward-in-t segments. We count loops using 'envelopes' one of which is pictured to the right of the path. The lower two segments of this envelope contribute positively to a density, the upper two negatively. The resulting 'density of right moving particles on the envelope' is pictured to the right. The density is 1 for $0 < t < 1$ because the envelope counts a single right moving particle on that interval. The density is -1 for $2 < t < 3$ because the envelope counts a single right-moving anti-particle in the interval. The density is 0 on the two remaining intervals because there are no right-moving particles on those intervals.

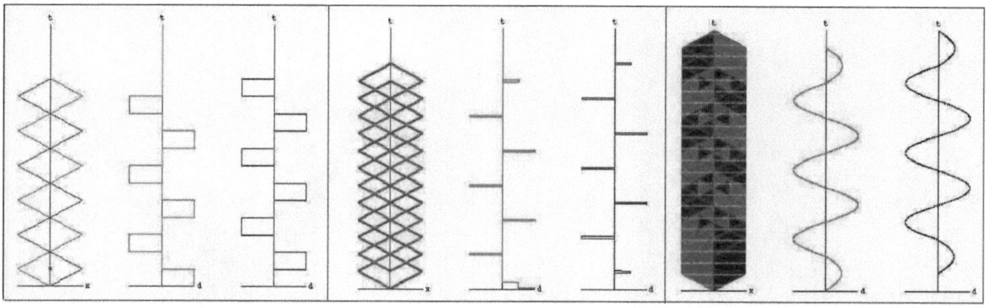

Figure 3. Evolution of the trigonometric functions. The left frame shows a single entwined path with the associated right and left densities. The second frame shows the path whose densities form a periodic array of alternating delta functions. The third frame shows the emergent trigonometric functions with the associated path. The details of the path are below the resolution of the figure at this point. In conventional quantum mechanics, the continuum language we use assumes the smooth version of the trigonometric functions on the right of the figure. That being the case, a path that forms the function would be below the resolution of the language itself.

single spacetime path is sketched in which a particle zig-zags in space while moving forward in time, and then zig-zags on the return journey to the origin in such a way that the forward and reversed paths are entwined. An observer moving forward in t would see this as a pair creation event at $t = 0$, an annihilation/creation event at $t = 2$ and an annihilation event at $t = 4$. However, since it is really particle pairs that always appear from closed loops, let us just count pair contributions by agreeing to count particles to the right of the line $x = 0$. We shall also count by spacetime direction. Since we shall have the perspective of an observer moving forward in time we shall associate a $+1$ with particles moving in a $+t$ direction and a -1 for particles moving in a $-t$ direction. Note that in the figure the density of right-moving particles alternates in sign. In Fig. 3 we first see a single path that has traversed the t-direction four times instead of just two times as in the previous figure. This time the regular geometry has been shifted so that the blank parts of the particle density have been filled in to produce square wave densities. Here both left and right moving densities are plotted. Note that the regular geometry forces the left and right densities to be automatically shifted by $1/4$ wavelength. We can repeat the construction shifting the trajectories in t slightly at each return in just such a way as to form periodic alternating delta functions as in the centre frame of the figure. Finally we can repeat the process to form the two trigonometric functions pictured in the last frame. Notice that although the path on the left looks complicated, it is still only a single path and the geometry is completely regular except for shifts of 'phase' at the origin. Notice also that if we are thinking of this as a model for the way in which a particle could write phase on spacetime, the smallest scale given by shifts of phase could be identified with the Planck scale. The wavelength of the trig function density that we wish to construct is the Compton length. The two densities are then the two components of e^{imt}.

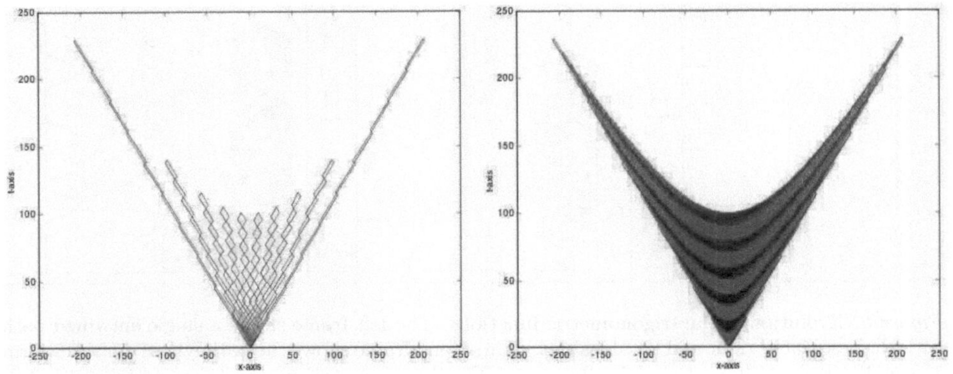

Figure 4. On the left, a few velocity eigen-paths are shown with velocities ranging from $-.9\ c$ to $+.9\ c$. These differ from a $v = 0$ eigenpath by a Lorentz transformation. On the right is a scan with a denser coverage of velocities. There is still a single path here but the details are below the resolution of the figure. Note the alternating bands corresponding to particle or antiparticle-rich areas. The oscillation between the regions is the origin of phase in this model.

Q. It does appear that you could construct the real and imaginary components of e^{imt} this way, but what relevance would this have to quantum mechanics?

A. In all current versions of quantum mechanics, our language *starts* in the continuum. In Fig. 5 we can imagine taking the continuum limit ('Planck length' goes to zero) so that the discrete approximations to the trig functions on the right approach their continuum version. The path on the left would then be infinitely long, and we would not be able to infer its presence with the language of the continuum. This is similar to the relationship between kinetic theory and thermodynamics. We can infer the latter from the former, but the language is wrong for going in the opposite direction. In this case if we start with the assumption of a continuum phase, we shall not be able to see that there is a way that nature could, in principle, approximate continuum phase using only a single continuous spacetime path for each physical 'particle' in the system.

Q. The process outlined gives a method for showing how we get e^{imt} from simple geometry and a continuum limit. How would we get a quantum propagator from this?

A. This was essentially the content of Feynman's 'Chessboard model' which started out as a problem in Feynman's book on path integrals [2], and subsequently developed into a small industry in a series of papers including [3, 4, 6, 5] and [7, 8]. All we really have to do is to notice that the paths we have generated correspond to a particle that has propagated with a macroscopic velocity $v = 0$. We can get the Dirac propagator from this model by applying a Lorentz transformation to get other velocities. We can still traverse all these 'velocity eigenpaths' with a single trajectory because they may all be joined serially at the origin. Fig. 4 shows a series of 'velocity eigenpaths' which are just our $v = 0$ path of the previous figure viewed from a moving frame. The dense scan to the right illustrates the hyperbolic geometry through the alternating light/dark patterns. It is the fluctuation of particle/antiparticle rich areas that imitates the phase of quantum mechanics and

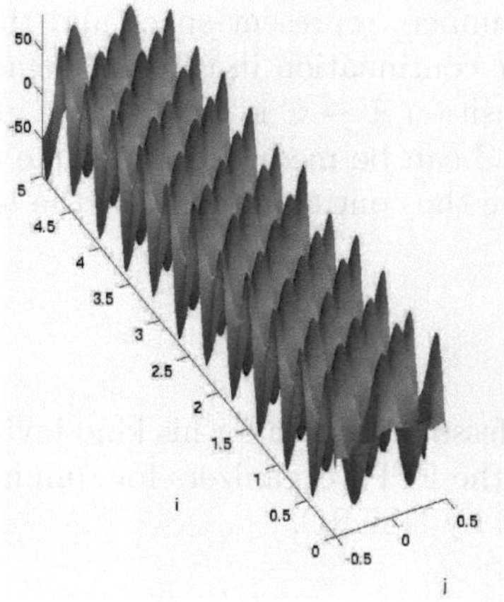

Figure 5. Here the velocity eigenpaths are chosen to correspond to the energy eigenstates of a particle-in-a-box. The initial condition is established by choosing the correct velocity eigenpaths, and moving the origin of the eigenpaths along the x-axis, shifting the t-coordinate in such a way that the propagator adds to the appropriate eigenfunction of the box. Here the typical standing wave pattern is still the manifestation of a single continuous trajectory. There is no canonical quantization, analytic continuation or wave mechanics in this picture.

allows us to construct the Dirac propagator from a single entwined path. Since the propagator can be constructed in this way we can simulate one-dimensional quantum phenomena with this ontological model. Fig. 5 is an illustration of this. The figure appears to be the standing-wave pattern corresponding to an excited state of a particle-in-a-box. Within numerical error this is exactly what it is. However it was generated by a single entwined path. The path is a concatenation of two velocity eigenpaths corresponding to the required energy, shifted in such a way that the path densities at $t = 0$ correspond to the eigenfunctions of the box. The single spacetime path then just propagates the pattern in t through its intrinsic geometry. Physically this is interesting because to get Fig. 5 we have not had to resort to quantization, analytic continuation or the introduction of waves. We are still solidly within the particle paradigm and the 'waves' appearing in the figure are ultimately a manifestation of simple spacetime geometry.

2 Conclusions

The thesis of this paper is that it is possible that the power and convenience of a language can sometimes be an impediment to its use as a tool to help understand nature. The use of real numbers as a representation of space and time is useful and intuitive. However, it may be the case that when we build the continuum limit into our language, we may not be able to see features hidden by that limit, and as a result we may have to build structures after the limit when they have a more natural appearance before the limit. We illustrated a case in point. Quantum mechanics typically assumes real numbers represent space and time and as a result has to propose a formal analytic continuation in the replacement of dynamical variables by operators. So the transition $\mathbb{R} \to \mathbb{C}$ is formal. The above demonstration shows that the transition $\mathbb{Z} \to \mathbb{C}$ can be mediated by simple spacetime geometry. This onlyrequires that we leave the continuum limit to the last in our construction.

3 Acknowledgments

I would like to thank Professor Siddarth for his kind invitation to the meeting. It is also a pleasure to thank the FFP6 organizers for running an excellent conference! This research was funded by NSERC.

References

[1] G. N. Ord, J. A. Gualtieri, and R. B. Mann .

[2] R. P. Feynman and A. R. Hibbs, *Quantum Mechanics and Path Integrals* (New York: McGraw-Hill, 1965).

[3] H. A. Gersch, Int. J. Theor. Physics **20**, 491 (1981).

[4] T. Jacobson and L. Schulman, J. Physics A **17**, 375 (1984).

[5] G. N. Ord, J. Stat. Phys. **66**, 647 (1992).

[6] B. Gaveau, T. Jacobson, M. Kac, and L. S. Schulman, Phys. Rev. Lett. **53**, 419 (1984).

[7] L. H. Kauffman and H. P. Noyes, Phys. Lett. A **218**, 139 (1996).

[8] G. N. Ord and J. A. Gualtieri, Phys. Rev. Lett **89** (2002).

CANTORIAN SPACE IN NATURE AND DYNAMICAL SYSTEMS

GERARDO IOVANE [a], PAOLA GIORDANO [a],
SAVERIO SALERNO [a]

[a] *Dipartimento di Ingegneria dell'Informazione e Matematica Applicata, Università di Salerno, via Ponte Don Melillo, 84084 Fisciano, Italy*

Abstract

In this paper we consider the motion of point-like massive systems on a Cantorian path. The results is that an harmonic material support with an external force is equivalent, under some assumptions, to a classical or quantum harmonic system on a continuous support. The idea that we want stress in this paper is that a Cantorian space could explain some relevant stochastic and quantum processes, if the space acts as an harmonic oscillating support, such as happen in Nature. This means that a quantum process, in some cases, could be explained as a classical one, but on a non continuous and fractal support. We consider the validity of this point of view, that in principle could be more realistic, since it describe the real nature of the matter and space, which does not only exist in Euclidean space or curved one, but in a Cantorian one.

1 Introduction

Nature shows us structures with scaling rules, where clustering properties from cosmological to nuclear objects reveals a form of hierarchy. Moreover, many systems shows an oscillatory behaviour. In the previous papers, the authors consider the compatibility of a Stochastic Self-Similar, Fractal Universe with the observation and the consequences of this model. In particular, it was demonstrated that the observed segregated Universe is the result of a fundamental self-similar law, which

B.G. Sidharth, F. Honsell and A. de Angelis. (eds.), Frontiers of Fundamental Physics, 183–190.

generalizes the Compton wavelength relation (see [1], [2], [3]). Starting from an universal scaling law, the author showed its agreement with the well–known Random Walk equation or Brownian motion relation [4], [5]. The main results by Iovane in [2], can be summarized as follows.

Theorem 1. *The structures of the Universe appears as if they were a classically self-similar random process at all astrophysical scales. The characteristic scale length has a self-similar expression*

$$R(N) = \frac{h}{Mc} N^{1+\phi} = \frac{h}{m_n c} N^{\phi},$$

where the mass M is the mass of the structure, m_n is the mass of a nucleon, N is the number of nucleons in to the structure and ϕ is the Golden Mean value.

In terms of Plankian quantities the scale length can be recast in

$$R_P(N) = \frac{l_P}{m_P}\sqrt{\frac{\hbar c}{G}} N^{(1+\phi)}.$$

The previous expression reflects the quantum memory of the Universe at all scales, which appears as hierarchy in the clustering properties.

Theorem 2. *The mass and the extension of a body are connected with its quantum properties, trough to the relation*

$$E_{E,N}(N) = E_P N^{1+\phi},$$

that links Plank's energy $E_P = h\nu$ and Einstein's one $E_E = mc^2$.
The quantum memory is reflected at all scales and it manifests itself trough a clusterization principle of the mass and extension of the body.

In the present paper we study some relevant force field on Cantorian space and analyze the differences with respect to the analogous case of a continuum, from a classical and quantum point of view. The idea that we want to stress in this paper is that a Cantorian space could explain some relevant stochastic and quantum processes, if the space acts as an harmonic oscillating support, such as happen in Nature. In other word, the vision is that an apparent indetermination, linked with a fractal support rather than a continuous one, can produce an indetermination on the motion of a physical object, which is explained via a stochastic or quantum process. This means that a quantum process, in some cases, could be explained as a classical one, but on a non continuous and fractal support. Consequently, an external observer looking at the motion of a particle under a fixed solicitation can measure an unusual behaviour with respect to a continuous material support, that is obvious with respect to the knowledge of the fractal support behaviour. In this case, he can make the hypothesis of an indetermination or a stochasticity in the process (motion), while there is just really only ignorance with respect to the support on which the motion take place.

Let us introduce some mathematical tools.

In descriptive set theory and the theory of polish spaces we find the following definitions [6].

Definition 1. *When a space $A^{\mathbb{N}}$ is viewed as the product of infinitely many copies of A with discrete topology and is completely metrizable and if A is countable, then the space is said to be polish.*

In particular, when $A = \{0,1\}, |A| = 2$, then we call $\mathcal{C} = 2^{\mathbb{N}}$ Cantor space. For $|A|^{-1}$ defined in an interval: $|A|^{-1} \in]0,1[$, then $C_F = A^{\mathbb{N}}$ is called a fuzzy Cantor space. If $|A|^{-1} = (\sqrt{5}-1)/2$ and $N = n-1$, where $-\infty \leq n \leq \infty$, then $C_F = \varepsilon^{(n)}$ is the E-infinity Cantorian space. Mohamed El Naschie in [7] showed the relationship between the Cantor space $\mathcal{C}$ and $\varepsilon^{(\infty)}$. As He reports: "the relationship comes from the cardinality problem of a Borel set in polish spaces Thus we call a subset of a topological space a Cantor set if it is homeomorphic to the Cantor space".

Let us consider a set of intervals $\mathcal{A} = \{\mathcal{A}^{(1)}, \mathcal{A}^{(2)}, ..., \mathcal{A}^{(n)}\}$ with

$$\begin{aligned}
\mathcal{A}^{(0)} &= [a,b] \qquad (1)\\
\mathcal{A}^{(1)} &= [a, a+(b-a)/3] \cup [a+2(b-a)/3, b]\\
&....\\
\mathcal{A}^{(n)} &= [a, a+(b-a)/3^n] \cup ... \cup [a+2(b-a)/3^n, b].
\end{aligned}$$

where a, b are real numbers.

If w is the level corresponding to $\mathcal{A}^{(w)}$ the number of extreme points is 2^w.

For $a = 0$ and $b = 1$ we obtain the Cantor space $\mathcal{C} = \bigcap_{n\in\mathbb{N}} \mathcal{C}_n$ with the following well know properties:

- $\mathcal{C}$ is compact, with null Reinmannian measure;
- There are no intervals in $\mathcal{C}$;
- $\mathcal{C}$ has the cardinality of continuum.

For our purpose we consider the set $\mathcal{A}$. In particular at the level w the length of a segment is $k_{w+1} = (b-a)/3^w$ with $w = 0, 1, 2, .., n-1$.

To evaluate the extremes for each a level w without using an iterative procedure we show the following algebraic method.

Let us introduce the vector $\mathbf{g}_m = (g_{m,1}, g_{m,2}, ..., g_{m,2^n})$ with $m = 1, ..., n-1$ and with 2^n components, whose values are:

$$g_{m,z} = \begin{cases} 2 & \text{if} \quad z = 2^m(1+2s)+h \\ 1 & \text{if} \quad z = 2^m(1+2s) \\ 0 & \text{elsewhere} \end{cases} \qquad (2)$$

with $s = 0, ..., 2^{n-m-1}-1$ and $h = 1, ..., 2^m - 1$. For $m = n$ we have a vector with all components equal to zero except the component at the place 2^n, that is equal to 1, i.e., $\mathbf{g}_n = (0, 0, ..., 0, 1)$. Consequently the coordinate of the extremes at a fixed level n is given by the following vector

$$\mathcal{A}_z^{(n)} = a + \sum_{p=0}^{n-1} k_{n-p} g_{p+1,z}. \tag{3}$$

Clearly appear how the previous results can be also used on the set $\mathcal{A}$, which generalizes the Cantor space on the interval $[a, b]$ instead of $[0, 1]$.

2 An elastic force field on Cantorian space

As we said a Cantorian space could explain some relevant stochastic and quantum processes, if the space acts as an harmonic oscillating support, such as happen in nature. In other words the vision is that an apparent indetermination, linked with a fractal support rather than a continuous one, can produce an indetermination on the motion of a physical object, which is explained thanks to stochastic or quantum process. This means that a quantum process in same case could be explained as a classical one, but on a non continuous and fractal material support.

To show this result we have to consider a classical harmonic oscillator on a fractal support and then we have to consider it not as the subject of our study, but as the scenario where a process could happen.

Let us suppose that an elastic force field $f(q) = -\alpha q$ is on the interval $\mathcal{A}^{(1)} = [a, b] = \left[\mathcal{A}_1^{(1)}, \mathcal{A}_2^{(1)}\right]$, moreover α is a nonnegative elastic constant; it is well known the motion equation for a material point of mass m coming from the Newtonian equation $f = ma$ is,

$$\ddot{q}^{(1)} + \left(\omega^{(1)}\right)^2 q = 0, \tag{4}$$

where $\omega^{(1)} = \alpha/m$ and with the solution

$$q^{(1)}(t) = \mathcal{A}_2^{(1)} \cos\left(\omega^{(1)} t\right), \tag{5}$$

where we have assumed the initial condition $q^{(1)}(t=0) = \mathcal{A}_2^{(1)} = b$ and $\dot{q}^{(1)}(t=0) = 0$. This can be seen as the level 1 of a recursive procedure, where we consider a fractal support. In fact, at the level 2 we have

$$\begin{aligned}
\mathcal{A}^{(2)} &= [a, a + (b-a)/3] \cup [a + 2(b-a)/3, b] \\
&= \left[\mathcal{A}_1^{(2)}, \mathcal{A}_1^{(2)} + k_2\right] \cup \left[\mathcal{A}_1^{(2)} + 2k_2, \mathcal{A}_4^{(2)}\right] \\
&= \left[\mathcal{A}_1^{(2)}, \mathcal{A}_2^{(2)}\right] \cup \left[\mathcal{A}_3^{(2)}, \mathcal{A}_4^{(2)}\right],
\end{aligned}$$

where $k_2 = (\mathcal{A}_2^{(1)} - \mathcal{A}_1^{(1)})/3 = (b-a)/3$; consequently we consider the force field

$$f^{(2)}(q) = \begin{cases} -\epsilon^{(2)}\alpha q & \forall q \in \left[\mathcal{A}_1^{(2)}, \mathcal{A}_2^{(2)}\right], \\ 0 & \forall q \in \left[\mathcal{A}_2^{(2)}, \mathcal{A}_3^{(2)}\right], \\ -\epsilon^{(2)}\alpha q & \forall q \in \left[\mathcal{A}_3^{(2)}, \mathcal{A}_4^{(2)}\right], \end{cases} \tag{6}$$

where $\epsilon^{(2)}$ is a parameter to take into account the reduction of the material support and to preserve the total energy. We will determine it in the following.

The total path of the level 2 can be see as composed by three sub-path: the first and the third with an oscillation motion, while the second where we find a uniform motion. In fact, it is easy to find for the third sub-path

$$q_3^{(2)}(t) = \mathcal{A}_4^{(2)} \cos\left(\omega^{(2)} t\right),$$

with $\left(\omega^{(2)}\right)^2 = \epsilon^{(2)} \left(\omega^{(1)}\right)^2$, where we have considered the initial conditions $q_3^{(2)}(t=0) = \mathcal{A}_4^{(2)} = b$ and $\dot{q}_3^{(2)}(t=0) = 0$. Analogously, we get for the second sub-path

$$q_2^{(2)}(t) = \beta_2^{(2)} t + \mathcal{A}_3^{(2)},$$

where we have used the following initial condition $q_3^{(2)}(t_1) = q_2^{(2)}(t_1) = \mathcal{A}_3^{(2)}$, and so $\mathcal{A}_3^{(2)} = \mathcal{A}_4^{(2)} \cos\left(\omega^{(2)} t_1\right) \rightarrow t_1 = \frac{1}{\omega^{(2)}} \cos^{-1}\left(\mathcal{A}_3^{(2)}/\mathcal{A}_4^{(2)}\right)$ and $\dot{q}_2^{(2)}(t=t_1) = \dot{q}_3^{(2)}(t=t_1) \equiv \beta_2^{(2)} = -\mathcal{A}_4^{(2)} \omega^{(2)} \sin\cos^{-1}\left(\mathcal{A}_3^{(2)}/\mathcal{A}_4^{(2)}\right)$. On the first path, we find again an oscillation motion

$$q_1^{(2)}(t) = \mathcal{A}_2^{(2)} \cos\left(\omega^{(2)} t + B_2^{(2)}\right),$$

with the following initial condition:
$q_2^{(2)}(t_2) = q_1^{(2)}(t_2) = \mathcal{A}_2^{(2)} \rightarrow t_2 = \left(\mathcal{A}_2^{(2)} - \mathcal{A}_3^{(2)}\right) / \beta_2^{(2)}$ and $\dot{q}_2^{(2)}(t=t_2) = \dot{q}_1^{(2)}(t=t_2)$ and so $B_2^{(2)} = \tan^{-1}\left(-\beta_2^{(2)}/\mathcal{A}_2^{(2)}\right) - \left(\mathcal{A}_2^{(2)} - \mathcal{A}_3^{(2)}\right) / \beta_2^{(2)}$. To summarize at the level 2 we obtain

$$q^{(2)}(t) = \begin{cases} q_1^{(2)}(t) = \mathcal{A}_2^{(2)} \cos\left(\omega^{(2)} t + B_2^{(2)}\right) & \forall q \in \left[\mathcal{A}_1^{(2)}, \mathcal{A}_2^{(2)}\right], \\ q_2^{(2)}(t) = \beta_2^{(2)} t + \mathcal{A}_3^{(2)} & \forall q \in \left[\mathcal{A}_2^{(2)}, \mathcal{A}_3^{(2)}\right], \\ q_3^{(2)}(t) = \mathcal{A}_4^{(2)} \cos\left(\omega^{(2)} t\right) & \forall q \in \left[\mathcal{A}_3^{(2)}, \mathcal{A}_4^{(2)}\right]. \end{cases} \tag{7}$$

With respect to the level 1 we have different amplitude, frequency and phase, but not casual ones since they are in agreement with the fractal model inside them.

Consequently, an external observer looking at the motion of a particle under a fixed solicitation can measure an unusual behaviour with respect to a continuous material support, that is obvious respect to the knowledge of the support behaviour. In this case he can make the hypothesis of an indetermination or a stochasticity in the process (motion), while there is just an ignorance respect to the material support on which the motion happen.

More in general at the level n the initial conditions for the velocity are

$$\beta_z^{(n)} = -\mathcal{A}_{z+2}^{(n)}\omega^{(n)} \sin\cos^{-1}\left(\mathcal{A}_{z+1}^{(n)}/\mathcal{A}_{z+2}^{(n)}\right), \text{ with } z = 0, ..., 2^n - 1. \tag{8}$$

and for the phases factor we can get

$$B_{z+1}^{(n)} = \tan^{-1}\left(-\beta_z^{(n)}/\mathcal{A}_{z+1}^{(n)}\right) - \left(\mathcal{A}_{z+1}^{(n)} - \mathcal{A}_z^{(n)}\right)\Big/\beta_z^{(n)} \tag{9}$$

Consequently, by taking into account Hausdorff $\aleph(\mathcal{C})$ measure the asymptotic behaviour gives us

$$\begin{array}{lcl} f^{(\infty)} & = & \lim\limits_{n\to\infty} f^{(n)} = -\frac{\alpha q}{\aleph(\mathcal{C})}, \\ k_\infty & = & \lim\limits_{n\to\infty} k_{n+1} = 0, \\ \left(\omega^{(\infty)}\right)^2 & = & \lim\limits_{n\to\infty} \left(\omega^{(n)}\right)^2 = \frac{\left(\omega^{(1)}\right)^2}{\aleph(\mathcal{C})} \\ \beta_z^{(\infty)} & = & \lim\limits_{n\to\infty} \beta_z^{(n)} = \beta_1^{(1)} = 0 \\ B_z^{(\infty)} & = & \lim\limits_{n\to\infty} B_z^{(n)} = 0 \\ A_z^{(\infty)} & \in & \mathcal{C} \end{array} \tag{10}$$

With respect to $\beta_1^{(1)}$, here we have used the initial condition $\beta_1^{(1)} = 0$, but it is obvious that there are no changes if $\beta_1^{(1)} = v_0$.

Thanks to the previous relation we obtain

$$q^{(\infty)}(t) = \begin{cases} q_{odd}^{(\infty)}(t) = \mathcal{A}_{odd+1}^{(\infty)} \cos\left(\omega^{(\infty)} t + B_{odd}^{(\infty)}\right), \\ q_{even}^{(\infty)}(t) = v_0 t + A_{even+1}^{(\infty)}. \end{cases} \tag{11}$$

The same considerations can be done for a $\epsilon^{(\infty)}$ El Naschie Cantorian space to obtain similar results.

It is interesting to note that if we have an external solicitation $F = \alpha q$ the motion equation on $\mathcal{C}$ becomes

$$\ddot{q} + 2\omega^2 q = F, \tag{12}$$

and so

$$\ddot{q} + \omega^2 q = 0, \tag{13}$$

that is the traditional motion equation for a massive point in an elastic force field. In other word, what we consider an external force for the support could be the classical elastic solicitation on a massive point moving on a continuous material support[1]. This example is just a toy model when we deal with macroscopic systems, since the frequency of oscillation of the material support could be very different with respect to the system. But this toy model could be very realistic with respect to microscopic and quantum processes and systems. For this reason in [8] we considered oscillating force fields in quantum mechanics to show how the Heisenmberg uncertainty principle can be translated from the processes and systems to the material support, where we have classical dynamics. We showed the validity of this point of view, that in principle could be more realistic, since it describe the real nature of the matter and space, which does not exist in Euclidean space or curved one, but in a Cantorian one. A more realistic model can be the linear chain (see [8]).

3 Conclusions

In this paper we studied the effect of a stochastic self-similar and fractal material support on some physical quantities and relations. In particular, we present an algebraic uniterative relation to find the extremes of a Cantor segmentation at any level of fragmentation. We studied the motion under an elastic force field on a Cantorian space, from classical point of view. The results of this work can be summarized as follows.

- A Cantorian space could explain some relevant stochastic and quantum processes, if the space acts as an harmonic oscillating support, such as happen in Nature (see [9]).

- An apparent indetermination, linked with a fractal support rather than a continuous one, can produce an indetermination on the motion of a physical object, which is explained thanks to stochastic or quantum process. This means that a quantum process, in some cases, could be explained as a classical one, but on a non continuous and fractal support. Consequently, an external observer looking at the motion of a particle under a fixed solicitation can measure an unusual behaviour with respect to a continuous material support, that is obvious with respect to the knowledge of the fractal support behaviour. In this case, he can make the hypothesis of an indetermination or a stochasticity in the process (motion), while there is just an ignorance respect to the material support on which the motion happen.

[1]To be more correct there is a change in the sign of the force, but this is linked with the subject of the problem (material support or massive point). In other words the sign minus of the elastic source becomes plus if we have it as an external source for the material support.

References

[1] G.Iovane, E.Laserra, F.S.Tortoriello, Stochastic Self-Similar and Fractal Universe, Chaos, Solitons and Fractals, 20, 3, 415, 2004.

[2] G.Iovane, Varying G, accelerating Universe, and other relevant consequences of a Stochastic Self-Similar and Fractal Universe, Chaos, Solitons and Fractals, 20, 4, 657, 2004.

[3] G.Iovane, E.Laserra, P.Giordano, Fractal Cantorian structures with spatial pseudo-spherical symmetry for a possible description of the actual segregated universe as a consequence of its primordial fluctuations, Chaos, Solitons & Fractals, 2004, 22, 4, 521.

[4] M.S. El Naschie, On the unification of the fundamental forces and complex time in the $\varepsilon^{(\infty)}$ space, Chaos, Solitons and Fractals, 11, 1149-1162, 2000.

[5] B.J.Sidharth, Chaos Solitons Fractals 11, 2155, 2000; Chaos Solitons Fractals 12, 795, 2001.

[6] T.Jech, Set Theory, Berlin: Springer, 2003.

[7] M.S. El Naschie, Quantum gravity, Clifford algebras, fuzzy set theory and fundamental constants of nature, Chaos, Solitons and Fractals, 20, 437, 2004.

[8] G.Iovane, P.Giordano, S.Salerno, Dynamical System on El Naschie's $\epsilon^{(\infty)}$ Cantorian space-time, accepted to Chaos Solitons & Fractals, 2004.

[9] G.Iovane, Self-Similar and Oscillating solutions of Einstein's equation and other relevant consequences of a Stochastic Self-Similar and Fractal Universe via El Naschie's $\epsilon^{(\infty)}$ Cantorian space-time, in press on Chaos, Solitons & Fractals, 2004.

MATHEMATICAL STRUCTURE OF INDIVIDUAL QUANTUM STATES

S.N. MAYBUROV [a]

[a] *Lebedev Institute of Physics, Leninsky Prospect 53, Moscow, 117924, Russia*

Abstract

The possible structure of quantum states is analyzed in relation to their preparation and the information available to observer; it's demonstrated that beside pure and mixed states the states of other kind - primordial states can exist. In particular, the states of environment are supposedly to be primordial, because they can't be related to the standard way of preparation admitted in QM.

1 Introduction

The development of quantum information systems revives the interest to the different aspects of Quantum Mechanics (QM) foundations. In this note we analyze the general structure of quantum states which can be especially important for the study of decoherence effects induced by the Environment interaction with the information system (detector) D [1]. Remind that QM deals with two sorts of states: an individual state ζ which characterizes the object state in the individual events representing so the most detailed object description, and the statistical state Ξ which describes the average properties of the objects ensemble. According to QM axiomatics any object, in principle, posseses the individual state which formally expressed by the time-dependent density matrix $\zeta = \rho(t)$ [2]. It's well known that the structure of states in QM depends of their production conditions; in particular, the pure state ψ_i of some object is prepared by the experimentalist O according to the particular preparation procedure PR_i. It defines unambiguously

B.G. Sidharth, F. Honsell and A. de Angelis. (eds.), Frontiers of Fundamental Physics, 191–194.

the produced state properties resulting in $\rho(t) = |\psi_i\rangle\langle\psi_i|$ which is the extremal positive functional (PF) on $\{Q\}$ set (algebra) of object observables [3]. The states mixture prepared by use of $\{PR_j\}$ set with the frequencies $\{P_j^p\}$, in this case at any t $\rho(t)$ is one of $|\psi_j\rangle\langle\psi_j|$ with the probabilities P_j^p; here and below it assumed that the different objects from the same ensemble don't intersect on t axe. Due to statistical character of QM measurements their results are normally averaged over large time interval T which are expressed via some functions of $\zeta(t)$. In particular, the ensemble state - Gemenge $W = \{P_i^p, \psi_i\}$ reproduces ζ statistical content, the statistical state (statistical operator) - $\Xi = \bar{\zeta} = \int \rho(t)dt$ averaged over T [4]. Meanwhile we observe around us many objects and systems which aren't prepared in the described way and a strict sense weren't prepared at all, yet they are undoubtly possess the individual quantum state of some kind, and here we attempt to describe their structure for an arbitrary system.

To illustrate our main hypothesis let's consider the toy model of particle production in cosmological Big Bang [5]. Despite its exotics, it is the simplest example of such unprepared system with high degree of symmetry. We consider the states of massive particles e with spin $\frac{1}{2}$ and assume that they are produced directly from the initial singularity. Obviously for any e production mechanism their statistical state Ξ_b possess the fundamental symmetries of Universe, in particular, the rotational invariance, so that $U_{\vec{n}}\Xi_b U_{\vec{n}}^{-1} = \Xi_b$, where $U_{\vec{n}}$ is the rotation in $\vec{n}$ direction. Assuming that the momentum $\vec{p}$ and spin $\vec{S}$ state components are factorized, then for spin component the only solution is $\Xi_b = \sigma_b$ where:

$$\sigma_b = \frac{1}{2}(|s_1\rangle\langle s_1| + |s_2\rangle\langle s_2|) \quad (1)$$

here $s_{1,2}$ are u, d e spinors. Because of the invariance of e production process relative to the time shift (at least at not very large time scale) we can assume that e individual state $\zeta_b = \rho_b(t)$ doesn't depend on t and therefore $\rho_b(t) = \sigma_b$ is constant. Such state called here the primordial state $\xi_b = \rho_b(t)$ isn't extremal PF on $\{Q\}$ as the pure states are, but in general is an arbitrary PF on $\{Q\}$. Because of the assumed time invariance it isn't the probabilistic mixture of pure states, and has completely uncertain spin $\vec{n}_s$ direction in each event. It can be characterized as the quantum state of third kind which is produced without participation of any observer. Meanwhile the same statistical state Ξ_b can be dispatched by $\rho(t)$ described by the mixture of pure states in case of the spontaneous breaking of rotational symmetry, For example, it can be the ensemble of pure states with random spin orientation $\rho(t_j) = |s(\vec{n}_r)\rangle$; yet currently no suitable randomization mechanism was proposed which can result in such effect [5]..

The same results can be derived for the collective system e production in $N \to \infty$ limit with total angular momentum $J^2 = 0$. It means that all N particles e_j are entangled and the collective state $\rho_c(t) = |\psi_c\rangle\langle\psi_c|$ where:

$$\psi_c = \sum_i^M a_i \prod_j^N |s_{l(1,i)}^1\rangle ... |s_{l(j,i)}^j\rangle ... |s_{l(N,i)}^N\rangle \quad (2)$$

here $M = \frac{4N!}{2^N}$, $l(j,i) = 1,2$ - the discrete function with $\sum_j^N l(j,i) = \frac{3}{2}N$, a_i are the complex parameters with $|a_i| = \frac{1}{M}$. For $N = 2$ it gives EPR state $\psi_c = \frac{1}{\sqrt{2}}(|s_1^1\rangle|s_2^2\rangle - |s_2^1\rangle|s_1^2\rangle)$ The partial e_i state for such system $\eta_i(t) = Tr\rho_c = \sigma_b$, where the trace taken over all $j \neq i$. The spin correlation for any particles pair in such state is: $\langle \vec{S}_j \vec{S}_i \rangle = -\frac{1}{4N}$ and is negligible for $N \to \infty$, consequently e states can be regarded as factorized in this limit and therefore are equivalent to the primordial state ξ_b. This example assumes that e primordial states can be produced in lab. conditions within some approximation.

Considering the states of environment E [1], it's plausible to assume that E states are also primordial states. Really, in some models E elements A_j^E are atoms (of atmospheric air, for example) which in lab conditions can't be prepared by O one by one in some pure state. Supposedly A^E interact with D independently of each other and therefore change its individual state ζ_D. If A^E atoms structure is similar to e, and there is no external influence on E state, then assuming that the rotational invariance holds also in lab. conditions, it follows that E statistical spin state $\Xi_E = \Xi_b$. If the lab. conditions are time invariant, then A^E individual spin states $\zeta_E = \rho_E(t)$ are equivalent to Big Bang e spin states $\zeta_E = \sigma_b$ so they are the primordial states $\xi_E = \zeta_E$. They differ from E individual states used in some decoherence models; in particular, in Zurek model it assumed that E states are pure and has the random orientation in space: $|E\rangle = |s(\vec{n}_r)\rangle$ [6]. This difference can lead to the important effects but we don't discuss them here.

In general E states properties can depend on some classical lab. parameters like the temperature τ, pressure $\mathcal{P}$, etc.; such ensembles are studied by Quantum Statistics (QS). According to its postulates the quantum ensemble of any objects is described by the statistical operator R [3]. For E embedded in thermostat its spin component is equal to:

$$R_E(s_i, s_j) = \sum_{\Lambda} e^{-\frac{F-\Lambda}{k\tau}} |\psi_\Lambda(s_i)\rangle\langle\psi_\Lambda^*(s_j)|$$

Here F, Λ is A^E free and total energy, ψ_Λ are its eigenstates. If spin and energy A^E state components are factorized then $R_E = \sigma_b$. In our approach the statistical state Ξ_E supposedly coincides with R_E, but in addition the individual E states ρ_E can be defined; from the mentioned t invariance it follows that $\rho_E(t) = R_E$ and A^E states are primordial states with $\xi_E = \rho_E(t)$. In this case also we don't see any dynamical mechanism which can make such individual states stochastic. Moreover if E atoms can interact with each other, then their collisions result in their states entanglement. Consequently, even if some A_j^E initial state was pure with the definite spin direction $\vec{n}_s$ it will become entangled with other A^E states and its partial state will approximate $\eta_i(t)$.

The primordial states has the analogy with the so called algebraic states φ arising in Algebraic QM formalism [3]. In this theory the set of states $\varphi \in \Omega$ is the linear space dual to Observable Algebra $\mathcal{U}$. The states φ, regarded as the individual states, are also PFs, but aren't necessarily extremal on $\mathcal{U}$ observables, and due to it their physical meanings is still questionable and put some doubts on

Algebraic QM consistency [2]. Now it seems that nonextremal φ can correspond to the primordial states ξ, at least for the situations regarded here. Therefore the introduction of primordial states supports Algebraic QM premises, and in particular, evidences in favor of the feasibility of algebraic states.

For the conclusion we argued here that the primordial quantum states, which differ from the pure and mixed states, can exists in the nature. Their appearance is defined by the fact that they are produced without direct involvement of any observer, in distinction from QM definition of the pure and mixed states preparation. It seems that the introduction of primordial states in Quantum Theory can help to fill the existing gap between QM and QS axiomatics.

References

[1] D.Guilini et al., 'Decoherence and Appearance of Classical World', (Springer-Verlag, Berlin, 1996)

[2] J.M.Jauch 'Foundations of Quantum Mechanics' (Adison-Wesly, Reading, 1968)

[3] G.Emch, 'Algebraic Methods in Statistical Physics and Quantum Mechanics', (Wiley, N-Y, 1972)

[4] (1996) P.Busch, P.Lahti, P.Mittelstaedt, 'Quantum Theory of Measurements' (Springer-Verlag, Berlin,1996)

[5] S.Dodelson 'Modern Cosmology' (Academic Press, London, 2003)

[6] W.Zurek, Phys Rev, D26,1862 (1982)

SPACE AND TIME PHYSICS WITH THE LORENTZ ETHER: THE CLOCK PARADOX

FRANCO SELLERI [a]

[a] *Università di Bari - Dipartimento di Fisica*
INFN - Sezione di Bari

Abstract

The description of natural phenomena by observers in motion is a problem that many consider solved by the Lorentz transformations, but that actually was left open. Consequences of my alternative "inertial" transformations are: (i) an explanation of the empirical data better than provided by the Theory of Special Relativity (TSR); (ii) the elimination of those features of the TSR giving rise to paradoxes thanks to the recovery of a privileged inertial frame in which the Lorentz ether is at rest. The example of the "clock paradox" is discussed and a complete resolution is obtained by giving an exhaustive unified description of all possible situations. Velocity (and nothing else) is thus seen to be responsible for the differential retardation effect.

1 The equivalent transformations

According to Poincaré [1], Reichenbach [2], Jammer [3], Mansouri and Sexl [4] the clock synchronization in inertial systems is conventional and the choice of the invariance of the one way velocity of light made in the TSR is only based on simplicity. Following the same line of thought, I introduced in the transformations of the time variable a suitable parameter e_1, describing synchronization [5]. The TSR is obtained for a particular nonzero value of e_1. In this way an infinite set of theories empirically equivalent to the TSR was developed.

Given the inertial frames and one can set up Cartesian coordinates and make the following standard assumptions:

B.G. Sidharth, F. Honsell and A. de Angelis. (eds.), Frontiers of Fundamental Physics, 195–208.

(i) Space is homogeneous and isotropic and time homogeneous, at least if judged by observers at rest in S_0;

(ii) In the isotropic system S_0 the velocity of light is "c" in all directions, so that clocks can be synchronized in S_0 and one way velocities relative to S_0 can be measured;

(iii) The origin of S, observed from S_0, is seen to move with velocity $v < c$ parallel to the $+x_0$ axis, that is according to the equation $x_0 = v\,t_0$;

(iv) The axes of S and S_0 coincide for $t = t_0 = 0$;

The system S_0 turns out to have a privileged status in all theories satisfying the assumptions (i) and (ii), with the exception of the TSR. Two further assumptions based on direct experimental evidence can be added:

(v) The two way velocity of light is the same in all directions and in all inertial systems;

(vi) Clock retardation takes place with the usual velocity dependent factor when clocks move with respect to S_0. This assumption is the same as A2 of the second section.

These conditions were shown [5] to lead to the following transformations of the space and time variables from S to S_0

$$\begin{cases} x = \dfrac{x_0 - v\,t_0}{R} \\ y = y_0 \qquad ; \qquad z = z_0 \\ t = R\,t_0 - e_1(x_0 - v\,t_0) \end{cases} \tag{1}$$

where

$$R = \sqrt{1 - v^2/c^2} \tag{2}$$

and e_1 is the synchronization parameter. By using again (ii) and Eq.s (1) one can find the one way velocity of light relative to the moving system S for light propagating at an angle θ from the velocity $\vec{v}$ of S relative to S_0 [5]:

$$c_1(\theta) = \frac{c}{1 + \Gamma \cos\theta} \tag{3}$$

with

$$\Gamma = e_1 R c + \frac{v}{c} \tag{4}$$

The TSR is a particular case, obtained for

$$e_1 = -\frac{v}{c^2 R} \tag{5}$$

giving $\Gamma = 0$ and $c_1(\theta) = c$ and reducing (1) to their Lorentz form.

These results are enough for the needs of the present paper. We should mention, however, that we also found that the choice $e_1 = 0$ is the only one allowing for a treatment of accelerations rationally connected with the physics of inertial systems S_0. The theory was applied to the rotating platform and to the Sagnac effect [6] with the result that only the choice $e_1 = 0$ could give a satisfactory explanation [7]. A consequence of this research was the discovery of a relativistic discontinuity between inertial systems and slowly accelerated systems. The discontinuity disappears only if $e_1 = 0$ [8]. Its existence in the TSR is the root of the difficulties met by Langevin [9], Post [10], Landau-Lifschitz [11], Anandan [12] in explaining the physics on a rotating platform.

There are several other good reasons to adopt $e_1 = 0$:

(a) Einstein in his 1905 article assumed that aberration depends on the relative velocity star-Earth, but this relativistic description works poorly, given that empirically the aberration angle is the same for all stars in the sky [13].

(b) The growing evidence for the existence of superluminal signals can easily be accommodated in the theory with $e_1 = 0$, while it is incompatible with relativity due to the presence of a causal paradox in which events belonging to the future of an observer can actively modify the past of the same observer [14].

(c) Practically all paradoxes of the special theory of relativity disappear in a theory based on the inertial transformations.

For these reasons a satisfactory theory of the physics of space and time has to be based on absolute simultaneity ($e_1 = 0$).

Assuming the inertial transformations, the S_0 system is initially considered to be privileged, and the velocity of light relative to it isotropic. Other inertial systems are described as "moving" and relative to them the observers detect an anisotropic velocity of light. In ref. [15] I described a resynchronization of clocks (ROC) and showed that it is uniquely determined by the new inertial frame S chosen to replace S_0 as "privileged" and by the requirement that absolute simultaneity should be preserved. Thus a weak form of relativity principle is restored. One can add, however, that from the point of view of the inertial transformations the validity of relativity appears accidental, more than fundamental. It would be enough to discover a very small noninvariance of the two way speed of light to make the whole game of ROC impossible.

2 The clock retardation formula

In 1905 Einstein [16] presented the clock retardation prediction as follows: Imagine one of the clocks which mark the time t_0 when at rest in the "stationary" inertial system S_0, and the time t when at rest in the "moving" inertial system S, to be located at the origin of the coordinates of S, where it marks the time t. What is the rate of this clock, if viewed from S_0? The quantities x_0, t_0 , and t, which refer to the position of the clock, satisfy $x_0 = v\,t_0$ and the Lorentz transformation of time

$$t = \frac{1}{R}(t_0 - v\,w_0/c^2)$$

where $R = \sqrt{1 - v^2/c^2}$. Therefore

$$t = t_0\,R = t_0 - (1 - R)t_0$$

whence it follows that the time t marked by the clock is slow by $1 - R$ seconds per second with respect to the S_0 time t_0.

From this Einstein deduced another consequence, which became known as "clock paradox". If at the points A and B of S_0 there are two synchronous stationary clocks; and if the clock at A is moved with the velocity v along the line AB to B, then on its arrival at B the two clocks no longer show the same time, but the clock moved from A to B lags behind the other which has remained at B by $(1 - R)t_0$, t_0 being the duration of the journey from A to B. Einstein considered evident that this result still holds good if the clock moves from A to B in any polygonal line, and also when the points A and B coincide. Furthermore, he assumed that the result obtained for a polygonal line holds also for a continuously curved line.

That is all. Clearly Einstein went beyond what he could say by considering only inertial systems, as he introduced also accelerated motions in the vertices of the polygonal line and along the continuous curve. He did so in an intuitive way, without solid foundations, and yet he was much closer to the correct result than with the 1918 paper (as we shall see).

Today, the retardation of moving clocks and its independence of acceleration are well established empirical facts. In a CERN experiment [17] muons with a velocity of $0.9994\,c$, corresponding to $R = 0.0346$, were circling in a ring with diameter of 14 m, with a centripetal acceleration $10^{18}g$. The lifetime τ_0 of the circling muons, measured in the laboratory, was in agreement with the formula $\tau = \tau_0 R$, where τ is the lifetime measured in the muon rest system. No effect of the huge acceleration on the lifetime was observed.

The experiment by Hafele and Keating [18] compared six synchronized caesium atomic clocks. Two were carried by ordinary commercial jets in an eastbound tour around the planet; another two were carried in a westbound tour; the last two remained on the ground. It was observed that with respect to the time shown by the latter clocks, those of the westbound trip had lost 59$\pm$10 ns, while those on the eastbound trip had advanced 273$\pm$7 ns. These results were in agreement with the usual formula $t = t_0 R$ if one used three different R's for the three pairs of clocks. The largest (smallest) R was that of eastward (westward) clocks, for which the Earth rotation velocity added to (subtracted from) the jet velocity. One had to include the effect of the Earth gravitational potential, variable with altitude, which modifies the rates of travelling clocks differently from those on the ground.

Similar conclusions have been obtained with the GPS (*Global Positioning System*) network of 24 satellites [19]. With an orbital radius of about four Earth radii and an orbital speeds of about 3.9 km/sec, each satellite has on board four atomic clocks marking time with an error of a few ns/day. The gravitational effect implies that the atomic clocks on board the satellites tick faster by about 45.900 ns/day because they are in a larger gravitational potential than atomic clocks on the Earth surface. The velocity effect makes atomic clocks moving at GPS orbital speeds tick slower by about 7.200 ns/day. Therefore the global prediction is a gain of about 38.700 ns/day. Rather than having clocks with such large rate differences, the satellite clocks were reset in rate before launch slowing them down by 38.700 ns/day. The rich data show that the on board atomic clock rates do indeed agree with ground clock rates.

The experimental evidence, in full agreement with Einstein's 1905 statements, points to the validity of the following clock retardation formula. If a clock U, marking the time t_0 when at rest in the isotropic inertial system S_0, is set in motion with arbitrarily oriented and possibly variable velocity $u(t_0)$ relative to S_0, the rate of the time marked by U at S_0 time t_0 is given by

$$d\tau = dt_0\sqrt{1 - u^2(t_0)/c^2} \tag{6}$$

This τ is exactly what an observer travelling with U reads on the clock itself. Therefore $d\tau$ is the "proper time" variation of U.

The idea behind Eq. (6) is that only instantaneous velocity (and not acceleration) fixes the clock rate, in agreement with observations. In physics one can recognize the cause of a phenomenon by varying it and verifying the existence of corresponding variations of the effect. Viceversa, if arbitrary variations of a physical quantity Q do not modify the effect E, one can exclude that Q is among the causes of E. Let us apply this criterion to Eq. (6). If $u(t_0)$ is varied a corresponding variation of the proper time rate arises: therefore velocity can be claimed to be a cause of the proper time rate variation. On the contrary, if the acceleration is modified at time t_0 while $u(t_0)$ remains the same $d\tau$ will not change. Therefore the acceleration has no effect on $d\tau$ and cannot be counted among the causes of its variation. This reasoning is important for determining the real physical roots of the "clock paradox".

3 Absolute motion resolution of the clock paradox

As stressed by S. Prokhovnik [20] the resolution of the clock paradox in terms of absolute motion was found by G. Builder [21] who showed that the differential retardation effect between two clocks which separate and reunite can be validly considered in respect to a single inertial reference frame. Such an effect (read on the clocks) appears obviously to be the same to observers in all states of motion, and in this sense is absolute ("invariant"). Builder concluded that the emergence of an absolute effect consequence of velocity implies the existence of a privileged inertial frame, in the sense that motion relative to this frame assumes an absolute significance and is quantitatively related to absolute effects.

Builder's paper is mostly qualitative and in some points difficult to understand.

My hopefully clearer reformulation of his argument is based on two assumptions:

A1. The velocity of light relative to an inertial system, S_0, is "c" in all directions, so that clocks can be synchronized in S_0 with the Einstein method and one way velocities relative to S_0 can be measured;

A2. A clock moving with speed $u(t_0)$ relative to S_0 during the S_0 time interval dt_0 marks a (proper) time increase $d\tau$ given by Eq. (6).

The TSR is well known to satisfy the above assumptions in all inertial systems. The theory of the "equivalent transformations" [5] accepts S_0 as the privileged system relative to which A1 and A2 are satisfied as well. Therefore all the consequences deduced below from A1 and A2 are valid both in the TSR and in the theory of the equivalent transformations.

In this paper we consider only one spatial dimension. According to A2 the proper time increase T marked by a clock moving from the point a at time t_{0a} to the point b at time t_{0b}, both fixed in S_0, (see Fig. 1) is

$$T = \int_{t_{0a}}^{t_{0b}} dt_0 \sqrt{1 - u^2(t_0)/c^2} \tag{7}$$

where

$$u(t_0) = \frac{dx_0}{dt_0} \tag{8}$$

and $x_0 = x_0(t_0)$ is the equation of motion of the clock on some "trajectory" in the (x_0, t_0) plane connecting the points a, b of Fig. 1. We consider a second ("varied") trajectory, very near to the original one, as follows

$$x_0(t_0) \qquad \rightarrow \qquad x_0(t_0) + \delta x_0(t_0) \tag{9}$$

with

$$\delta x_0(t_{0a}) = \delta x_0(t_{0b}) = 0 \tag{10}$$

According to Eq. (10) the clock on the varied trajectory occupies the points a and b at the same times t_{0a} and t_{0b} as on the unvaried trajectory. This clearly corresponds to the situation in which two clocks separate at point a at time t_{0a}, follow different trajectories and reunite again in point b at time t_{0b}.

Also the velocity undergoes a variation

$$u(t_0) \qquad \rightarrow \qquad u(t_0) + \delta u(t_0) \tag{11}$$

with $u(t_0)$ given by Eq. (8) and

$$\delta u(t_0) \equiv \frac{d}{dt_0} \delta x_0(t_0) \tag{12}$$

The proper time integral will correspondingly become

$$T + \delta T = \int_{t_{0a}}^{t_{0b}} dt_0 \sqrt{1 - \left[u(t_0) + \delta u(t_0)\right]^2 / c^2} \tag{13}$$

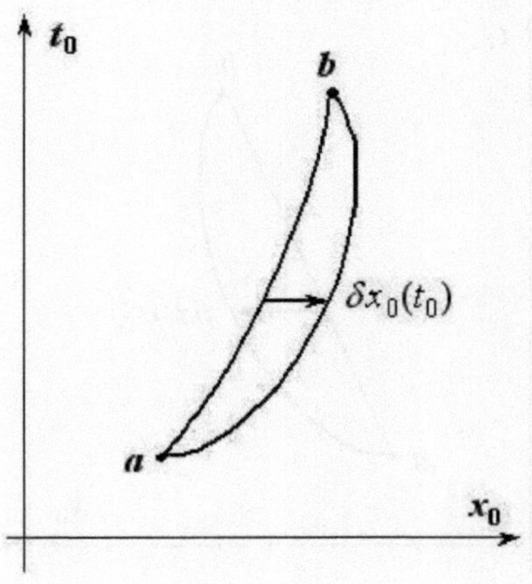

Figure 1. A space and time diagram showing a "trajectory" between two points, and a second varied trajectory between the same points.

From (12) and (13) it follows, for small variations

$$\delta T = -\frac{1}{c^2}\int_{t_{0a}}^{t_{0b}} dt_0 \frac{u(t_0)}{\sqrt{1-u(t_0)^2/c^2}}\frac{d}{dt_0}\delta x_0(t_0) \tag{14}$$

Integrating by parts one gets

$$\delta T = -\frac{1}{c^2}\left\{\left[\frac{u(t_0)\delta x_0(t_0)}{\sqrt{1-u(t_0)^2/c^2}}\right]_{t_{0a}}^{t_{0b}} - \int_{t_{0a}}^{t_{0b}} dt_0 \frac{d}{dt_0}\left[\frac{u(t_0)}{\sqrt{1-u(t_0)^2/c^2}}\right]\delta x_0(t_0)\right\} \tag{15}$$

Due to (10) the first term in the right hand side vanishes. The derivative in the second term gives

$$\delta T = \frac{1}{c^2}\int_{t_{0a}}^{t_{0b}} dt_0 \left[\frac{u'(t_0)}{\sqrt{1-u(t_0)^2/c^2}}\right]^{3/2}\delta x_0(t_0) \tag{16}$$

where $u'(t_0) = du(t_0)/dt_0$. Clearly, $\delta T = 0$ for arbitrary $\delta x_0(t_0)$ satisfying (10) if and only if $u'(t_0) = 0$ at all times. This is like saying that the extremum proper time T of all motions is the uniform one with the constant velocity

$$u_1 = \frac{x_{0b} - x_{0a}}{t_{0b} - t_{0a}} \tag{17}$$

for which the proper time integral (7) takes the value

$$T_1 = (t_{0b} - t_{0a})\sqrt{1 - u_1^2/c^2} \tag{18}$$

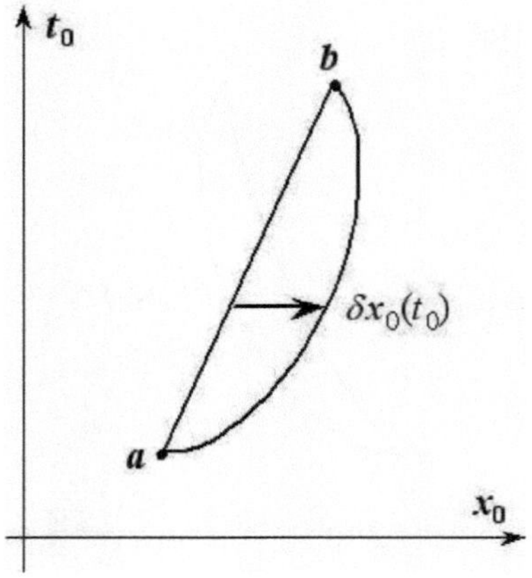

Figure 2. Space and time diagram showing a constant velocity connection between two points a and b, and a second varied trajectory between the same points.

Among all motions connecting a and b this extremum is unique, as it can be obtained for $u'(t_0) = 0$ only. Therefore it gives either the maximum or the minimum proper time of all possible motions from a to b, not only of those obtained with an infinitesimal deformation of the straight line.
That the extremum is actually a maximum can be seen as follows. One has

$$\sqrt{1-(u_1+\delta u)^2/c^2} \cong \sqrt{1-u_1^2/c^2} - \frac{1}{c^2}\frac{u_1\delta u}{\sqrt{1-u_1^2/c^2}} - \frac{1}{2c^2}\frac{(\delta u)^2}{\left(1-u_1^2/c^2\right)^{3/2}} \tag{19}$$

But u_1 is constant and δu satisfies (12) so that, after integration, the first order variation of T_1 arising from (19) vanishes due to (10). The second order variation is

$$\delta^2 T = -\frac{1}{2c^2}\int_{t_{0a}}^{t_{0b}} dt_0 \frac{\left[\delta u(t_0)\right]^2}{\left[1-u_1^2/c^2\right]^{3/2}} \tag{20}$$

Clearly, $\delta^2 T < 0$ for all possible velocity variations. Therefore, moving away from the constant velocity line of Fig. 2 to a different line connecting a and b implies in all cases a decrease of the elapsed proper time. Then the found extremum is a maximum. Comparing the motion with velocity u_1 with any different motion from a to b with velocity $u_2(t_0)$ one has

$$\Delta T = T_1 - T_2 = \int_{t_{0a}}^{t_{0b}} dt_0 \left[\sqrt{1-u_1^2/c^2} - \sqrt{1-u_2(t_0)^2/c^2}\right] > 0 \tag{21}$$

As a difference of proper times, ΔT is exactly what two observers who traveled with the clocks find by direct comparison of the clocks readings.

The clock moving with rectilinear uniform motion can be considered at rest in a different inertial system. Therefore the previous argument can be taken to describe all possible "clock paradox" situations from the point of view of the particular inertial system S_0 we have considered. The description is the same in the TSR and in all other theories satisfying the assumptions A1 and A2 above. But the principle of relativity, the metrics of Minkowski space and/or the gravitational potential of the fictitious forces have nothing to do with the essence of the matter. Only velocities are able to influence the working of a clock. In a forthcoming paper we will show that the previous argument can be extended to three dimensional space [22].

4 Most general clock retardation formula

We now generalize the obtained results by substituting assumption A2 with the following one: If during the S_0 time interval dt_0 a clock is moving with velocity $u(t_0)$ relative to S_0, it marks a (proper) time change $d\tau$ given by

$$d\tau = dt_0 F\big[u(t_0)\big] \tag{22}$$

where F is an arbitrary function of its argument. Thus the proper time T spent by a clock moving between any two points a, b fixed in S_0 is given by

$$T = \int_{t_{0a}}^{t_{0b}} dt_0 F\big[u(t_0)\big] \tag{23}$$

where

$$u(t_0) = \frac{dx_0}{dt_0} \tag{24}$$

and $x_0 = x_0(t_0)$ is the equation of motion of the clock on some "trajectory" connecting the points a, b of Fig. 1 in the (x_0, t_0) plane. We consider a variation of this trajectory as in eq. (9), with the conditions (10) satisfied. Thus the proper time integral becomes

$$T + \delta T = \int_{t_{0a}}^{t_{0b}} dt_0\, F\big[u(t_0) + \delta u(t_0)\big] \tag{25}$$

From (12) and (25) it follows, for small variations

$$\delta T = \int_{t_{0a}}^{t_{0b}} dt_0\, F'\big[u(t_0)\big] \frac{d}{dt_0} \delta x_0(t_0) \tag{26}$$

where F' indicates the u derivative of F. An integration by parts now gives

$$\delta T = \Big\{F'\big[u(t_0)\big]\, \delta x_0(t_0)\Big\}_{t_{0a}}^{t_{0b}} - \int_{t_{0a}}^{t_{0b}} dt_0\, \frac{d}{dt_0} \left\{F'\big[u(t_0)\big]\right\} \delta x_0(t_0) \tag{27}$$

Due to (10) the first term in the right hand side vanishes and from the second term one gets

$$\delta T = -\int_{t_{0a}}^{t_{0b}} dt_0\, F''\big[u(t_0)\big] u'(t_0)\, \delta x_0(t_0) \tag{28}$$

Clearly, $\delta T = 0$ for arbitrary $\delta x_0(t_0)$ if, at all times t_0

$$F''\big[u(t_0)\big]u'(t_0) = 0 \tag{29}$$

Thus we see that at least one of the possible solutions is $u'(t_0) = 0$ at all times, the uniform motion. Whether it corresponds to a maximum or to a minimum has to be seen case by case. Many functions F give rise to a maximum of the proper time integral for $u'(t_0) = 0$, e.g.

$$F\big[u(t_0)\big] = \exp\left\{-\frac{1}{2}\frac{u^2}{c^2}\right\} \tag{30}$$

Therefore there is nothing typical of a relativistic theory in the maximum of the proper time integral provided by the rectilinear uniform motion.

5 Clock retardation in general relativity

The 1905 formulation of the clock paradox had implications that probably Einstein did not like. The delay is an absolute effect, as all observers agree that the clock moving with variable velocity marks a smaller time. They disagree, however, on the numerical value of this variable velocity. In relativity all potential observers (forming an infinite set) are completely equivalent, so that, in a sense, one can say that the clock velocity assumes at any time all conceivable values. But a quantity having infinitely many values is totally undefined. In this way the presumed cause of the differential retardation effect seems to vanish into nothingness. But the cause of a real physical effect should be concrete as well, in spite of the evasive description given by the theory. Therefore velocity itself should be well defined, that is, relative to a physically active reference background (ether) defining at the same time the privileged reference frame.

It is no surprise, then, that the original formulation was completed with a later one based on the theory of general relativity (TGR) [23], whose essential points we will now review. Let S be an inertial reference system. Further, let U_1 and U_2 be two exactly similar clocks working at the same rate when at rest in S. If one of the clocks - let us say U_2 - is in a state of uniform translatory motion relative to S, then, according to the TSR it works more slowly than U_1, which is at rest in S. At this point Einstein adds a remark: "This result seems odd in itself. It gives rise to serious doubts when one imagines the following thought experiment." In the thought experiment A is the origin of S, and B a different point of the positive x-axis. The two clocks, initially at rest at A, work at the same rate and their readings are the same. Next, a constant velocity in the direction $+x$ is imparted to U_2, so that it moves towards B. At B the velocity is reversed, so that U_2 returns towards A. When it arrives at A its motion is stopped, so that it is again at rest near U_1. Since U_2 works more slowly than U_1 during its motion along the line AB, U_2 must be behind U_1 on its return.

Now comes the problem, says Einstein. According to the principle of relativity the whole process must surely take place in exactly the same way if it is considered in a reference frame S' sharing the movement of U_2. Relatively to S' it is U_1 that executes the to-and-fro movement while U_2 remains at rest throughout. From this it would seem to follow that, at the end of the process, U_1 must be behind U_2, which contradicts the former result.

But, Einstein adds, the TSR is inapplicable to the second case, as it deals only with inertial frames, while S is at times accelerated. Only the TGR deals with accelerated frames. From the point of view of the TGR, one can use the coordinate system S' just as well as S. But in describing the whole process, S and S' are not equivalent as the following comparison shows.

Description relative to the S Reference System

1. The clock U_2 is accelerated by an external force in the direction $+x$ until it reaches the velocity v. U_1 is at rest, now as in all the subsequent steps;
2. U_2 moves with constant velocity v to the point B on the $+x$-axis;
3. U_2 is accelerated by an external force in the direction $-x$ until it reaches the velocity v in the direction $-x$;
4. U_2 moves with constant velocity v in the direction $-x$ back to the neighbourhood of U_1;
5. U_2 is brought to rest by an external force very near to U_1.

Description relative to the S′ Reference System

1. A gravitational field, oriented along $-x$, appears, in which the clock U_1 falls with an accelerated motion until it reaches the velocity v. When U_1 has reached the velocity v the gravitational field vanishes. An external force applied to U_2 prevents U_2 from falling in the gravitational field;
2. U_1 moves with constant velocity v to a point B' on the x-axis. U_2 remains at rest;
3. A homogeneous gravitational field in the direction $+x$ appears which accelerates U_1 in the direction $+x$ until it reaches the velocity v, whereupon the gravitational field vanishes. U_2 is kept at rest by an external force;
4. U_1 moves with constant velocity v in the direction $+x$ into the neighbourhood of U_2. U_2 remains at rest;
5. A gravitational field in the direction $-x$ appears, which brings U_1 to rest. The gravitational field then vanishes. U_2 is kept at rest by an external force.

The second description is based on the principle of equivalence between fictitious and gravitational forces. According to both descriptions, at the end of the process the clock U_2 is retarded by a definite amount compared with U_1. With reference to S' this is explained by noticing that during the stages 2 and 4, the clock U_1, moving with velocity v, works more slowly than U_2, which is at rest. But this retardation is overcome by the quicker working of U_1 during stage 3. For, according to the GTR, a clock works faster the higher is the gravitational potential in the point where it is placed, and during stage 3 U_1 is indeed placed in a region of higher gravitational potential than U_2. A calculation made with instantaneous acceleration shows that the consequent advancement amounts to exactly twice as much as the retardation during stages 2 and 4 [24]. After this conclusion Einstein states: "This completely clears up the paradox."

The prediction of the TGR that a clock works faster the larger is the gravitational potential ϕ in the region in which it is placed is fully confirmed by the experiments in the gravitational field of the Earth. At first sight the 1918 reasoning could seem to be a consequence of the empirical facts. The mathematical treatment of the clock paradox situation given by the TGR leads to the right result by describing the retardation of U_2 as a consequence of the action of ϕ on U_1 [24]. Yet the theory shows its weakness in several ways.

It is unreasonable that something happens to U_1, as the effect should be objective and during the (short) times of acceleration some change in U_1 should be seen also by some independent observers.

For the sake of clarity let us compare two different experiments, E1 and E2 below, both starting in the same way, as follows. Let A be a point of the inertial system S. Let U_1 be the clock constantly at rest in A and U_2 the mobile clock, initially at rest in A. At time $t = 0$ we give U_2 a constant acceleration in the direction $+x$ until it reaches the constant velocity v as in Einstein's thought experiment. After this U_2 moves with velocity v until, at time $t = t_1$, two alternative developments can start.

E1. In the time interval (t_1, t_2), with $t_2 > t_1 > 0$, U_2 experiences a constant acceleration, exactly reversing its velocity. The 1918 formulation introduces a gravitational potential ϕ in the rest system of U_2, during (t_1, t_2). The role of ϕ is very important as it must give rise to a variation in the time marked by U_1 opposite and twice as large as that arising in the long stretches of uniform motion during which U_1 is delayed with respect to U_2.

E2. In the time interval (t_1, t_2) U_2 continues its rectilinear and uniform motion in the direction $+x$. The two clocks do not reunite anymore. No gravitational potential can arise, as there is no acceleration. Therefore in this second case there is no retardation of U_1 with respect to U_2.

In comparing E1 and E2, we note that the observers at rest with respect to U_1 see in both cases the same identical situation, namely no modification of the rate of time keeping of U_1. This can also be checked from a distance by observers in arbitrary states of motion, e.g. by monitoring U_1 on a TV screen. The fundamental fact is always the same: nothing ever happens to U_1, in particular nothing happens in the time interval (t_1, t_2), when U_2 accelerates. Therefore ϕ cannot be the cause of any change of U_1, as in passing from E1 to E2 the presumed cause varies from a position dependent ϕ to $\phi =$ const. without any variation of the effect.

The criticism becomes even stronger if one considers not just one but several clocks

$$U_1\,,\, U_1'\,,\, U_1''\,,\, \ldots$$

at rest in different points of the line AB of the inertial reference frame S. When U_2 accelerates these clocks should be influenced differently, as ϕ depends on distance from U_2. But in reality nothing happens: observers can check in particular that the delays with which a light signal originated near U_1 touches U_1', U_1'', ... are the same before and after the time interval during which U_2 was accelerated.

We can conclude that the gravitational potential of the fictitious forces exerts no action on the clocks, contrary to Einstein's 1918 opinion. The gravitational fields in the accelerated systems are not ordinary static fields, but arise from the accelerations of bodies [25]. Einstein assumed that these fields had on clocks the same action as ordinary fields, but we can now conclude that on this particular point he was not right, in spite of the very probable correctness of the general idea of equivalence between fictitious and gravitational forces. One finds an analogy in the magnetic field, which can be considered a dynamical manifestation of the electric field, but has quite different interaction properties.

We can claim, finally, that concerning the nature of the differential retardation of separating and reuniting clocks Einstein was much nearer to the truth in 1905 than in 1918.

References

[1] H. Poincaré, *La mesure du temps*, Rev. Métaphys. Morale 6, 1-13 (1898).

[2] H. Reichenbach, *The Philosophy of Space & Time*, Dover, New York, (1958).

[3] M. Jammer, *Some fundamental problems in the special theory of relativity*, in: G. Toraldo di Francia, ed., *Problems in The Foundations of Physics*, pp. 202-236, Società Italiana di Fisica, Bologna and North Holland, Amsterdam, (1979).

[4] R. Mansouri and R. Sexl, *A Test Theory of Special Relativity, I, II and III*, General Relat. Gravit., 8, 497-513, 515-524, 809-814 (1977).

[5] F. Selleri, *Noninvariant One-Way Velocity of Light*, Found. Phys. 26, 641-664 (1996); *Noninvariant One-Way Velocity of Light and Particle Collisions*, Found. Phys. Lett. 9, 43-60 (1996).

[6] M. G. Sagnac, *Effet tourbillonnaire optique. La circulation de l'éther lumineux dans un interférographe tournant*, J. de Phys. 4, 177 (1914).

[7] F. Selleri, *Sagnac effect: end of the mystery*, in *Relativity in rotating frames*, pp. 57-77, G. Rizzi and M. L. Ruggiero, eds., Kluwer, Dordrecht, (2004).

[8] F. Selleri, *Zero acceleration discontinuity and absolute simultaneity*, to be published in the book: *Absolute simultaneity*, W. L. Craig, ed., Routledge (2005).

[9] P. Langevin, *Sur l'expérience de Sagnac*, Compt. Rend. 205, 304 (1937).

[10] E. J. Post, *Sagnac effect*, Rev. Mod. Phys. 39, 475 (1967).

[11] L. D. Landau and E. M. Lifshitz, *The classical theory of fields*, Butterworth-Heinemann, Oxford (1996), §§ 82-87.

[12] J. Anandan, *Sagnac effect in relativistic and nonrelativistic physics*, Phys. Rev. D 24, 338 (1981).

[13] G. Puccini and F. Selleri, *Doppler effect and aberration of light from the point of view of absolute motion*, Nuovo Cim. B, 117, 283 (2002).

[14] F. Selleri, *Superluminal signals require absolute simultaneity*, Bull. Soc. Sciences et Lettres de Lodz, LII, 63 (2002)

[15] F. Selleri, *The inertial transformations and the relativity principle*, in print on Found. Phys. Letters (2005).

[16] A. Einstein, *Zur Elektrodynamik bewegter Körper*, Ann. der Phys. 17, 891-921 (1905). English translation: *On the electrodynamics of moving bodies*, in A. Einstein, H. A. Lorentz, ..., *The principle of relativity*, pp. 37-65, Dover, New York, (1952).

[17] J. Bailey, et al., *Measurements of relativistic time dilatation for positive and negative muons in a circular orbit*, Nature 268, 301-304 (1977).

[18] J. C. Hafele and R. E. Keating, *Around the World Atomic Clocks*, Science 177, 166-170 (1972).

[19] T. Van Flandern, *What the Global Positioning System Tells us about Relativity*, in: F. Selleri, ed., *Open questions in relativistic physics*, pp. 81-90, Apeiron, Montreal, (1998).

[20] S. J. Prokhovnik, *An introduction to the neo-Lorentzian relativity of Builder*, Spec. Science and Techn. 2, 225-229 (1979).

[21] G. Builder, *The resolution of the clock paradox*, Aust. J. Phys. 10, 246-262 (1957).

[22] F. Selleri, *Absolute velocity resolution of the clock paradox*, to appear in the book: *The Aether: Poincaré and Einstein*, V. Dvoeglazov & C. Roy Keys, eds., (2005).

[23] A. Einstein, *Dialog über Einwände gegen die Relativitätstheorie*, Die Naturwissenschaften 6, 697-702 (1918).

[24] C. Møller, *The Theory of Relativity*, Clarendon Press, Oxford (1957), § 98.

[25] D. W. Sciama, *On the Origin of Inertia*, Monthly Notices Roy. Astron. Soc., 113, 34-42 (1953). A. Ghosh, *Origin of Inertia*, Apeiron, Montreal, (2000).

Part III

Nuclear and High-Energy Particle Physics and Astrophysics

CLUSTERS OF MATTER AND ANTIMATTER - A MECHANISM FOR COLD COMPRESSION

WALTER GREINER [a]

[a] *Frankfurt Institute for Advanced Studies, Johann Wolfgang Goethe Universität, D–60054 Frankfurt am Main, Germany*

Abstract

In this talk I first present the vacuum for the e^+-e^- field of QED and show how it is modfied for baryons in nuclear environment. Then I discuss the possibility of producing new types of nuclear systems by implanting an antibaryon into ordinary nuclei. The structure of nuclei containing one antiproton or antilambda is investigated within the framework of a relativistic mean-field model. Self-consistent calculations predict an enhanced binding and considerable compression in such systems as compared with normal nuclei. I present arguments that the life time of such nuclei with respect to the antibaryon annihilation might be long enough for their observation.

It is generally accepted that physical vacuum has nontrivial structure. This conclusion was first made by Dirac on the basis of his famous equation for a fermion field which describes simultaneously particles and antiparticles. The Dirac equation in the vacuum has a simple form

$$(i\gamma^\mu \partial_\mu - m)\Psi(x) = 0\,, \tag{1}$$

where $\gamma^\mu = (\gamma^0, \boldsymbol{\gamma})$ are Dirac matrices, m is the fermion mass and $\Psi(x)$ is a 4-component spinor field. For a plane wave solution $\Psi(x) = e^{-ipx}u_p$ this equation is written as

$$(\widehat{p} - m)u_p = 0\,, \tag{2}$$

B.G. Sidharth, F. Honsell and A. de Angelis. (eds.), Frontiers of Fundamental Physics, 211–220.

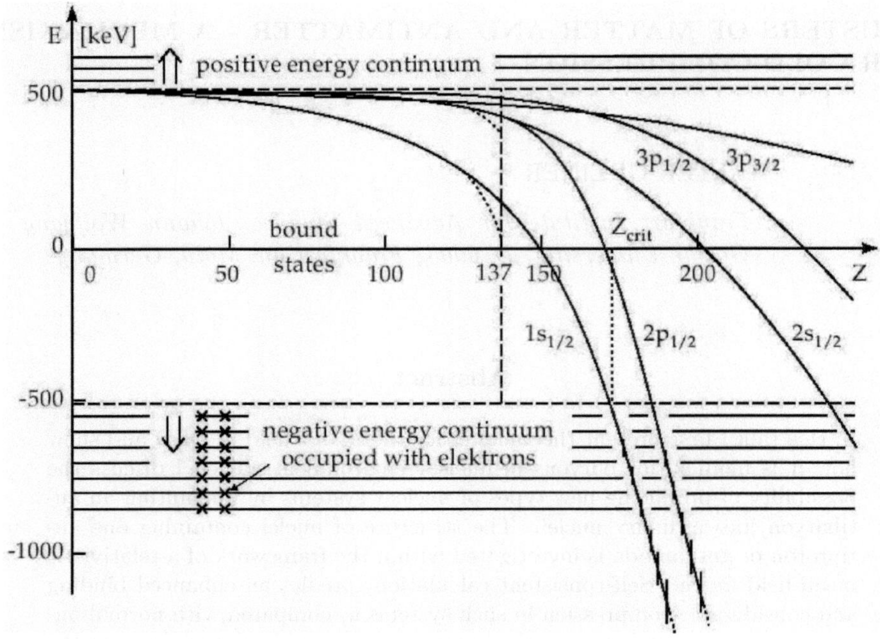

Figure 1. Lowest bound states of the Dirac equation for nuclei with charge Z. While the Sommerfeld fine-structure energies (dashed line) for $\xi = 1$ (s states) end at $Z = 137$, the solutions for extended Coulomb potentials (full line) can be traced down to the negative-energy continuum reached at the critical charge Z_{cr} for the 1s state. The bound states entering the continuum obtain a spreading witdth as indicated.

where $\widehat{p} = \gamma^0 E - \boldsymbol{\gamma}\mathbf{p}$. Multiplying by $(\widehat{p} + m)$ and requiring that $u_p \neq 0$ one obtains the equation $E^2 - \mathbf{p}^2 - m^2 = 0$ which has two solutions

$$E^{\pm}(\mathbf{p}) = \pm\sqrt{\mathbf{p}^2 + m^2}\,. \tag{3}$$

Here the + sign corresponds to particles with positive energy $E_N(\mathbf{p}) = E^+(\mathbf{p})$, while the − sign corresponds to solutions with negative energy. To ensure stability of the physical vacuum Dirac has assumed that these negative-energy states are occupied forming what is called now the Dirac sea. Then the second solution of eq. (3) receives natural interpretation: it describes holes in the Dirac sea. These holes are identifed with antiparticles. Their energies are obviously given by $E_{\overline{N}}(\mathbf{p}) = -E^-(\mathbf{p}) = \sqrt{\mathbf{p}^2 + m^2}$. Unfortunately, the Dirac sea brings divergent contributions to physical quantities such as energy density, and one should introduce a proper regularization scheme to rid off these divergences. This picture has received numerous cofirmations in quantum electrodynamics and other fields.

One of the most fascinating aspects is the structure of the vacuum in QED and its change into charged vacuum states under the influence of strong (supercritical) electric fields [1]. I shortly remind of this phenomenon.

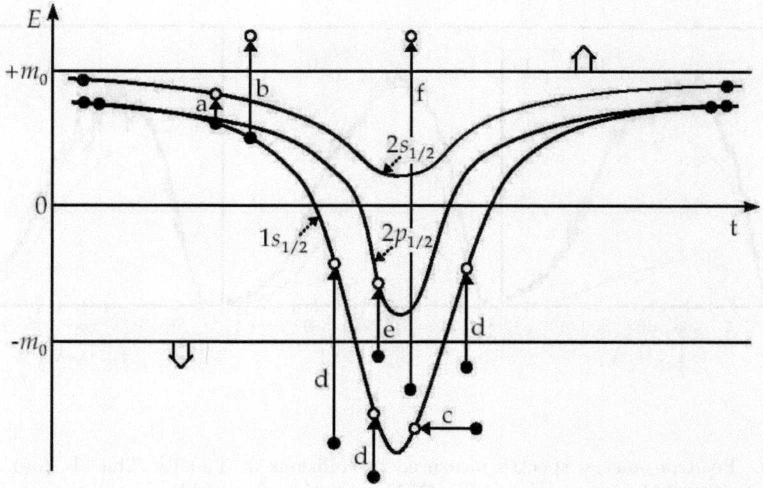

Figure 2. Time dependence of the quasi-molecular energy levels in a supercritical heavy ion collision. The arrows denote various excitation processes which lead to the production of holes and positrons.

Fig. 1 shows the diving of the deeply bound states into the lower energy continuum of the Dirac equation. In the supercritical case the dived state is degenerate with the (occupied) negative electron states. Hence spontaneous e^+e^- *paircreation* becomes possible, where an electron from the Dirac sea occupies the additional state, leaving a hole in the sea which escapes as a positron while the electron's charge remains near the source. This is a fundamentally new process, whereby the neutral vacuum of QED becomes unstable in supercritical electrical fields. It decays within about $10^{-19}s$ into a charged vacuum. The charged vacuum is now stable due to the Pauli principle, that is the number of emitted particles remains finite. The vacuum is first charged twice because two electrons with opposite spins can occupy the $1s$ shell. After the $2p_{1/2}$ shell has dived beyond $Z_{cr} = 185$, the vacuum is charged four times, etc. This change of the vacuum structure is not a perturbative effect, as are the radiative QED effects (vacuum polarization, self-energy, etc.).

The time-dependence of the energy levels in a supercritical heavy-ion collision is depicted in Fig. 2. An electron (or hole) which was in a certain molecular eigenstate at the beginning of the collision can be transfered with a certain probability into different states by the dynamics of the collision. This can lead to the hole production in an inner shell by excitation of an electron to a higher state and/or hole production by ionization of an electron to the continuum. Further possibilities are induced positron production by excitation of an electron from the lower continuum to an empty bound level and direct pair production [2].

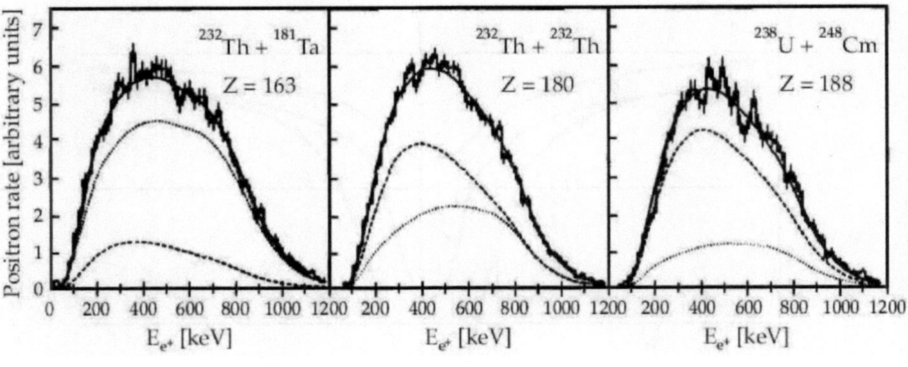

Figure 3. Positron energy spectra measured in collisions of Th+Ta, Th+Th, and U+Cm at energies of about 6 MeV per nucleon. The QED predictions (dashed lines) and the experimentally determined background from nuclear pair conversion (dotted lines) add up to the full lines which are in close agreement with experiment.

A comparison of the theoretical predictions and expectations and experimental data is shown in figure 3. Sharp positron peeks can be expected if there were a mechanism in the heavy ion collision leading to a time delay. This may be caused by a pocket in the potential between the two ions. Spontaneous pair production should then be enhanced in supercritical systems. Until now, however, the situation remains inconclusive [2].

It has been noticed already many years ago (see e. g. ref. [3]) that nuclear physics may provide a unique laboratory for investigating the Dirac picture of vacuum. The basis for this is given by relativistic mean-field models which are widely used now for describing nuclear matter and finite nuclei. Within this approach nucleons are described by the Dirac equation coupled to scalar and vector meson fields. Scalar S and vector V potentials generated by these fields modify plane-wave solutions of the Dirac equation as follows

$$E^{\pm}(\mathbf{p}) = V \pm \sqrt{\mathbf{p}^2 + (m - S)^2} \tag{4}$$

Again, the + sign corresponds to nucleons with positive energy $E_N(\mathbf{p}) = V + \sqrt{\mathbf{p}^2 + (m - S)^2}$, and the − sign corresponds to antinucleons with energy $E_{\overline{N}}(\mathbf{p}) = -E^-(-\mathbf{p}) = -V + \sqrt{\mathbf{p}^2 + (m - S)^2}$. It is remarkable that changing sign of the vector potential for antinucleons is exactly what is expected from the G-parity transformation of the nucleon potential. As follows from eq. (4), in nuclear environment the spectrum of single-particle states of the Dirac equation is modified in two ways. First, the mass gap between positive- and negative-energy states, $2(m - S)$, is reduced due to the scalar potential and second, all states are shifted upwards due to the vector potential. These changes are illustrated in Fig. 4.

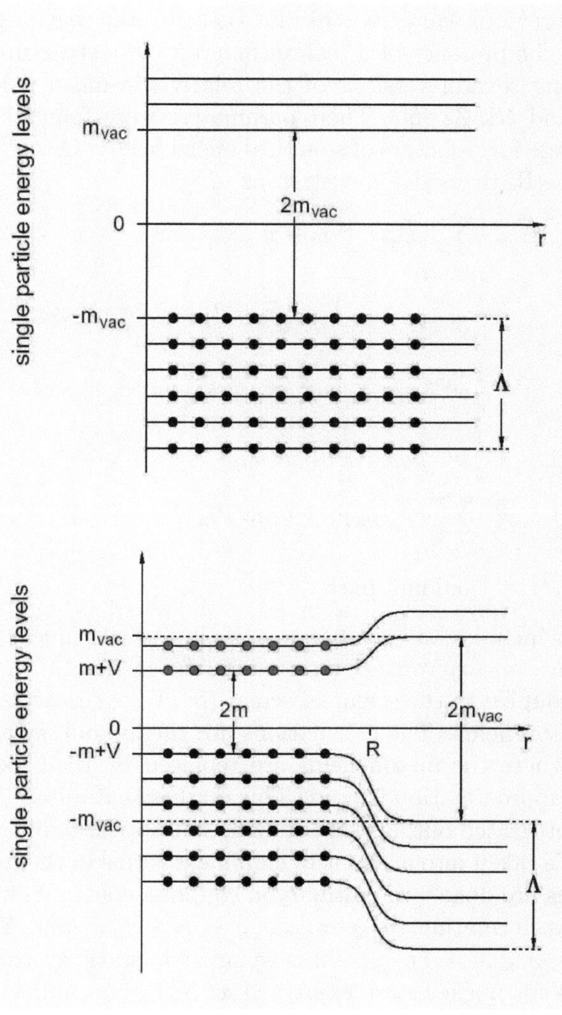

Figure 4. Schematic spectrum of Dirac equation in vacuum (upper panel) and in a nucleus of radius R (lower panel). A divergent contribution of negative-energy states is often regularized by introducing a cut-off momentum Λ.

It is well known from nuclear phenomenology that good description of nuclear ground state is achieved with $S \simeq 350$ MeV and $V \simeq 300$ MeV so that the net potential for nucleons is $V - S \simeq -50$ MeV. Using the same values one obtains for antinucleons very a deep potential, $-V - S \simeq -650$ MeV. Such a potential would produce many strongly bound states in the Dirac sea. However, if these states are occupied they are hidden from the direct observation. Only creating a hole in this sea, i.e. inserting a real antibaryon into the nucleus, would produce an observable effect. If this picture is correct one should expect the existence of strongly bound states of antinucleons with nuclei. Below I report on our recent study of antibaryon doped nuclear systems [4].

Unlike some previous works, we take into account the rearrangement of nuclear structure due to the presence of a real antibaryon. The structure of such systems is calculated using several versions of the relativistic mean field model (RMF): TM1 [5], NL3 and NL-Z2 [6]. Their parameters were found by fitting binding energies and charge form-factors of spherical nuclei from ^{16}O to ^{208}Pb. The general Lagrangian of the RMF model is written as

$$\begin{aligned}
\mathcal{L} = & \sum_{j=B,\overline{B}} \overline{\psi}_j (i\gamma^\mu \partial_\mu - m_j)\psi_j \\
& + \frac{1}{2}\partial^\mu \sigma \partial_\mu \sigma - \frac{1}{2} m_\sigma^2 \sigma^2 - \frac{b}{3}\sigma^3 - \frac{c}{4}\sigma^4 \\
& - \frac{1}{4}\omega^{\mu\nu}\omega_{\mu\nu} + \frac{1}{2} m_\omega^2 \omega^\mu \omega_\mu + \frac{d}{4}(\omega^\mu \omega_\mu)^2 \\
& - \frac{1}{4}\vec{\rho}^{\mu\nu}\vec{\rho}_{\mu\nu} + \frac{1}{2} m_\rho^2 \vec{\rho}^\mu \vec{\rho}_\mu \\
& + \sum_{j=B,\overline{B}} \overline{\psi}_j (g_{\sigma j}\sigma + g_{\omega j}\omega^\mu \gamma_\mu + g_{\rho j}\vec{\rho}^\mu \gamma_\mu \vec{\tau}_j)\psi_j \\
& + \text{Coulomb part}
\end{aligned} \tag{5}$$

Here summation includes valence baryons B, in fact the nucleons forming a nucleus, and valence antibaryons $\overline{B}$ inserted in the nucleus. They are treated as Dirac particles coupled to the scalar-isoscalar (σ), vector-isoscalar (ω) and vector-isovector ($\vec{\rho}$) meson fields. The calculations are carried out within the mean-field approximation where the meson fields are replaced by their expectation values. Also a "no-sea" approximation is used. This implies that all occupied states of the Dirac sea are "integrated out" so that they do not appear explicitly. It is assumed that their effect is taken into account by nonlinear terms in the meson Lagrangian. Most calculations are done with antibaryon coupling constants which are given by the G-parity transformation ($g_{\sigma\overline{N}} = g_{\sigma N}, g_{\omega\overline{N}} = -g_{\omega N}$) and $SU(3)$ flavor symmetry $g_{\sigma\overline{\Lambda}} = \frac{2}{3}g_{\sigma\overline{N}}, g_{\omega\overline{\Lambda}} = \frac{2}{3}g_{\omega\overline{N}}$). In isosymmetric static systems the scalar and vector potentials for nucleons are expressed as $S = g_{\sigma N}\sigma$ and $V = g_{\omega N}\omega^0$.

Following the procedure suggested in Ref. [7] and assuming the axial symmetry of the nuclear system, we solve effective Schrödinger equations for nucleons and an antibaryon together with differential equations for mean meson and Coulomb fields. We explicitly take into account the antibaryon contributions to the scalar and vector densities. It is important that antibaryons give a negative contribution to the vector density, while a positive contribution to the scalar density. This leads to increased attraction and decreased repulsion for surrounding nucleons. To maximize attraction, nucleons move to the center of the nucleus, where the antiproton has its largest occupation probability. This gives rise to a strong local compression of the nucleus and leads to a dramatic rearrangement of its structure.

Results for the ^{16}O nucleus are presented in Fig. 5 which shows 3d plots of nucleon density distributions. The calculations show that inserting an antiproton into the ^{16}O nucleus leads to the increase of central nucleon density by a factor 2–4 depending on the parametrization. Due to a very deep antiproton potential

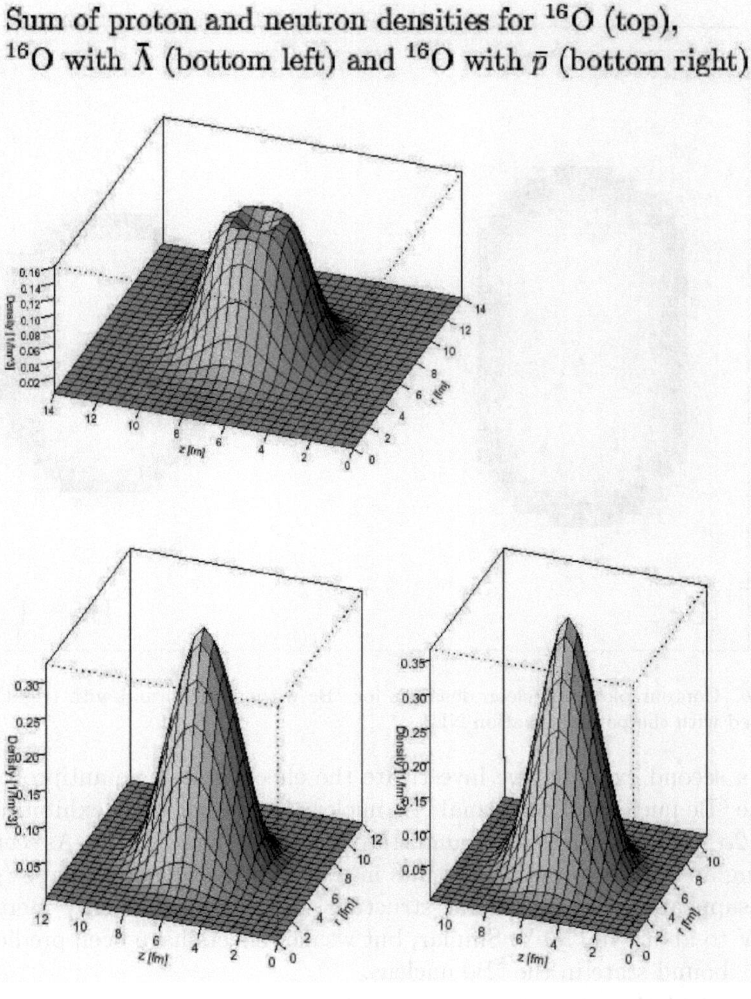

Figure 5. Sum of neutron and proton densities for ^{16}O (top), ^{16}O with $\overline{p}$ (bottom right) and ^{16}O with $\overline{\Lambda}$ (bottom left) calculated with the parametrization NL-Z2.

the binding energy of the whole system is increased significantly as compared with 130 MeV for normal ^{16}O. The calculated binding energies of the $\overline{p}$–^{16}O system are 830, 1050 and 1160 MeV for the NL-Z2, NL3 and TM1, respectively. Due to this anomalous binding we call such systems super bound nuclei (SBN). In the case of antilambdas we rescale the coupling constants with a factor 2/3 that leads to the binding energy of 560÷700 MeV for the $\overline{\Lambda}$–^{16} system.

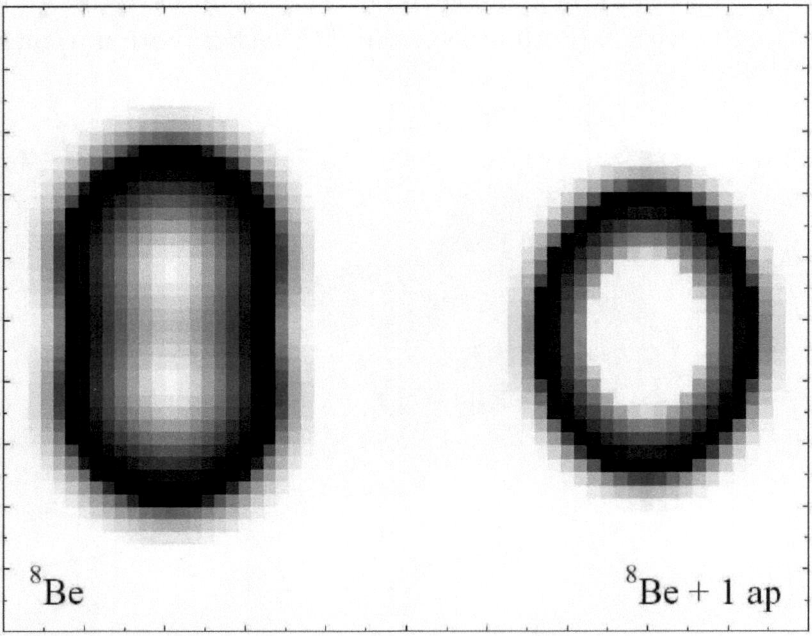

Figure 6. Contour plot of nucleon densities for ^{8}Be without (left) and with (right) antiproton calculated with the parametrization NL3.

As a second example, we investigate the effect of a single antiproton inserted into the ^{8}Be nucleus. The normal ^{8}Be nucleus is not spherical, exhibiting a clearly visible 2α structure with the ground state deformation $\beta_2 \simeq 1.20$. As seen in Fig. 6, inserting an antiproton in ^{8}Be results in a much less elongated shape ($\beta_2 \simeq 0.23$) and disappearance of its cluster structure. The binding energy increases from 53 MeV to about 700 MeV. Similar, but weaker effects have been predicted [8] for the K^- bound state in the ^{8}Be nucleus.

The calculations have been performed also with reduced antinucleon coupling constants as compared to the G-parity prescription. We have found that the main conclusions about enhanced binding and considerable compression of $\overline{p}$-doped nuclei remain valid even when coupling constants are reduced by factor 3 or so.

The crucial question concerning possible observation of the SBNs is their life time. The main decay channel for such states is the annihilation of antibaryons on surrounding nucleons. The energy available for annihilation of a bound antinucleon equals $Q = 2m_N - B_N - B_{\overline{N}}$, where B_N and $B_{\overline{N}}$ are the corresponding binding energies. In our case this energy is at least by a factor 2 smaller as compared with the vacuum value of $2m_N$. This should lead to a significant suppression of the available phase space and thus to a reduced annihilation rate in medium. We have performed detailed calculations assuming that the annihilation rates into different channels are proportional to the available phase space. All intermediate states with heavy mesons like ρ, ω, η as well as multi-pion channels have been considered. Our conclusion is that decreasing the Q value from 2 GeV to 1 GeV may lead to the reduction of total annihilation rate by factor 20÷30. Then we estimate the SBN life times on the level of 5-25 fm/c which makes their observation feasible. This large margin in the life times is mainly caused by uncertainties in the overlap integral between antinucleon and nucleon scalar densities. Longer life times may be expected for SBNs containing antihyperons. The reason is that instead of pions more heavy kaons must be produced in this case. We have also analyzed multi-nucleon annihilation channels (Pontecorvo-like reactions) and have found their contribution to be less than 40% of the single-nucleon annihilation.

We believe that such exotic nuclear states can be produced by using antiproton beams of multi-GeV energy, e.g. at the future GSI facility. It is well known that low-energy antiprotons annihilate on the nuclear periphery (at about 5% of the normal density). Since the annihilation cross section drops significantly with energy, a high-energy antiproton can penetrate deeper into the nuclear interior. Then it can be stopped there in an inelastic collision with a nucleon, e.g. via the reaction $A(\overline{p}, N\pi)_{\overline{p}}A'$, leading to the formation of a $\overline{p}$-doped nucleus. Reactions like $A(\overline{p}, \Lambda)_{\overline{\Lambda}}A'$ can be used to produce a $\overline{\Lambda}$-doped nucleus. Fast nucleons or lambdas can be used for triggering such events. In order to be captured by a target nucleus final antibaryons must be slow in the lab frame. Rough estimates of the SBN formation probability in a central $\overline{p}A$ collision give the values $10^{-5} - 10^{-6}$. With the $\overline{p}$ beam luminocity of $2 \cdot 10^{32}$ cm^{-2}s^{-1} planned at GSI this will correspond to the reaction rate of a few tens of desired events per second.

Several signatures of SBNs can be used for their experimental observation. First, annihilation of a bound antibaryon can proceed via emission of a single photon, pion or kaon with an energy of about 1 GeV (such annihilation channels are forbidden in vacuum). So one may search for relatively sharp lines, with width of 10÷40 MeV, around this energy, emitted isotropically in the SBN rest frame. Another signal may come from explosive disintegration of the compressed nucleus after the antibaryon annihilation. This can be observed by measuring radial collective velocities of nuclear fragments.

It is interesting to look at the antibaryon-nucleus system from somewhat different point of view. An antibaryon implanted into a nucleus acts as an attractor for surrounding nucleons. Due to the uncompensated attractive force these nucleons acquire acceleration towards the center. As the result of this inward collective motion the nucleons pile up producing local compression. If this process would be completely elastic it would generate monopole-like oscillations around the compressed SBN state. The maximum compression is reached when the attractive potential energy becomes equal to the compression energy. Simple estimates show that local baryon densities up to 5 times the normal nuclear density may be obtained in this way. It is most likely that the deconfinment transition will occur at this stage and a high-density cloud containing an antibaryon and a few nucleons will appear in the form of a multi-quark-antiquark cluster. One may speculate that the whole ^{4}He or even ^{16}O nucleus can be transformed into the quark phase by this mechanism. As shown in ref. [9], an admixture of antiquarks to cold quark matter is energetically favorable. The problem of annihilation is now transferred to the quark level. But the argument concerning the reduction of available phase space due to the entrance-channel nuclear effects should work in this case too. Thus one may hope to produce relatively cold droplets of the quark phase by the inertial compression of nuclear matter initiated by an antibaryon.

I am grateful to T. Bürvenich, I.N. Mishustin and L.M. Satarov for fruitful discussions and help in the preparation of this talk.

References

[1] W. Greiner, B. Müller, J. Rafelski, *Quantum electrodynamics of strong fields*, Springer Verlag, 2nd edition, December 1985

[2] J. Reinhardt and W. Greiner, *Quantum electrodynamics*, Springer Verlag, 3rd edition, February 2003

[3] N. Auerbach, A.S. Goldhaber, M.B. Johnson, L.D. Miller, A. Picklesimer, Phys. Lett. **B182** (1986) 221.

[4] T. Bürvenich, I.N. Mishustin, L.M. Satarov, H. Stöcker, W. Greiner, Phys. Lett. **B542** (2002) 261.

[5] Y. Sugahara and H. Toki, Nucl. Phys. **A579** (1994) 557.

[6] M. Bender, K. Rutz, P.–G. Reinhard, J.A. Maruhn, and W. Greiner, Phys. Rev. **C60** (1999) 34304.

[7] G. Mao, H. Stöcker, and W. Greiner, Int. J. Mod. Phys. **E8** (1999) 389.

[8] Y. Akaishi and T. Yamazaki, Phys. Rev. **C65** (2002) 044005.

[9] I.N. Mishustin, L.M. Satarov, H. Stoecker, W. Greiner, Phys. Rev. **C 59** (1999) 3343.

UNDERSTANDING THE NUCLEON SPIN

FRANCO BRADAMANTE [a,b]

[a] *Dipartimento di Fisica, Università degli Studi di Trieste, via A. Valerio 2, 34100 Trieste, Italy*
[b] *INFN, Sezione di Trieste, via A. Valerio 2, 34100 Trieste, Italy*

Abstract

A major effort is undergoing at DESY, at CERN, and at BNL to investigate the spin structure of the nucleon. Following the original EMC discovery in 1988 that the quarks contribute only a small amount to the nucleon spin, a direct determination of ΔG, the gluon polarization in a polarized nucleon, has become the primary goal of many presently running experiments. After reviewing our present knowledge of the spin of the nucleon, I briefly describe COMPASS, a new fixed target experiment at CERN, and present some preliminary physics results.

1 The nucleon spin puzzle

The original EMC discovery [1] that the singlet axial vector current matrix element $\Delta\Sigma$ is smaller than the value predicted by the Ellis-Jaffe sum rule is by now firmly established.

In the simple quark model the spin of the proton is carried by its three valence quarks, so that $\Delta\Sigma = \Delta u + \Delta d = \frac{4}{3} - \frac{1}{3} = 1$. In the usual notation, the quantities

$$\Delta q = \int_0^1 \left\{ \left(q(x)^{\downarrow\uparrow} + \bar{q}(x)^{\downarrow\uparrow}\right) - \left(q(x)^{\uparrow\uparrow} + \bar{q}(x)^{\uparrow\uparrow}\right) \right\} \, dx = \int_0^1 \Delta q(x) \, dx$$

are the differences of the integrated quark densities of a given flavour q, for quark spin anti-parallel or parallel to the proton direction. In general terms one writes

B.G. Sidharth, F. Honsell and A. de Angelis. (eds.), Frontiers of Fundamental Physics, 221–230.

the spin equation for the proton (or neutron) as

$$\frac{1}{2} = \frac{1}{2}\Delta\Sigma + \Delta G + < L_z >$$

where $\Delta\Sigma = \Delta u + \Delta d + \Delta s$ is the contribution of the quarks spins, ΔG is the contribution of the gluons, and $< L_z >$ is a possible contribution from the gluons and quarks angular momentum.

Since fifteen years, i.e. since the EMC discovery, we know that this picture does not correspond to reality, and that the contribution of the quarks to the spin of the nucleons is much smaller.

Polarized lepton - polarized nucleon deep inelastic scattering (DIS) provided a way to access the polarized quark densities Δq. It can be shown [2] that the cross-section difference $\Delta\sigma$ for parallel and antiparallel spins depends on two structure functions, g_1 and g_2, $\Delta\sigma = a{\cdot}g_1(x,Q^2)+b{\cdot}g_2(x,Q^2)$, very much like the unpolarized cross-section depends on two structure functions, $F_1(x,Q^2)$ and $F_2(x,Q^2)$. The coefficients a and b are kinematical factors, and can be calculated in QED. In the case of longitudinally polarized beam and target the quantity b is small, and a measurement of $\Delta\sigma$ allows $g_1(x,Q^2)$ to be measured. Conversely, if the target spin is perpendicular to the beam direction, a is smaller than b and $g_2(x,Q^2)$ can be measured.

In the quark and parton model the structure function g_1 and F_1 are simply related to the quark densities $\Delta q(x,Q^2)$:

$$g_1(x,Q^2) = \tfrac{1}{2}\sum e_q^2 \cdot \Delta q(x,Q^2); \qquad F_1(x,Q^2) = \tfrac{1}{2}\sum e_q^2 \cdot \left\{q(x,Q^2) + \bar{q}(x,Q^2)\right\}.$$

The summations extend over the quark flavours. By measuring $\Delta\sigma$ over a large x and Q^2 domain one can derive the first moment of g_1, i.e.

$$\Gamma_1^{p(n)}(Q^2) = \int_0^1 g_1^{p(n)}(x,Q^2)dx\,.$$

To leading order in QCD we can neglect scaling violation, identify $\Delta q(x,Q^2)$ with $\Delta q(x)$, and write the moment $\Gamma_1^{p(n)}$ as a linear combination of Δq:

$$\Gamma_1^{p(n)}(Q^2) = +(-)\tfrac{1}{12}a_3 + \tfrac{1}{36}a_8 + \tfrac{1}{9}a_0.$$

In this expression the a_i are the diagonal combinations in the SU(3) nonet of the axial matrix elements, and are related to the quantities Δq by the relations:

$$a_3 = \Delta u - \Delta d = g_A, \;\; a_8 = \Delta u + \Delta d - 2\Delta s, \;\; a_0 = \Delta u + \Delta d + \Delta s = \Delta\Sigma.$$

From the identity $\Delta q \cdot 2MS^\mu =< P,S|\bar{\Psi}_q\gamma^\mu\gamma_5\Psi_q|P,S >$, where S^μ is the spin of the proton, one can see that the quantities Δq are also related to the weak decay of the baryons. Using the values of a_3 and a_8 derived from the neutron and hyperon decay, and the measured value of Γ_1 coming from polarized DIS one can derive $\Delta\Sigma$. When this was done by the EMC Collaboration [1] the measured value of Γ_1^p implied that $\Delta\Sigma = 0.12 \pm 0.09 \pm 0.14$, i.e. only a small part (compatible with zero!!) of the spin of the proton is carried by the quarks.

In order to understand the picture which emerged from the EMC experiment, a number of new experiments were executed (SMC at CERN [3], E142, E143, E154 and E155 at SLAC [4], and HERMES at DESY [5], the only one which is still running), and a much larger number of theoretical papers were dedicated to the subject.

On the experimental side, the finding of EMC for the proton was confirmed, and a similar result was obtained for the neutron. Today the compilation of all the available measurements for g_1 for the proton, the deuteron and the neutron is impressive [6].

On the theoretical side, much progress was done to go beyond the quark-parton model and analyze the polarized DIS data within the framework of QCD. Perturbative QCD analysis in next-to-leading order have been performed by several groups to determine the polarized parton distribution functions Δq and ΔG using all the available data sets.

By now it is understood that

i) the singlet axial charge a_0 receives a contribution from ΔG due to the Adler - Bell - Jackiw anomaly, thus the $\Delta\Sigma$ and ΔG contributions to a_0 cannot be separately determined;

ii) the results depend on the renormalisation scheme which is used;

iii) although in principle ΔG could be determined from the Q^2 evolution of g_1 (given the large span of energies of the incident lepton in the various experiments, a broad range of Q^2 is covered by the data in most bins of x-Bjorken), it turns out that the present inclusive DIS data do not put a strong constraint on the polarized gluon distribution.

The results of various fits suggest a large positive value (1 - 2) for ΔG (see f.i. [7]), but the uncertainties in these determinations are $\sim 100\%$.

A direct measurement of ΔG seems essential to progress in this field.

2 The spin puzzle, present activities

To understand the nucleon spin today three lines of attack are being pursued:

i) a direct measurement of ΔG. This is the main goal of both the COMPASS [8] experiment at CERN and of the RHIC spin program [9] at BNL, which all have recently started data taking.

ii) to measure transversity. These parton distributions represent a new territory and there is a broad interest in their properties. COMPASS and RHIC have definite plans to measure them, and at lower energy transversity is an important part of the physics objectives of the HERMES experiment at DESY and of the experiments at JLab.

iii) on a longer term there are ideas to derive $< L_z >$ from measurements of Generalized Parton Distributions (GPDs) via the Ji sum rule [10]. First measurements of exclusive reactions (in particular deeply virtual Compton scattering) have been done by HERMES and at JLab, but the full program is terribly ambitious and anyway necessitates of a new high energy and high intensity CW electron machine.

Particularly interesting is the physics case for transversity. Large transverse spin effects at high energy have been discovered since many years, but their explanation in terms of QCD has always been lacking. Well known examples are the Λ polarization in NN scattering, the impressive C_{nn} values in elastic pp scattering at 90° c.m. measured at BNL, and the asymmetry of pion production in polarized proton-proton scattering measured at FNAL. Undoubtedly the awareness of these phenomena played a major role in launching the polarized proton collider project at BNL and has been an important stimulus in developing a theory of transversity.

Although the basic ideas are almost 30 years old, the rigorous treatment of transversity is relatively recent, and has gained a high momentum only the past few years. As originally shown by Jaffe and Ji [11], to completely specify the quark state at the twist-two level, to the momentum distribution $q(x)$ and to the helicity distribution $\Delta q(x)$ one has to add the transverse spin distributions $\Delta_T q(x)$. Its measurement gives access to new information related to relativistic effects for bound quark states, the study of new evolution in QCD, the knowledge of the tensor charge of the nucleon, and predictions for other processes involving transversity.

The transversity distributions $\Delta_T q(x)$ have never been measured, since they are chirally-odd functions and do not contribute to inclusive deep inelastic scattering. They may instead be extracted from measurements of the spin asymmetries in cross-sections for semi-inclusive deep inelastic scattering between leptons and transversely polarized nucleons, in which a hadron is also detected in the final state. In such processes the measurable asymmetry is due to the combined effect of $\Delta_T q(x)$ and another chirally-odd function, which contributes to the fragmentation of the transversely polarized quark. In the case in which the observed final hadron is a pion or, in general, a scalar particle, this new fragmentation function is the so-called Collins function $\Delta_T D_q^h$ that describe the hadronization of a transversely polarized quark q in a hadron h [12], as yet unmeasured, which in its own right merits serious study. In the case in which the observed final particle is for example a Λ°, the chirally-odd function is a transverse fragmentation function, also unknown and interesting. Other channels for accessing $\Delta_T q$ require the detection of a vector particle or two pions in the final state. An important side-product of the study of transversity is, therefore, the determination of a certain number of fragmentation functions.

3 Spin physics with COMPASS

Inclusive DIS does not allow to solve the spin puzzle, and to progress a direct determination of ΔG is necessary. A new experimental approach is required, namely semi-inclusive DIS with a full reconstruction of the hadronic current jet. A flavour tagging procedure allows to identify the struck quark in the DIS process. A suggestion to isolate the photon-gluon fusion process and directly measure ΔG was put

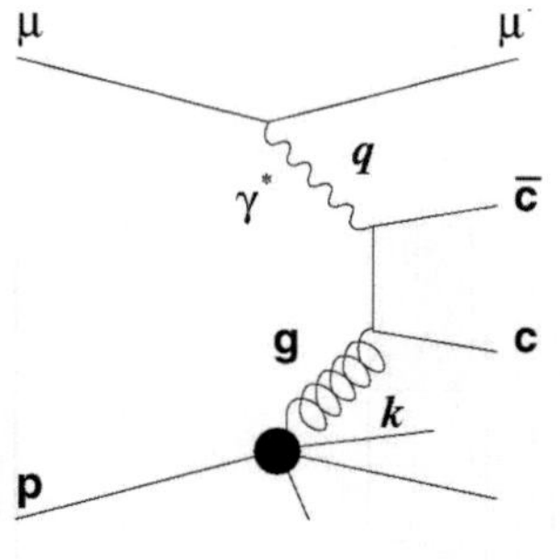

Figure 1. The photon-gluon fusion diagram, dominant mechanism for charm production at COMPASS energies.

forward already in 1988 [13, 14], and implied measuring the cross-section asymmetry of open charm in DIS. A new experiment, with full hadron identification and calorimetry, seemed to be necessary: COMPASS (COmmon Muon and Proton Apparatus for Structure and Spectroscopy) was proposed in 1996 and approved one year later.

The main goal of COMPASS is indeed a direct measurement of ΔG by measuring the cross-section asymmetry $A_{\mu N}^{c\bar{c}}$

$$A_{\mu N}^{c\bar{c}} = \frac{\Delta\sigma^{\mu N \to c\bar{c}X}}{\sigma^{\mu N \to c\bar{c}X}}. \qquad (1)$$

At COMPASS energies the production of charm goes predominantly via photon-gluon fusion (PGF), according to the diagram shown in Fig. 1, and the quantities $\sigma^{\mu N \to c\bar{c}X}$ and $\Delta\sigma^{\mu N \to c\bar{c}X}$ can be expressed as a convolution of the elementary photon-gluon cross-section with the gluon distributions G and ΔG.

The most promising additional way to measure ΔG in COMPASS uses the asymmetry of charged hadron pairs at high p_t [15]. Originally developed for the COMPASS experiment, the method has been recently applied also to the HERMES data [16]. The basic diagram is still the PGF, $\gamma g \to q\bar{q} \to h^+h^-X$, and the hardness of the process is guaranteed by the large p_t. The background from the leading order process $\gamma q \to q$, and the QCD-Compton process, $\gamma q \to \gamma q$, is in general dominating the PGF creation of a light $q\bar{q}$ pair, but suitable kinematic cuts can enhance considerably this process and allow for a statistically precise measurement.

Apart from ΔG, the COMPASS spectrometer is measuring Δq and $\Delta_T q$ from the relevant identified hadron asymmetries, in semi-inclusive polarized muon - polarized nucleon DIS, in the longitudinal and in the transversal mode respectively.

COMPASS has also an important physics programme using hadron beams, in particular to address the long standing question of exotic states. A first pilot run will take place at the end of the 2004 run, and the measurement will be resumed after the technical stop of the CERN accelerators in 2005.

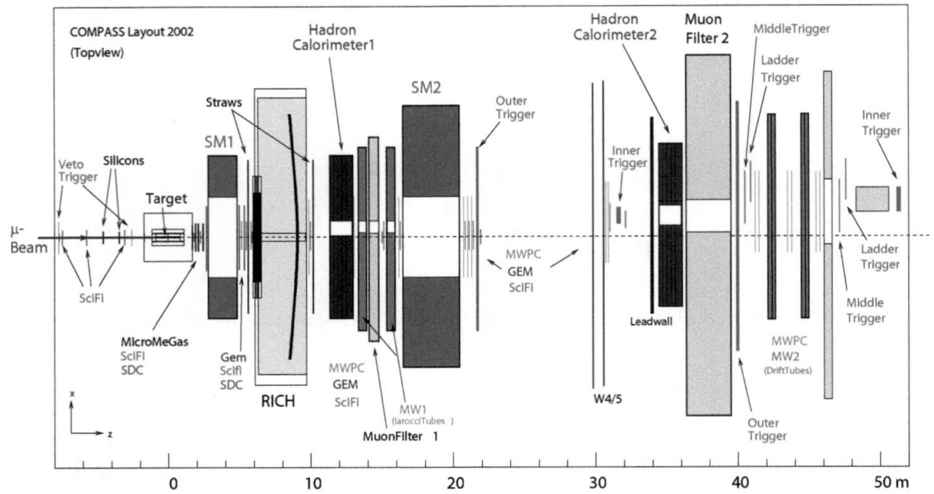

Figure 2. Top view of the lay-out of the spectrometer for the COMPASS experiment in 2002. The labels and the arrows refer to the major components of the tracking, trigger, and PID systems.

4 The COMPASS spectrometer

A common requirement of all the measurements foreseen by COMPASS is the detection and identification of particles over a large angular (±200 mrad) and dynamical (up to $\sim$ 150 GeV) range. To perform these measurements an International Collaboration of 31 Institutes, presently consisting of more than 200 physicists from 11 different Countries, including a strong group from Trieste, has built COMPASS, a new state-of-the-art spectrometer, capable of standing beam intensities of up to $2 \cdot 10^8$ particles/spill. The COMPASS spectrometer comprises two magnetic stages. Both stages are complemented with charged particle identification with fast RICH detectors, electromagnetic calorimetry, hadronic calorimetry, and muon identification via filtering through thick absorbers. The lay-out of the spectrometer which was on the floor in the year 2002 is shown in Fig. 2.

The experiment has been run at a muon energy of 160 GeV. The beam is naturally polarized by the $\pi-$decay mechanism. The beam polarization was measured by the SMC Collaboration to be about 80% at 190 GeV. The triggering system and the tracking system of COMPASS have been designed to stand the associated rate of secondaries, and use state-of-the-art detectors. Also, fast front-end electronics, multi-buffering, and a large and fast storage of events are essential.

We use the polarized target system of the SMC experiment, which allows for two oppositely polarized target cells, 60 cm long each. The PT magnet can provide both a solenoidal field (2.5 T) and a dipole field (0.5 T), for adiabatic spin rotation. The target polarization can then be oriented either longitudinally or transversely to the beam direction. Use of two different target materials, NH_3 as proton target and ^{6}LiD as deuteron target, is foreseen. Polarizations of 85 % and 50 % have been reached, respectively. In so far we have used ^{6}LiD: its favourable dilution factor of 0.4 is of the utmost importance for the measurement of ΔG.

To match the expected particle flux in the various locations along the spectrometer, COMPASS uses very different tracking detectors, in particular the experiment has been the first to use on a large scale novel detectors like Micromegas's and triple-GEM detectors.

The charged particle identification relies on the RICH technology. Presently, only RICH1 (the RICH in the first magnetic spectrometer) exists. As VUV photon detectors we use a novel technique, developed at CERN, i.e. MWPC's with a CsI photocathode (segmented in 8×8 mm^2 pads) which detect photons with wave length shorter than 200 nm, i.e. in the far UV domain. The total active area of the photon detectors is 5.6 m^2 and the total number of pads is about 80,000. The front-end electronics uses a modified version of the Gassiplex chip, and the read-out cards constitute a major project of the Trieste ICTP Microprocessor Laboratory, which utilises hundreds of DSP's.

The readout system uses a modern concept, involving highly specialized integrated circuits. The readout chips are placed close to the detectors and the data are concentrated at a very early stage via high speed serial links. At the next level high bandwidth optical links transport the data to a system of readout buffers. The event building system is based on PCs and Gigabit or Fast Ethernet switches and is highly scalable. This high performance network is also used to transfer the assembled data to the computer center for database formatting, reconstruction, analysis and mass storage. The data are sent via an optical link from the Hall 888 directly to the Computer building for Central Data Recording (CDR).

To handle the huge amount of data (the collected raw data size is ~300 TB/year) we used Objectivity/DB until the end of 2002, and Oracle since. The power needed to process COMPASS data is about 100 kSI2k. In the off-line farm, the data servers handle the network traffic from the CDR, distribute the raw data to the CPU clients (where they are put in the data base), receive them back from the PCs, and finally send them to a hierarchical storage manager (HSM) system. In parallel, the data servers receive the data to be processed from the HSM, send them to the PCs for processing, collect the output (DST or mDST), and send it to the HSM. Data processing is performed on the farm at CERN while DST's data analysis is done on satellite farms in the major home institutes (including Trieste).

A major effort was devoted to writing from scratch the off-line programs (CORAL, the new COmpass Reconstruction and AnaLysis program) using object-oriented technology and C++ language.

5 First physics results

Over the 80 days of the 2002 run, a total of about 6000 millions of events have been collected, corresponding to a data size of 260 TB. Similar numbers have been taken in the 2003 run and are presently being collected in the 2004 run. Many different physics channels are presently being investigated,

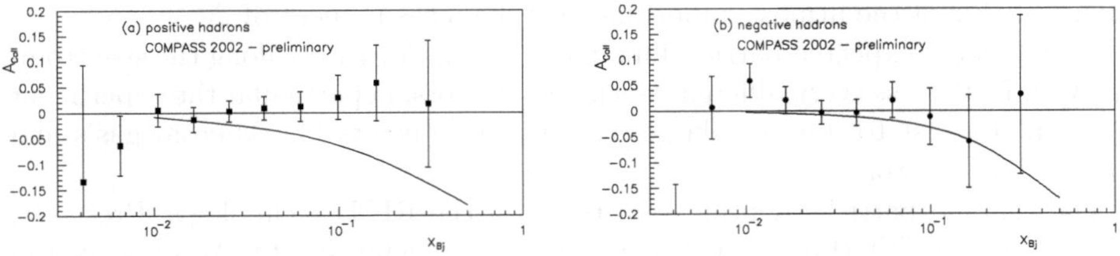

Figure 3. Preliminary results for the asymmetry A_{UT} for positive (a) and negative (b) leading hadrons produced by 160 GeV μ^+ on a transversely polarized deuterium target. The data refer to the year 2002 COMPASS run, while the curves are from Ref. [17].

- ΔG from open charm and high p_T hadron pair,
- A_{LL} to extract g_1,
- vector meson (ρ, ϕ, J/Ψ) production to test S-channel helicity conservation,
- Λ physics,
- transversity (single hadron, hadron pairs, Λ),
- Chan asymmetries,
- search of exotics (Θ^+, Ξ^{--}, ...).

A flavour of the physics results is given below.

5.1 COLLINS ASYMMETRY AND TRANSVERSITY

Semi-Inclusive Deep Inelastic Scattering (SIDIS) provides the possibility to measure the transverse polarized parton distribution function $\Delta_T q(x)$ via the azimuthal dependence of the momentum of the leading hadron. This measurement requires to operate with a transversely polarized target: about 20% of the COMPASS running time has been devoted to this measurement. Figure 3 shows preliminary values of the first ever measured Collins asymmetry on a deuteron target, separately for positive and negative leading hadrons. The data are compared with a model calculation [17] of the asymmetry A_{UT}, which includes the transversity distribution function $\Delta_T q(x)$ through the following linear combination of quark flavours:

$$A_{UT}(x) = \frac{\Sigma e_q^2 \cdot \Delta_T q(x) \cdot \Delta_T D_q^h}{\Sigma e_q^2 \cdot q(x) \cdot D_q^h}.$$

The small values of the measured asymmetries at all x might imply either a cancellation between the proton and the neutron asymmetries, or a small Collins effect in the fragmentation.

5.2 $\Delta G/G$ FROM CHARM PHOTO-PRODUCTION AND HIGH p_T HADRON PAIRS

As mentioned in Section 3, the gluon polarization will be obtained from the measured cross-section asymmetry of open charm events. Open-charm events are

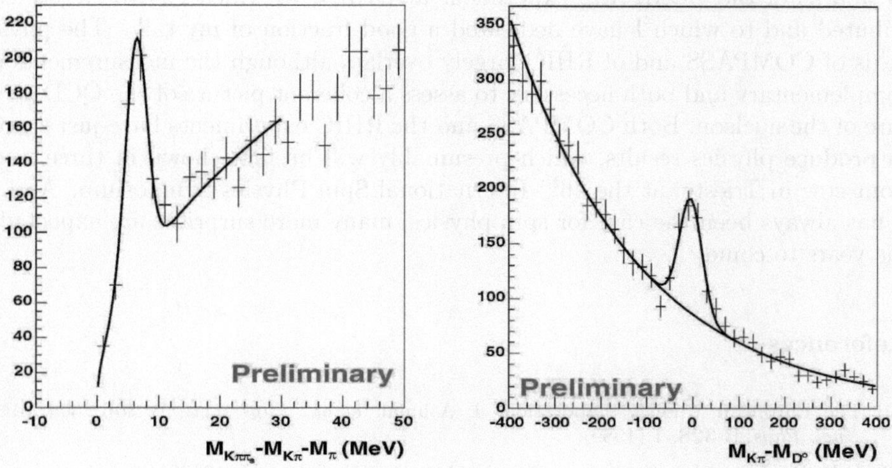

Figure 4. Left: D^* produced by requiring the invariant mass of the $K\pi$ pair to be in a 60 MeV window around the D^0 peak and a soft pion being detected. Cutting the events around the D^* mass, the D^0 peak in the invariant mass spectrum of $K\pi$ is very clear (figure at the right).

identified by reconstructing D°, $\bar{D}^\circ$, and $D^{*\pm}$ mesons from they decay products, i.e. $D^\circ \rightarrow K^-\pi^+$ and $D^{*+} \rightarrow D^\circ\pi^+ \rightarrow K^-\pi^+\pi^+$ and charge conjugate. In the first case, cuts on the K direction in the D° rest frame ($|cos(\theta^*_K)| < 0.5$) and on the D° energy fraction ($z_D = E_D/E_{\gamma^*} > 0.25$) are needed to reduce the background contamination. Preliminary signals of the D mesons are shown in fig. 4 Kaon-pion pairs are selected by asking: $z_D > 0.2$; $|cos(\theta^*_K)| < 0.85$; $10 < p_K < 35$ GeV in order to be in the RICH K identification region. A soft pion (< 10 GeV) is also required. This measurement is statistically limited. The D° signal in our data is at the level of 10^{-7}, and other decay channels are also being investigated presently. From the present analysis the projected error on $\Delta G/G$ using the data from 2002, 2003 and 2004 should be about 0.24. From the same data, using unidentified hadron pairs at high p_T, as mentioned in Section 3, we should be able to estimate $\Delta G/G$ with an error of 0.16 when requiring $Q^2 > 1$. Releasing the cut on Q^2 (as HERMES did) the hadron pair statistics increases by one order of magnitude, both the theoretical interpretation of the result becomes more difficult.

6 Conclusions and outlook

I have briefly summarized our present understanding of the nucleon spin. Following the EMC discovery in 1988, the importance of this issue has been broadly recognized, and the field has enjoyed a true "Renaissance". In this context one has witnessed the approval and the starting up of the most ambitious project spin physics could ever dream of, namely the polarized proton-proton collider at BNL, as well as of the COMPASS experiment at CERN, to which I have largely contributed and to which I have dedicated a good fraction of my talk. The physics goals of COMPASS and of RHIC largely overlap, although the measurements are complementary and both necessary to assess a coherent picture of the QCD structure of the nucleon, Both COMPASS and the RHIC experiments have just started to produce physics results, which presumably will be first shown in three weeks from now in Trieste, at the 16th International Spin Physics Symposium. And, as it has always been the case for spin physics, many more surprises are expected in the years to come.

References

[1] The European Muon Collaboration, J. Ashman *et al.*, *Phys. Lett.* B **206**, 364 (1988); *Nucl. Phys.* B **328**, 1 (1989).

[2] V.H. Hughes and J. Kuti, Ann. Rev. Nuclear Particle Science 331983611.

[3] The Spin Muon Collaboration, B. Adeva *et al.*, *Phys. Rev.* D **58**, 112001 (1998); *Phys. Rev.* D **60**, 072004 (1999).

[4] SLAC E-142, D.L. Anthony *et al.*, *Phys. Rev. Lett.* **71**, 959 (1993); SLAC E-143, K. Abe *et al.*, *Phys. Rev. Lett.* **75**, 25 (1995); SLAC E-154, K. Abe *et al.*, *Phys. Rev. Lett.* **79**, 26 (1997); SLAC E-155, P.L. Anthony *et al.*, *Phys. Lett.* B **436**, 339 (1999).

[5] The HERMES Collaboration, K. Ackerstaff *et al.*, *Phys. Lett.* B **404**, 383 (1997).

[6] Particle Data Group, *Phys. Rev.* D **66**, 010001 (2002).

[7] The Spin Muon Collaboration, B. Adeva *et al.*, *Phys. Rev.* **D58**, 112002 (1998).

[8] COMPASS, A Proposal for a COmmon Muon and Proton Apparatus for Structure and Spectroscopy, CERN/SPSLC 96-14, SPSLC/P297, 1 March 1996.

[9] see, for instance, G. Bunce, N. Saito, J. Soffer, and W. Vogelsang, *Annual Rev. Nucl. Part. Sci.* **50**, 525 (2000).

[10] X. Ji, *Phys. Rev. Lett.* **78**, 610 (1997);*Phys. Rev.* **D55**, 7114 (1997).

[11] R.L. Jaffe and X. Ji, *Phys. Rev. Lett.* **67**, 552 (1991).

[12] J. Collins, *Nucl. Phys.* B **396**, 161 (1993).

[13] G. Altarelli and G.G. Ross, *Phys. Lett.* B **212**, 391 (1988).

[14] M. Glück and E. Reya, *Z. Phys.* C **39**, 569 (1988).

[15] A. Bravar, D. von Harrach, and A. Kotzinian, *Phys. Lett.* B **421**, 349 (1998).

[16] A. Airapetian *et al.*, DESY-99-07, July 1999.

[17] A.V. Efremov *et al. Eur. Phys. J.* **C32**, 346 (2003)

THE MASS AND SPIN OF THE MESONS, BARYONS, AND LEPTONS

E. L. KOSCHMIEDER

Center for Statistical Mechanics,
The University of Texas at Austin, Austin, TX 78712, USA
e-mail: koschmieder@mail.utexas.edu

Abstract

The rest masses of the stable mesons and baryons and the rest masses of their antiparticles, as well as the rest masses of the $\mu^{\pm}$ and $\tau^{\pm}$ leptons can be explained, within 1% accuracy, with the standing wave model, which uses only photons, neutrinos, charge and the weak nuclear force. And we can explain the spin of the stable mesons and baryons and the spin of the $\mu^{\pm}$ and $\tau^{\pm}$ leptons without any additional assumption. We can also determine the rest masses of the e, μ and τ neutrinos.

1 Introduction

When elementary particles are discussed usually the so-called "Standard Model" is invoked and it is implied that this solves the problem. However, it is an obvious fact that in the forty years since the introduction of the quark concept we have not arrived at precise theoretical values for the masses of the mesons and baryons and for the masses of the leptons. That means that neither the mass of the fundamental proton nor the mass of the fundamental electron have been explained so far.

Many other attempts to explain the elementary particles have been made. For example, El Naschie [1] has proposed a topological theory from which the masses of the proton and the electron, as well as of the pions, can be determined with great accuracy. And, as another example, Sidharth [2] has suggested a model of the π^0 meson in which an electron and a positron circle their center of mass.

B.G. Sidharth, F. Honsell and A. de Angelis. (eds.), Frontiers of Fundamental Physics, 231–237.

In the following I will show that one can explain the masses of the mesons, baryons and leptons with the "Standing Wave Model". This model uses only photons, neutrinos, charge and the weak nuclear force. We can also explain the spin of the mesons and baryons without any additional assumption. A complete description of the standing wave model has been posted on the internet [3].

2 The integer multiple rule of the particle masses

The spectrum of the so-called "stable" mesons and baryons consists of the γ-branch and the ν-branch as we have shown in [4]. The masses of the $\pi^0, \eta, \Lambda, \Sigma^0, \Xi^0$, $\Omega^-, \Lambda_c^+, \Sigma_c^0$, Ξ_c^0, and Ω_c^0 particles, i.e. the masses of the γ-branch, are in a first approximation integer multiples of the mass of the π^0 meson. A least square analysis shows them to follow the formula

$$\mathrm{m}(N)/\mathrm{m}(\pi^0) = 1.0065\,N - 0.0043 \qquad N \geq 1, \tag{1}$$

with the correlation coefficient 0.999. The letter N stands for the integer number closest to the ratio of any particle of the γ-branch to the mass of π^0 meson. The principal decays, i.e. decays with a percentage >1%, of the γ-branch particles occur always *with* the emission of γ-rays or π^0 mesons, that means *without* the emission of neutrinos.

The neutrino branch, or ν-branch, consists of the $\pi^\pm$, $\mathrm{K}^{\pm,0}$, n, p, $\mathrm{D}^{\pm,0}$ and $\mathrm{D}_s^\pm$ particles. Their masses follow the formula

$$\mathrm{m}(N)/0.853\mathrm{m}(\pi^\pm) = 1.000\,N + 0.00575 \qquad N \geq 1, \tag{2}$$

with a correlation coefficient 0.998. The principal decays of the ν-branch particles occur always *with* the emission of neutrinos.

That means: The masses of the γ-branch particles are, in a first approximation, integer multiples of the mass of π^0. The masses of the ν-branch particles are, in a first approximation, integer multiples of $\pi^\pm$ times a factor ≈ 0.85.

3 The γ-branch particles

The characteristic particle of the γ-branch is the π^0 meson which is created in the process $\gamma + \mathrm{p} \to \pi^0 + \mathrm{p}$ and decays via $\pi^0 \to \gamma\gamma$ (98.8%). A γ-ray impinges on a proton and creates a π^0 meson. Fourier analysis dictates that a continuum of frequencies of electromagnetic waves must be present in the wave packet created by the high energy collision. The wave packet is the π^0 meson. We assume that the wave packet is organized in a *cubic lattice*, because the continuous Fourier frequency spectrum of the high energy collision can be absorbed by the continuous frequency spectrum of the lattice oscillations. The relevance of cubic lattices for particle theory was first suggested by Wilson [5]. According to lattice theory the lattice oscillations are standing waves. A rest mass can only be formed by *standing waves.*

The frequencies of longitudinal standing waves in a cubic lattice are according to Born and v. Karman [6] given by

$$\omega = \pm\, 2\sqrt{\alpha/\mathrm{m}}\,\sin(\phi/2)\,, \tag{3}$$

where α is the constant of the force acting between the particles of mass m within the lattice, and ϕ is given by $\phi = 2\pi a/\lambda$. There is a continuum of frequencies with a cutoff at $\phi = \pi$. However, the increase of the frequencies is limited by the requirement that the group velocity cannot exceed the velocity of light. That means that the frequencies of the oscillations must follow the formula

$$\nu = \nu_0\,\phi\,. \tag{4}$$

The rest mass of the packet of standing waves is the sum of the energy of all oscillations. The second mode of the oscillations has frequencies twice the frequencies of the basic mode and twice as many oscillations, that means $4\times$ as much energy as the basic mode. That confirms the integer multiple rule: The particles of the γ-branch consist of the basic mode (the π^0 meson), or of higher modes (the η meson $\mathrm{m}(\eta) \cong 4\mathrm{m}(\pi^0)$), or of superpositions of modes, for example two higher modes make the Λ baryon $\mathrm{m}(\Lambda) \cong 2\mathrm{m}(\eta)$.

Summing up: The π^0 meson and the other particles of the γ-branch are like cubic black bodies filled with standing electromagnetic waves. A conservative explanation of the γ-branch particles which does not use hypothetical particles. It appears that the π^0 meson can be derived from the theory of electromagnetic waves, that means from first principle.

4 The ν-branch particles

The characteristic particles of the ν-branch are the $\pi^\pm$ mesons which can be created in the reaction $\gamma + \mathrm{p} \to \pi^+ + \pi^- + \mathrm{p}$. A γ-ray impinges on a proton and is converted into a π^+ and a π^- meson. The $\pi^\pm$ mesons decay as $\pi^\pm \to \mu^\pm + \nu_\mu(\bar\nu_\mu)$ (99.98%), followed by the decay of the $\mu^\pm$ mesons, e.g. $\mu^+ \to e^+ + \bar\nu_\mu + \nu_e$ ($\approx$100%). It appears that the $\pi^\pm$ mesons consist of $\nu_\mu, \bar\nu_\mu, \nu_e, \bar\nu_e$ neutrinos which are held together in a cubic lattice by the weak nuclear force. The existence of the neutrinos and of the antineutrinos and of the weak nuclear force is unquestionable. Since the weak nuclear force has a range of only 10^{-16} cm and the radius of the $\pi^\pm$ mesons is $\approx 0.8{\cdot}10^{-13}$ cm, the weak force can hold the $\pi^\pm$ mesons together only if the neutrinos and antineutrinos are arranged in a lattice. The lattice points must oscillate because the lattice has been created in a high energy collision. The size of the nucleon and the range of the weak nuclear force determine the number of the lattice points. From the measured radius of the proton $\mathrm{r}_p = 0.88{\cdot}10^{-13}$ cm and the lattice constant which is equal to the range of the weak nuclear force $\mathrm{r}_w = 1{\cdot}10^{-16}$ cm follows that the number of the lattice points is

$$\mathrm{N} = 2.854 \cdot 10^9\,. \tag{5}$$

The energy in the rest mass of the $\pi^\pm$ mesons is the energy of all standing lattice oscillations plus the energy in the rest masses of all neutrinos in the lattice

$$\mathrm{m}(\pi^\pm)\mathrm{c}^2 = \mathrm{E}_\nu(\pi^\pm) + \Sigma\,\mathrm{m(neutrinos)c}^2\,. \tag{6}$$

Using Eq.(4) we can calculate the energy

$$\mathrm{E}_\nu(\pi^\pm) = 67.82\,\mathrm{MeV} \approx 0.5\,\mathrm{E}(\mathrm{m}(\pi^\pm))$$

of the lattice oscillations. If the number of the neutrinos and their rest masses is also known, then $\mathrm{m}(\pi^\pm)$ is known. Assuming that $\mathrm{m}(\nu_e) \ll \mathrm{m}(\nu_\mu)$ we obtain from Eq.(6) a first approximation for the mass of the muon-neutrino

$$\mathrm{m}(\nu_\mu) \approx 50\,\mathrm{milli\,eV/c^2}\,, \tag{7}$$

Summing up: The $\pi^\pm$ mesons can be explained by an oscillating cubic lattice of $\nu_\mu, \bar{\nu}_\mu, \nu_e, \bar{\nu}_e$ neutrinos plus an electric charge.

The primary decay of the $\mathrm{K}^\pm$ mesons $\mathrm{K}^\pm \rightarrow \mu^\pm + \nu_\mu(\bar{\nu}_\mu)$ (63.5%) leads to the same products as the decay of the $\pi^\pm$ mesons $\pi^\pm \rightarrow \mu^\pm + \nu_\mu(\bar{\nu}_\mu)$ (99.98%). Therefore it appears that the $\mathrm{K}^\pm$ mesons should consist at least partially of the same $\nu_\mu, \bar{\nu}_\mu, \nu_e, \bar{\nu}_e$ neutrinos as the $\pi^\pm$ mesons. However the $\mathrm{K}^\pm$ mesons with a mass of $0.8843 \cdot 4\mathrm{m}(\pi^\pm)$ cannot be solely the second mode of the $\pi^\pm$ mesons because the second mode of $\pi^\pm$ does not have the energy 4 times of $\pi^\pm$, because the rest masses of the neutrinos which make up part of the energy of $\pi^\pm$ do not change when the energy of the lattice oscillations is increased fourfold in the second mode of the oscillations. The difference between the second mode of $\pi^\pm$ and the real mass of the $\mathrm{K}^\pm$ mesons is made up by a π^0 meson, whose presence in $\mathrm{K}^\pm$ is confirmed by the appearance of π^0 among the decay products of $\mathrm{K}^\pm$, in particular in the decay $\mathrm{K}^\pm \rightarrow \pi^\pm\pi^0$ (21.13%). So the $\mathrm{K}^\pm$ mesons are the state $(2.)\pi^\pm + \pi^0$. The K^0 and $\overline{\mathrm{K}^0}$ mesons are the state $(2.)\pi^\pm + \pi^\mp$, they consist of neutrinos only.

The neutron is a superposition of a K^0 and a $\overline{\mathrm{K}^0}$ meson and must have a neutrino lattice. The structure of the proton follows from the β-decay of the neutron. The proton must have a neutrino lattice as well.

The characteristic feature of the particles of the ν-branch is their neutrino lattice of $\nu_\mu, \bar{\nu}_\mu, \nu_e, \bar{\nu}_e$ neutrinos. We do not use hypothetical particles.

5 The mass of the $\mu^\pm$ mesons

Surprisingly the mass of the $\mu^\pm$ mesons can be explained with the standing wave model as well. The $\mu^\pm$ mesons are leptons which do not interact strongly with the mesons and baryons. Leptons make up 1/2 of the number of stable elementary particles. The $\mu^\pm$ mesons originate from the decay $\pi^\pm \rightarrow \mu^\pm + \nu_\mu(\bar{\nu}_\mu)$ (99.98%). From the experiments we learn that

$$\mathrm{m}(\mu^\pm)/\mathrm{m}(\pi^\pm) = 0.757027 = 1.00936 \cdot 3/4 \cong 3/4\,. \tag{8}$$

With $m(\nu_\mu)$ from Eq.(7) or from Eq.(12) we find that

$$m(\pi^\pm) - m(\mu^\pm) \approx N/4 \cdot m(\nu_\mu)\,, \tag{9}$$

where N is the number of all neutrinos in the $\pi^\pm$ lattice as given by Eq.(5). We find also that the oscillation energies E_ν are

$$E_\nu(\pi^\pm) \approx E_\nu(\mu^\pm)\,. \tag{10}$$

Eq.(9) says that the energy in the rest masses of all muon (respectively) anti-muon neutrinos is consumed in the $\pi^\pm$ decay. Eq.(10) says that the oscillation energy of all neutrinos in the $\pi^\pm$ mesons is conserved in the $\pi^\pm$ decay. Consequently there is *no other source* for the kinetic energy of the decay products of the $\pi^\pm$ mesons than the energy in the rest masses of either all muon or anti-muon neutrinos in $\pi^\pm$.

That means: The $\mu^\pm$ mesons have a neutrino lattice which differs from the neutrino lattice of the $\pi^\pm$ mesons by the absence of the muon or anti-muon neutrinos.

This contradicts the common belief that the μ mesons are *point particles.* However, since neutrinos do not interact, in a good approximation, with neither mass nor charge the finite size of the neutrino lattice of the μ mesons cannot be detected by conventional scattering experiments.

Summing up: The mass of the $\mu^\pm$ mesons can be explained with a neutrino lattice which contains the remains of the $\pi^\pm$ lattice and an electrical charge. Since either all muon neutrios or all antimuon neutrinos have been removed from the $\pi^\pm$ lattice to form the $\mu^\pm$ mesons it must be that $m(\mu^\pm) \approx 3/4\ m(\pi^\pm)$ as the experiments found. The mass of the $\tau^\pm$ mesons can be explained along similar lines.

6 The neutrino masses

The rest mass of the electron neutrino can be determined from the energy released in the decay of the neutron $n \rightarrow p + e^- + \bar{\nu}_e$. Likewise the rest mass of the anti-electron neutrino follows from the decay of the antineutron. We find that

$$m(\nu_e) = m(\bar{\nu}_e) = 0.365\,\text{meV}/c^2\,. \tag{11}$$

Inserting Eq.(11) into Eq.(6) we obtain an accurate value for the mass of the muon neutrino

$$m(\nu_\mu) = m(\bar{\nu}_\mu) = 49.91\,\text{meV}/c^2\,. \tag{12}$$

Since the same considerations apply for either the π^+ or π^- meson it must be that $m(\nu_\mu) = m(\bar{\nu}_\mu)$. The ratio of the neutrino masses is then

$$m(\nu_\mu)/m(\nu_e) = 136.74\,, \tag{13}$$

or 99.8% of the inverse of the fine structure constant $\alpha_f = 1/137.036$. It does not seem likely that this is just a coincidence.

We also find from the decay of the $D_s^{\pm}$ mesons that the mass of the τ neutrino is

$$m(\nu_\tau) = m(\bar{\nu}_\tau) \cong 0.54\,\mathrm{eV/c^2}\,. \tag{14}$$

Since $N/4 \cdot m(\nu_e) = 0.51\,m(e^\pm) \cong 0.5\,m(e^\pm)$ we arrive with Eq.(13) at

$$m(\mu^\pm)/m(e^\pm) \cong 3/2\alpha_f + 2\,, \tag{15}$$

whereas the empirical formula for the ratio of the mass of the $\mu^\pm$ mesons to the mass of the electron is $m(\mu^\pm)/m(e^\pm) = 3/2\alpha_f + 1 = 206.55$. The experimental ratio is 206.768. Similarly we arrive at

$$m(\pi^\pm)/m(e^\pm) \cong 2/\alpha_f + 2\,, \tag{16}$$

whereas the empirical formula is $m(\pi^\pm)/m(e^\pm) = 2/\alpha_f - 1 = 273.07$. The experimental ratio is 273.13. Finally we find for the long sought for ratio of the mass of the proton to the mass of the electron the value

$$m(p)/m(e^-) \cong 0.9426[14/\alpha_f + 14] = 1821.5\,, \tag{17}$$

which is 99.2% of the experimental value 1836.16. For details see [3].

Summing up: We can explain the rest masses of the $\mu^\pm$ and $\tau^\pm$ mesons and we can determine the rest masses of the e, μ, τ neutrinos. In other words, we have found the masses of all leptons exempting the electron. We will deal with the electron later.

7 The spin of the particles

It appears to be a necessary condition for the validity of a model of the particles that the model can also explain the *spin* of the particles. We have proposed an explanation of the spin of the particles in [7].

The spin, or the intrinsic angular momentum, of a particle is, of course, the sum of the angular momentum vectors of all components of the particle. The π^0 and η mesons do not have spin because their $O(10^9)$ standing electromagnetic waves must be linearly polarized and hence do not have an angular momentum. The longitudinal oscillations of the π^0 or η lattices do not have an angular momentum either because for longitudinal oscillations $\vec{r} \times \vec{p} = 0$. So $s(\pi^0, \eta) = 0$ as it must be.

The Λ baryon appears to be the superposition of two η mesons, $m(\Lambda) = 1.019 \cdot 2\,m(\eta)$. The superposition of two perpendicular oscillations of the same amplitude and frequencies produces a circular oscillation whose angular momentum is $\hbar/2$, see [7]. The angular momentum vector of each of the $O(10^9)$ circular oscillations is canceled by the angular momentum vector of the circular oscillation at the mirror position in the lattice. Only the angular momentum $\hbar/2$ of the circular oscillation at the *center* is not canceled. That means that the intrinsic angular momentum of the Λ baryon is $\hbar/2$, or $s(\Lambda) = 1/2$ as it must be. The other baryons of the γ-branch are composites of the Λ baryon with spin 1/2 and one or two π mesons which do not have spin. Consequently their spin is 1/2, exempting the Ω^- particle.

The explanation of the spin of the ν-branch particles is different because each of the O(10^9) neutrinos which are in the ν-branch particles has spin 1/2. However, the spin vectors of the neutrinos in the lattice cancel because the spin vector of each neutrino has an opposite spin vector at the mirror position in the lattice, but for the central neutrino. That makes the intrinsic angular momentum of the neutrino lattice of the $\pi^{\pm}$ mesons equal to $\hbar/2$, however s($\pi^{\pm}$) is zero. This problem is solved by the spin of the electric charge $e^{\pm}$ carried by the $\pi^{\pm}$ mesons. The spin 1/2 carried by one of the electric charges cancels the spin 1/2 of the neutrino lattice, so $s(\pi^{\pm}) = 0$ as it must be. Similar considerations apply for $K^{\pm}$ with $s(K^{\pm}) = 0$.

The explanation of $s(K^0) = 0$ is more difficult because there is no charge to cancel the spin of the neutrino lattice. However, the K^0 lattice does not contain single neutrinos, but neutrino pairs, because the K^0 and $\overline{K^0}$ mesons are superpositions of the second mode of the $\pi^{\pm}$ mesons and the first mode of the $\pi^{\mp}$ mesons, $K^0 = (2.)\pi^{\pm} + \pi^{\mp}$. The spin of the neutrino pairs is zero, therefore the spin of the entire neutrino lattice is zero. The K^0 meson also contains a pair of opposite electrical charges whose spin cancels likewise. That means that the intrinsic angular momentum of K^0 is zero.

The spin of the neutron whose mass is $\approx 2\,m(K^0)$ is $s(n) = 1/2$. The spin of the neutron is caused by the angular momentum of circular oscillations which result from the superposition of two perpendicular lattice oscillations from a K^0 and a $\overline{K^0}$ meson, similar to the superposition of two η mesons in the Λ baryon. For details see [7]. The spin of the proton $s(p) = 1/2$ is determined by the decay of the neutron, details are in [7].

8 Conclusions

The standing wave model solves a number of problems for which an answer heretofore has been hard to come by. Only photons, neutrinos, charge and the weak nuclear force are needed to explain the masses of the stable mesons and baryons and of the leptons. We can also explain the spin of the mesons and baryons and of the $\mu^{\pm}$ and $\tau^{\pm}$ mesons.

References

[1] El Naschie MS. Chaos,Sol.Frac. **22**,494 (2004).

[2] Sidharth BJ. The Chaotic Universe p.49, Nova Science Publishers (2001).

[3] Koschmieder EL. http://arXiv.org/physics/0408070 (2004).

[4] Koschmieder EL. Chaos,Sol.Frac. **18**,1129 (2003).

[5] Wilson KG. Phys.Rev.D **10**,2445 (1974).

[6] Born M. and v.Karman Th. Phys.Z. **13**,297 (1912).

[7] Koschmieder EL. Hadr.J. **26**,471 (2003).

EQUALITY AND IDENTITY AND (IN)DISTINGUISHABILITY IN CLASSICAL AND QUANTUM MECHANICS FROM THE POINT OF VIEW OF NEWTON'S NOTION OF STATE

PETER ENDERS [a]

[a] *Fischerinsel 2, D-10179 Berlin, Germany*

Abstract

The notion of state is a central notion for all branches of physics. Surprisingly enough, Newton's notion differs from the nowadays notion. Our review of the benefits of Newton's notion comprises Gibbs's paradox, Einstein's derivation of the classical and quantum distribution laws from the energetic spectrum (serving to remove anthropomorphic elements), the difference between 'identical' and 'indistinguishable' (being a property of states rather than of particles), a new physical content of $|\psi(x,t)|$ (the invariance of $|\psi(x,t)|$ rather than $\psi(x,t)$ against permutations yields not only fermions and bosons, but also anyons), and a novel classification of forces (leading eventually to a derivation of the Maxwell-Lorentz equations from classical mechanics).

1 Introduction

Classical mechanics is the safest (if not the *only* safe) ground we can move on. For this, we will analyze the implications of Newton's notion of state differing considerably from the contemporary one for the notions 'equality', 'identity' and '(in)distinguishability' playing a paramount role in statistics and in quantum mechanics. Newton's notion allows for considering them within classical point mechanics, what frees the discussion from anthropomorphic elements. Bach's (1997) fundamental results are obtained within an *elementary dynamical* framework.

B.G. Sidharth, F. Honsell and A. de Angelis. (eds.), Frontiers of Fundamental Physics, 239–245.

2 Newton's notion of state

Newton's First Law "Every body perseveres in its state of being at rest or of moving uniformly straight forward, except insofar as it is compelled to change its state by forced impressed." –
Newton's state corresponds to nowadays' stationary state.

Newton's Second Law "A change in motion is proportional to the motive force impressed and takes place along the straight line in which that force is impressed." –
$\Delta\vec{p} \sim \vec{F}$: Newton's state variable is the momentum (the conserved quantities).

Laplace's demon "A sufficiently powerful intelligence knowing all loci and velocities in a mechanical systems at one time is able to calculate the loci and velocities at all later times." –
Laplace's state is nowadays' state, including both stationary and non-stationary ones, the state variables are a complete set of independent dynamical variables.

Advantages of Newton's notion:

- Easily generalized to classical and quantum systems (not tied to orbits);
- Amount of conserved quantities ˜ amount of quantum (state) numbers;
- Symmetry of state = symmetry of state function → gauge symmetry, geometric phases, ...

Advantages of modern (Laplace's) notion:

- Complete description of motion;
- Identification of phase space points differing only by interchanging equal bodies → multiply connected spaces → appropriate topology for anyons.

Disadvantages of modern (Laplace's) notion:

- The state changes even in absence of causes;
- State at rest is ignored;
- Interchange of equal bodies changes state → Gibbs's paradox;
- Inapplicable to quantum systems.

3 Equal bodies in Newtonian states

1) Equal / identical bodies / particles

Equality We call two classical bodies or quantum particles equal, when their interchange does not change the properties / state / motion of a system. (*cf* Helmholtz, §10)

Identity "Particles are called identical, if they agree in all their intrinsic (i.e. state independent) properties." (Bach, p.15)
Remark: The restriction to the *intrinsic* properties circumvents the conflict with the logical notion 'identical' (= equal in *all* properties).

2) Permutation symmetry of Newtonian states

All conserved quantities of a classical-mechanical system: total energy / momentum / angular momentum / ..., are invariant w.r.t. the permutation of equal parameters, *ie*, w.r.t. the permutation of (labels of) equal bodies, *ie*, w.r.t. bodies with equal properties concerning the system considered, ⇒

- A Newtonian state is invariant against interchanging (labels of) equal bodies,
- Equal bodies cannot be distinguished or identified by means of the conserved quantities (Newtonian/ stationary-state variables),
- Anthropomorphic arguments like 'particle can be marked or not' are not relevant points of view (interchanging or marking two resting red balls in a snooker game does not interfere the game);
- Indistinguishable classical particles have no trajectories (Bach provides probabilistic proof).

3) Comparison with Laplace's notion of state (continued)

The locus of a body is that part of space it occupies. Euler's exclusion principle (not to be interchanged with Pauli's exclusion principle!) states, that no body can occupy more than one locus, and no part of space can be occupied by more than one body. ⇒

- Equal bodies can, at least in principle, always be distinguished and identified by means of their locus.
- Equal bodies are distinguishable within Laplace's notion of state.
- *(In)distinguishability is a property of states, not of particles/bodies.* (*cf* Bach, p.15)

4 Classical and quantum distribution laws

Einstein (1907) has derived the classical and quantum distribution laws using just the energetic spectra of a classical (continuous spectrum) and a quantum harmonic oscillator (discrete spectrum):

$$\langle E\rangle_{class} = \frac{\int_0^\infty E e^{-\frac{E}{kT}} dE}{\int_0^\infty e^{-\frac{E}{kT}} dE} = kT \tag{1}$$

$$\langle E\rangle_{quant} = \frac{\sum_{n=0}^\infty \hbar\omega n e^{-\frac{\hbar\omega n}{kT}}}{\sum_{n=0}^\infty e^{-\frac{\hbar\omega n}{kT}}} = \frac{\hbar\omega}{e^{\frac{\hbar\omega}{kT}} - 1} = \langle E\rangle_{class} \frac{\frac{\hbar\omega}{kT}}{e^{\frac{\hbar\omega}{kT}} - 1} \tag{2}$$

(In)distinguishability does not play any role (*cf* Bach's "Bose-Einstein statistics as invented by Boltzmann").

5 New physical content of $|\psi(x,t)|$

FFP5: Quantum-mechanical systems are conservative systems which may assume configurations for which $V(x) > E$

⇒ There is an *effective* potential energy

$$V_{nkl}(x) = V_{E_{nkl}}(x) = F_{E_{nkl}}(x) \cdot V(x) \leq E_{nkl}; \quad -\infty < x < +\infty \tag{3}$$

$F_{E_{nkl}}(x) \sim |\psi_{E_{nkl}}(x)|^2$ is a limiting function such, that $V_{nkl}(x) \leq E_{nkl}$ even if $V(x) > E_{nkl}$ (→ tunnel effect demystified).

Progress since FFP5:

- Common principles of state change for classical and quantum systems → Eulerean derivation of time-dependent Schrödinger equation from time-independent one;
- $F_{E_{nkl}}$ is dimensionless ⇒ $F_{E_{nkl}} = F_{E_{nkl}}(x/x_0)$: *All* quantum systems exhibit characteristic length x_0;
- $F_{E_{(nkl)}}(x/x_0) \sim |\psi_E(x)|^2$ → gauge symmetry;
- Novel classification of fields:
  A) Fields being related to total energy, E (accelerating fields like electric field, $\vec{E}$),
  B) Fields being related not to E, but to x_0 (refracting fields like magnetic field, $\vec{B}$),
  C) Fields being related neither to E, nor to x_0 (gauge fields, see Aharonov-Bohm effect);
  Remark: The *Maxwell-Lorentz equations* turn out to be just *compatibility conditions* for such fields $\vec{E}$ and $\vec{B}$.

6 Newtonian state variable $|\psi(x,t)|$

1) By virtue of their definition, limiting functions are invariant against permutations of equal particles (omitting x_0):

$$F_E(x_2, x_1) = F_E(x_1, x_2) \geq 0 \tag{4}$$

$\Rightarrow$ the most general representation of $F_E(x_1, x_2)$ reads (m_E entire)

$$F_E(x_1, x_2) = \left|\psi_E^{\gamma,m}(x_1, x_2)\right|^2; \tag{5}$$

$$\psi_E^{\gamma,m}(x_1, x_2) = \frac{1}{\sqrt{2}} e^{i\gamma_E(x_1,x_2)} \left[\tilde{\psi}_E(x_1, x_2) + e^{im_E\pi}\tilde{\psi}_E(x_2, x_1)\right] \tag{6}$$

2) Wigner's theorem:
If $\psi_E^{\gamma,m}(x_1, x_2)$ is eigenfunction, then

$$\hat{R}\psi_E^{\gamma,m}(x_1, x_2) = e^{i\rho_E^{(1,2)}} \psi_E^{\gamma,m}(x_2, x_1) \tag{7}$$

$$= e^{i\rho_E^{(1,2)}} e^{im_E\pi} e^{i[\gamma_E(x_2,x_1) - \gamma_E(x_1,x_2)]} \psi_E^{\gamma,m}(x_1, x_2) \tag{8}$$

is also eigenfunction $\Rightarrow$ $[\gamma_E(x_2, x_1) - \gamma_E(x_1, x_2)]$ can be absorbed into $\rho_E^{(1,2)}$.

3) 2nd permutation:

$$\hat{R}^2\psi_E^m(x_1, x_2) = \hat{R}e^{i\rho_E^{(1,2)}} \psi_E^{\gamma,m}(x_2, x_1) = e^{i\rho_E^{(1,2)}} e^{i\rho_E^{(2,1)}} \psi_E^m(x_1, x_2) \tag{9}$$

Standard case:

$$\hat{R}^2 = e^{i\rho_E^{(1,2)}} e^{i\rho_E^{(2,1)}} = 1; \quad \rho_E^{(2,1)} = -\rho_E^{(1,2)}; \quad \gamma_E(x_1, x_2) = 0 \tag{10}$$

$\Rightarrow$ Wave functions are either symmetric (bosons, +) or anti-symmetric (fermions, $-$):

$$\psi_E^{\pm}(x_1, x_2) = \frac{1}{\sqrt{2}} \left[\tilde{\psi}_E(x_1, x_2) \pm \tilde{\psi}_E(x_2, x_1)\right] \tag{11}$$

Non-standard, 'anyonic' case:

$$\hat{R}^2 = e^{i\rho_E^{(1,2)}} e^{i\rho_E^{(2,1)}} \neq 1; \quad \rho_E^{(2,1)} \neq -\rho_E^{(1,2)}; \quad \gamma_E(x_2, x_1) - \gamma_E(x_1, x_2) \neq 0 \tag{12}$$

$\Rightarrow$ If there are topologically inequivalent paths, the wave function is neither symmetric, nor anti-symmetric, but can exhibit any intermediate behaviour $\rightarrow$ anyons (the clue to the fractional quantum Hall effect).

7 About the meaning of 'identical' and 'indistinguishable'

1) Some rigorous definitions

Equal means 'equal in some well-defined properties such as mass, density, shape, charge ..., but not in all'
Example: the 2 electrons in the ground state of He (they differ in s_z)
'equal' depends on the view, *ie*, which properties one is looking at.

Congruent means 'equal in all essential (geometric) properties, but not in locus'
Example: 2 red snooker balls of high quality

Identical means 'one and the same', *ie*, equal in all properties (strictly speaking, *no* exception at all)
Example: 2 squares of equal side length on the same place of a sheet

Indistinguishable means, that there is *no* mean (no one differing property/attribute) for discrimination.

$\Rightarrow$ Indistinguishable things are identical (Leibniz)

2) Questions

If we weaken the definition of 'identical', there may be a weakening of 'distinguishable' to 'identifiable', so that we are led to the question

- Are there non-classical indistinguishable bodies/particles not being identical?
- Are there principally distinguishable (*ie*, not identical) bodies/particles not being identifiable?

3) Observations

- Quantum particles are identical w.r.t. intrinsic properties and 'almost identical' w.r.t. state properties: All electrons (protons, ...) exhibit the same mass at rest, electrical charge, modulus of spin, etc.;
- Pauli's exclusion principle: 2 electrons differ in at least one quantum number – however: it does *not* say, which electron is in which state (entanglement);
- The quanta occupying an oscillator loose their individuality: Say, 12 quanta in state E_{12} occupy all together the *one* 12-quanta state, not 12 single-quantum states, $\Rightarrow$ they have got no individual properties (parameter values) (in contrast to electrons, these quanta – Planck's "energy elements" – occupy not single-particle, but *single-system* states); – nevertheless: these 12 quanta are not one and the same (one thing) as we are thinking them as 12 particles.
- Cluster law: Wave functions of distinct systems need not to be entangled $\leftrightarrow$ There are distinguishable equal quantum particles.

4) Conclusions

- The notions 'identical' or 'distinguishable' as used in logics play almost no role, in fact: the actual physical meaning of 'distinguishability' as property of states is the *identifiability*;
- 'Identical' is meaningful in the sense of Bach (including only intrinsic properties, *eg*, spin s – but not s_z);
- There is no principal difference between classical bodies, bosons and fermions w.r.t. these properties.

8 Summary

Newton's notion of state is an addition to, though not a complete replacement of Laplace's notion. Our treatment of equality, identity and (in)distinguishability accounting for Newton's notion of state reveals the following advantages and new results.

- Common treatment of classical bodies and quantum particles;
- Non-probabilistic classification of (bodies/particles in) states;
- The limiting function F_E gives $|\psi_E|$ a new physical meaning as relative space occupation; both exhibit the same symmetry as Newtonian (stationary-) state variables (quantum numbers) w.r.t. external fields ($\rightarrow$ gauge invariance) and permutations ($\rightarrow$ fermions, bosons, *and anyons*);
- Distribution functions can be related to energetic spectra and occupation $\Rightarrow$ they are independent of (in)distinguishability;
- The meaning of 'identity' and 'indistinguishability' in physics is partly at variance with their meaning in logics, hence, careful restrictions are necessary;
- There are particles exhibiting different extrinsic properties (*eg*, spin direction in EPR), but not being identifiable: Equal bodies in symmetric states cannot be identified, be there different attributes or not.

References

[1] A. Bach, *Indistinguishable Classical Particles* (Lecture Notes in Physics m44) Springer, Berlin Heidelberg (1997)

[2] P. Enders, *From Classical to Quantum Physics. A Deductive Approach*, Springer, Berlin Heidelberg (2005) (in German)

[3] H. v. Helmholtz, *Introduction to the Lectures on Theoretical Physics*, Barth, Leipzig (1903) (in German)

NUMERICAL MODELLING OF QUANTUM STATISTICS IN HIGHENERGY PHYSICS

OLEG V. UTYUZH [a], GRZEGORZ WILK [b], ZBIGNIEW WŁODARCZYK [b]

[a] *The Andrzej Sołtan Institute for Nuclear Studies, Nuclear Theory Department, ul. Hoża 69; 00-681 Warszawa, Poland; emails: utyuzh@fuw.edu.pl, wilk@fuw.edu.pl*

[b] *Institute of Physics, Świętokrzyska Academy, ul. Świętokrzyska 15, 25-406 Kielce, Poland, e-mail: wlod@pu.kielce.pl*

Abstract

Numerical modelling of quantum effects caused by bosonic or fermionic character of secondaries produced in high energy collisions of different sorts is at the moment still far from being established. In what follows we propose novel numerical method of *modelling* Bose-Einstein correlations (BEC) observed among identical (bosonic) particles produced in such reactions. We argue that the most natural approach is to work directly in the momentum space of produced secondaries in which the Bose statistics reveals itself in their tendency to bunch in a specific way in the available phase space. Fermionic particles can also be treated in similar fashion.

The multiparticle production processes consist substantial part of the high energy collisions and are of considerable theoretical interest. Unfortunately their description is so far available only by means of numerical Monte Carlo codes based to some extent on modern theoretical ideas but otherwise remaining purely phenomenological [1]. They are build in such manner as to describe as close as possible the complicated final state of such reaction, cf., Fig. 1. However, using classical (positive defined) probabilities as their basic tool such MC codes cannot directly describe

B.G. Sidharth, F. Honsell and A. de Angelis. (eds.), Frontiers of Fundamental Physics, 247–254.

some observed features, which are connected with Bose-Einstein (BE) (or Fermi-Dirac (FD)) statistics of produced secondaries (in case they are identical and occupy almost the same parts of the phase space defined by the uncertainty relation, see Fig. 1). When two (or more) identical particles of the same kind are observed, their common wave function should be symmetrized (for BE statistics) or antisymmetrized (for FD statistics) what results in characteristic shapes of (two particle, for example) correlation function $C_2(Q = |p_1 - p_2|) = N(p_1, p_2)/[N(p_1) \cdot N(p_2)]$, cf., Fig. 2.

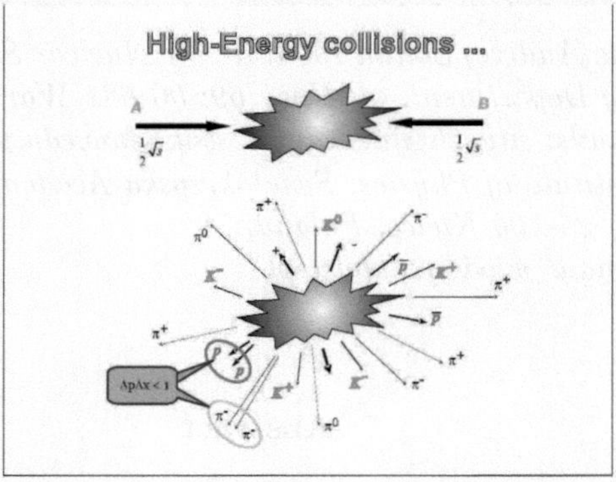

Figure 1. Schematic view of high energy collision resulting in production of many particles of different statistics.

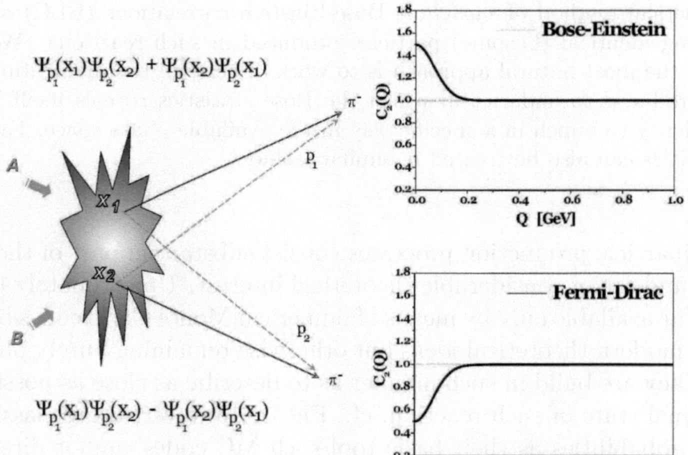

Figure 2. Example of BE and FD statistics (left panel) and two particle correlation functions they lead to (right panels).

Referring for details of Bose-Einstein correlations (BEC) to the literature (cf., for example, [2] and references therein) let us concentrate here directly on the problem of their *proper numerical modelling*, i.e., such in which the bosonic character of secondaries produced in hadronization process are going to be accounted for from the very beginning. This problem was so far considered only in [3] (using statistical approach based on information theory approach, cf., however, also [4]). All other approaches, which claim to model BEC numerically [5], simply add to the outcomes of existing MC codes [1] some *afterburners*, which modify them in a suitable way to be able fit the BEC data. Such approach inevitably leads to such unwanted features as violation of energy-momentum conservation or changes in the original (i.e., obtained directly form MC code) multiparticle spectra.

In [6] we have proposed afterburner free from such unwanted effects. It was based on different concept of introducing quantum mechanical (QM) effects in the otherwise purely probabilistic distributions then those proposed in [7]. Namely, each MC code provides us usually with a given number of particles, each one endowed with either (+) or (−) or (0) charge and with well defined spatio-temporal position and energy-momentum. But experiment provides us information on only the first and last characteristics. The spatio-temporal information is not available directly. In fact, the universal hope expressed in [2, 5] is that precisely this information can be deduced from the previous two via the measured BEC. Our reasoning was as follows: (*i*) BEC phenomenon is of QM origin therefore one has to introduce in the otherwise purely classical distributions provided by MCG the new element mimicking QM uncertainties; (*ii*) it cannot be done with energy-momenta because they are measured and therefore fixed; (*iii*) the next candidate, i.e., spatio-temporal characteristics can be changed but it was already done in [7, 5]; (*iv*) one is thus left with charges and in [6] we have simply assigned (on event-by-event basis) new charges to the particles from MCG conserving, however, the original multiplicities of $(+/-/0)$. This has been done in such way as to make particles of the same charge to be located maximally near to each other in the phase space exploring for this natural fluctuations in spatio-temporal and energy-momentum characteristic of the outcome of MCG. The advantages of such approach are: (*a*) energy-momentum is automatically conserved and multiparticle distributions are not modified and (*b*) it is applicable already on the level of each event provided by MCG (not only, as some of propositions of [5] only to all events). However, the new assignment of charges introduces a profound change in the structure of the original MCG. Generally speaking (cf. [6] for details) it requires introduction of bunchings of particles of the same charge.

This observation will be the cornerstone of our new proposition. Let us first remind that idea of bunching of particles as quantum statistical (QS) effect is not the new one [8]. It was used in connection with BEC for the first time in [9] and then was a basis of the so called *clan model* of multiparticle distributions leading in natural way to their negative binomial (NB) form observed in experiment [10]. It was then again introduced in the realm of BEC in [11] and [3, 4]. Because our motivation comes basically from [3] let us outline shortly its basic points. It deals with the problem of how to distribute in a least biased way a given number of bosonic secondaries, $\langle n\rangle = \langle n^{(+)}\rangle + \langle n^{(-)}\rangle + \langle n^{(0)}\rangle$, $\langle n^{(+)}\rangle = \langle n^{(-)}\rangle = \langle n^{(0)}\rangle$. Using information theory approach (cf., [12]) their rapidity distribution was obtained in form of grand partition function with temperature T and chemical potential μ. In addition, the rapidity space was divided into *cells* of equal size δy each (it was fitted parameter). It turned out that whereas the very fact of existence of such cells was enough to obtain reasonably good multiparticle distributions, $P(n)$, (actually, in the NB-like form), their size, δy, was crucial for obtaining the characteristic form of the 2−body BEC function $C_2(Q = |p_i - p_j|)$ (peaked and greater than unity at $Q = 0$ and then decreasing in a characteristic way towards $C_2 = 1$ for large values of Q, see Fig. 2) out of which one usually deduces the spatio-temporal characteristics of the hadronization source [2] (see [3] for more details). The outcome was obvious: to get C_2 peaked and greater than unity at $Q = 0$ and then decreasing in a characteristic way towards $C_2 = 1$ for large values of Q one must have particles located in cells in phase space which are of nonzero size. It means then that from C_2 one gets not the size of the hadronizing source but only size of the emitting cell, in [3] $R \sim 1/\delta y$, cf. [13]. In the quantum field theoretical formulation of BEC this directly corresponds to the necessity of replacing delta functions in commutator relations by a well defined peaked functions introducing in this way same dimensional scale to be obtained from fits to data [14]. This fact was known even before but without any phenomenological consequences [15].

Let us suppose now that we have mass M and we know that it hadronizes into $N = \langle n \rangle$ bosonic particles (assumed to be pions of mass m) with equal numbers of $(+/-/0)$ charges and with limited transverse momenta p_T. Let the multiplicity distribution of these pions follows some NB-like form, broader than Poissonian one. Suppose also that the two-particle correlation function of identical particles, $C_2(Q)$, has the specific BEC form mentioned above. How to model such process from the very beginning, i.e., in such way that bosonic character of produced particles is accounted for from the very beginning and not imposed at the end? We propose the following steps (illustrated by comparison to some selected LEP e^+e^- data [16]):

(1) Using some (assumed) function $f(E)$ select a particle of energy $E_1^{(1)}$ and charge $Q^{(1)}$. The actual form of $f(E)$ should reflect somehow our *a priori* knowledge of the particular collision process under consideration. In what follows we shall assume that $f(E) = \exp(-E/T)$, with T being parameter (playing in our example the role of "temperature").

(2) Treat this particle as seed of the first *elementary emitting cell* (EEC) and add to it, until the first failure, other particles of the same charge $Q^{(1)}$ selected according to distribution $P(E) = P_0 \cdot f(E)$, where P_0 is another parameter (actually it plays here the role of "chemical potential" $\mu = T \cdot \ln P_0$). This assures that the number of particles in this first EEC, k_1, will follow geometrical (or Bose-Einstein) distribution, and in precisely this way one accounts for the bosonic character of produced pions. This results in $C_2(Q) > 1$ but only *at one point*, namely for $Q = 0$.

(3) To get the observed spread out of $C_2(Q)$ one has to allow that particles in this EEC have (slightly) different energies from energy of the particle being its seed. To do it allow that each additional particle selected in point (2)

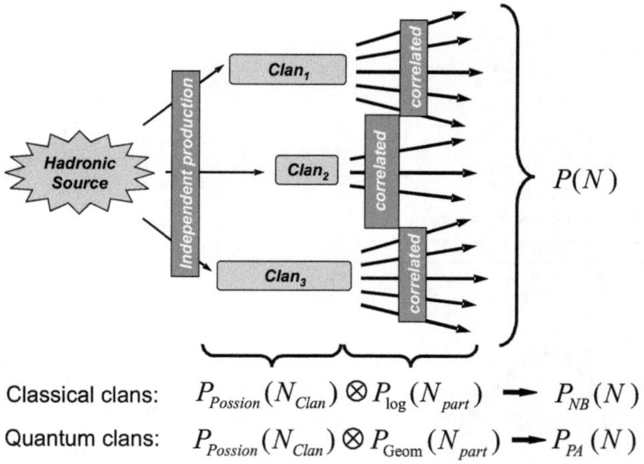

Figure 3. Schematic view of our algorithm, which leads to bunches of particles (*clans*). Whereas in [10] these clans could consist of any particles distributed logarithmically in our case they consist of particles of the same charge and (almost) the same energy and are distributed geometrically to comply with their bosonic character.

above have energy $E_i^{(1)}$ selected from some distribution function peaked at $E_1^{(1)}$, $G\left(E_1^{(1)} - E_i^{(1)}\right)$.

(4) Repeat points (1) to (2) as long as there is enough energy left. Correct in every event for every energy-momentum nonconservation caused by selection procedure and assure that $N^{(+)} = N^{(-)}$.

As result we get a number of EECs with particles of the same charge and (almost) the same energy, which we regard as being equivalent to *clans* in the [10] (see Fig. 3). These clans are distributed in the same way as the particles forming the seeds for those EEC, i.e., according to Poisson distribution (see Fig. 4, upper-left panel). On the other hand, as was already said, particles in each EEC will be distributed according to geometrical distribution (see Fig. 4, upper-right panel). As a result the overall distribution of particles will be of the so called Pòlya-Aeppli distribution [17]. It fits our examplatory data reasonable well. It is interesting to notice at this point that to get NB distribution resulting from the classical clan model of [10] one should have logarithmic rather than geometrical distribution of particles in EEC, which would then not account for the bosonic character of produced secondaries. In this respect our model differs from this *classical* clan model and we see that what we have obtained is indeed its *quantum* version, therefore its proposed name: *quantum clan model.*

The first preliminary results presented in Fig. 4 are quite encouraging (especially when one remembers that so far effects of resonances and all kind of final state interactions to which C_2 is sensitive were neglected here). It remains now to be checked what two-body BEC functions for other components of the momentum differences and how they depend on the EEC parameters: T, P_0 and σ. So far the

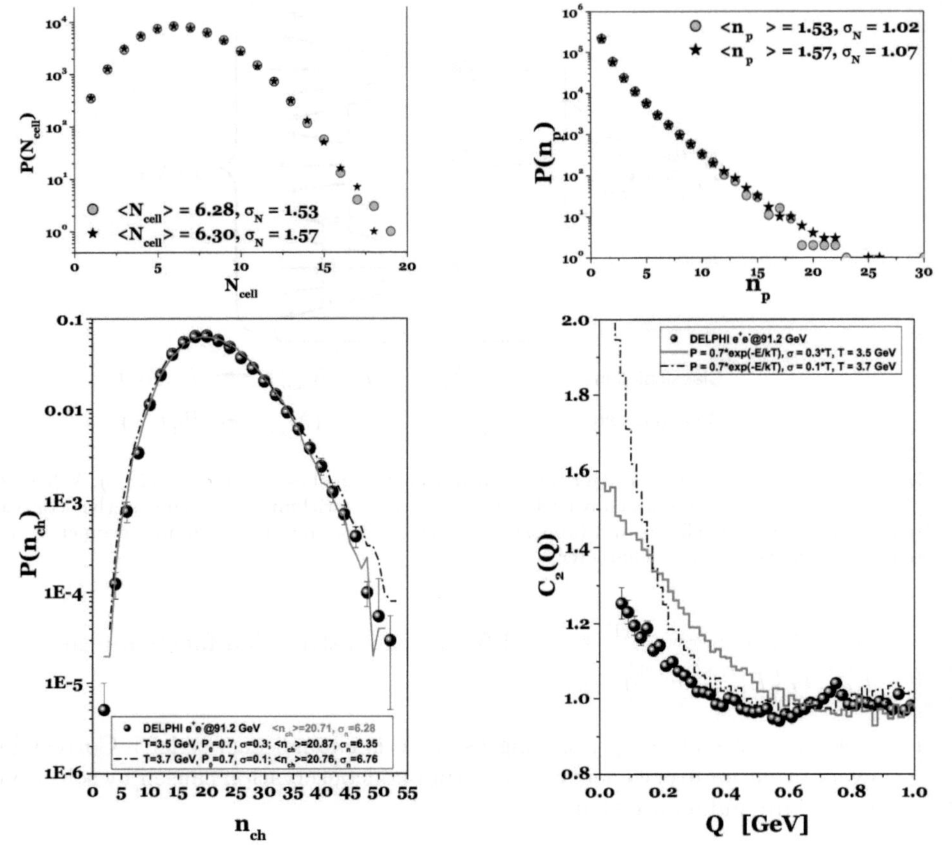

Figure 4. Upper panels: distribution of cells and particles in a given cell. Lower-left panel: the corresponding summary $P(n)$ which is convolution of both $P(n_{cell})$ and $P(n_p)$. Lower-right panel: examples of the corresponding corresponding correlation functions $C_2(Q)$. Two sets of parameters were used. Data are from [16].

main outcome is that BEC are due to EEC's only and therefore provide us mainly with their characteristics (it is worth to mention at this point that essentially this type of approach has been also proposed to simulate Bose-Einstein condensate phenomenon in [18]). This should clear at least some of many apparently "strange" results obtained from BEC recently (see Quark Matter 2004 proceedings, especially [19]). The most intriguing is the fact that apparently the "size" of the hadronizing source deduced from the BEC data does not vary very much with energy and with the size of colliding objects as has been naively expected [2]. In our approach this has simple explanation, see Fig. 5. The point is that BEC are mainly sensitive to the correlation length, which in our case is dimension of the emitting cell, not to dimension of the "fireball" in which hadronization process takes place. The size of this fireball depends mainly on the number of produced secondaries [19], which in our case is given by the "partition temperature" pa-

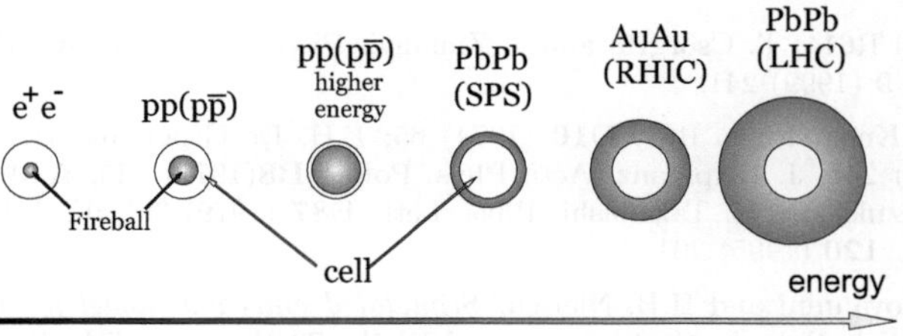

Figure 5. Schematic view of how in our approach the size of the EEC compares with the size of hadronizing fireball at different energies and for different types of projectiles.

rameter T and by the "chemical potential" parameter $\mu = T \cdot \ln P_0$. They are changing with mass M (and therefore with the energy of reaction and the type of projectile). On the other hand dimension of EEC is given entirely by parameter σ describing the spread of energy of particles belonging to this EEC, which is only weakly depending on energy (if at all). We shall close with mentioning that our approach accounts also for multiparticle BEC and, because of this, by intermittence effects seen in data (at least to some extend) [6].

Acknowledgments

Partial support of the Polish State Committee for Scientific Research (KBN) (grant 2P03B04123 (ZW) and grants 621/E-78/SPUB/CERN/P-03/DZ4/99 and 3P03B05724 (GW)) is acknowledged.

References

[1] K.J. Eskola, Nucl. Phys. **A698** (2002) 78; S.A.Bass et al. Prog. Part. Nucl. **41** (1998) 225.

[2] U.A. Wiedemann and U. Heinz, Phys. Rep. **319** (1999) 145; R.M. Weiner, Phys. Rep. **327** (2000) 249; T. CsörgHo, in *Particle Production Spanning MeV and TeV Energies*, eds. W.Kittel et al., NATO Science Series C, Vol. 554, Kluwer Acad. Pub. (2000), p. 203 (see also: hep-ph/0001233); G. Baym, Acta Phys. Polon. **B29** (1998) 1839; W. Kittel, Acta Phys. Polon. **B32** (2001) 3927; G. Alexander, Rept. Prog. Phys. **66** (2003) 481.

[3] T. Osada, M. Maruyama and F. Takagi, Phys. Rev. **D59** (1999) 014024.

[4] W. Zajc, Phys. Rev. **D53** (1987) 3396; R.L. Ray, Phys. Rev. **C57** (1998) 2352; J.G.Cramer, *Event Simulation of High-Order Bose-Einstein and Coulomb Correlations*, Univ. of Washington preprint (1996, unpublished).

[5] L. Lönnblad and T. Sjöstrand, Eur. Phys. J. **C2**(1998) 165; K. Fiałkowski and R. Wit, Eur. Phys. J. **C2**(1998) 691; K. Fiałkowski, R. Wit and J. Wosiek, Phys. Rev. **D58**(1998) 094013; T.Wibig, Phys. Rev. **D53** (1996) 3586. B. Andersson, Acta Phys. Polon. **B29** (1998) 1885; J.P. Sullivan et al., Phys. Rev. Lett. **70**, 3000 (1993); K. Geiger, J. Ellis, U. Heinz and U.A. Wiedemann, Phys. Rev. **D61**(2000) 054002.

[6] O.V. Utyuzh, G. Wilk and Z. Włodarczyk, Phys. Lett. **B522** (2001) 273 and Acta Phys. Polon. **B33**(2002) 2681.

[7] A. Białas and A. Krzywicki, Phys. Lett. **B354**, (1995) 134; H. Merlitz and D. Pelte, Z. Phys. **A351**(1995) 187 and **A357** (1997) 175; U.A. Wiedemann et al., Phys. Rev. **C56**

(1997) R614; T. CsörgHo and J. Zimányi, Phys. Rev. Lett. **80** (1998) 916 and Heavy Ion Phys. **9** (1999) 241.

[8] W.J. Knox, Phys. Rev. **D10** (1974) 65; E.H. De Groot and H. Satz, Nucl. Phys. **B130** (1977) 257; J. Kripfganz, Acta Phys. Polon. **B8**(1977) 945; A.M. Cooper, O. Miyamura, A. Suzuki and K. Takahashi, Phys. Lett. **B87** (1979) 393; F. Takagi, Prog. Theor. Phys. Suppl. **120** (1995) 201.

[9] A. Giovannini and H.B. Nielsen, *Stimulated emission model for multiplicity fluctuations*, The Niels Bohr Institute preprint, NBI-HE-73-17 (unpublished) and *Stimulated emission effect on multiplicity distribution* in: Proc. of the IV Int. Symp. on Multip. Hydrodynamics, Pavia 1973, Eds. F.Duimio, A.Giovannini and S.Ratii, p. 538.

[10] A. Giovannini and L. Van Hove, Z. Phys. **C30** (1986) 381; see also: P. Carruthers and C.S.Shih, Int. J. Mod. Phys. **A2** (1986) 1447.

[11] M. Biyajima, N. Suzuki, G. Wilk and Z. Włodarczyk, Phys. Lett. **B386** (1996) 297.

[12] O.V. Utyuzh, G. Wilk and Z. Włodarczyk, *Multiparticle production processes from the nonextensive point of view*, hep-ph/0410341 and references therein.

[13] W.A. Zajc, *A pedestrian's guide to interferometry*, in "Particle Production in Highly Excited Matter", eds. H.H.Gutbrod and J.Rafelski, Plenum Press, New York 1993, p. 435.

[14] G.A. Kozlov, O.V. Utyuzh and G. Wilk, Phys. Rev. **C68** (2003) 024901 and Ukr. J. Phys. **48** (2003) 1313.

[15] K. Zalewski, Lecture Notes in Physics **539**(2000) 291.

[16] P.Abreu et al. (DELPHI Collab.), Phys. Lett. **B286** (1992) 201 and Phys. Lett. **B247** (1990) 137.

[17] J. Finkelstein, Phys. Rev. **D37** (1988) 2446; Ding-wei Huang, Phys. Rev. **D58** (1998) 017501.

[18] R.Kutner and M.Regulski, Comp. Phys. Com. **121-122** (1999) 586; cf. also: R.Kutner, K.W.Kehr, W.Renz and R.Przeniosło , J.Phys. **A28** (1995) 923.

[19] J.T. Mitchell, J. Phys. **G30** (2004) S819; H. Appelhäuser, *ibidem*, S935.

FRONTIERS OF HIGH ENERGY COSMIC RAYS

MÁRIO PIMENTA [a,b]

[a] *LIP, Av. Elias Garcia, 14-12, 1000-149 Lisboa, Portugal*
[b] *IST, Av. Rovisco Pais,1049-001 Lisboa, Portugal*

Abstract

Ultra high energy cosmic rays are a mystery and a challenge. Its theoretical and experimental status, as well as the expected performance of the future generation of experiments, are briefly reviewed.

Cosmic rays were always a frontier domain. In this short lecture I have no time to give the due credit to the many people involved and to the many breakthroughs of the already 100 years of history. I would nevertheless refer two events: the discovery by Victor Hess in 1912, with pioneer balloon flights, of a strange radiation coming from the space and not correlated with the sun [1]; the discovery of Ultra High Energy Cosmic Rays (UHECR), with an energy greater than 10^{19} eV, by John Linsley in 1962 at Volcano Ranch [2].

The energy spectrum of cosmic rays extends over more than 11 decades of energy, following an almost perfect power law (see figure 1) [3]. The fluxes are quite high at low energies (up to 1 $particle/m^2/s$ for energies of the order of the GeV), but extremely small at the highest energies (1 $particle/km^2/century$ at 10^{20} eV). Under a closer inspection, changes in the slope can be observed around 10^{15} eV (the knee) and 10^{18} eV (the ankle). There is a great deal of interest these two regions. This review will however be dedicated to the far end of the spectrum (figure 2), where just a few events have been observed, mainly in two experiments: AGASA [4] and HiRes [5]. While AGASA seems to be compatible with no slope change in the spectrum at these extreme energies, HiRes seems to suggest a dramatic drop. The statistical significance of the difference between the two experiments is however slightly above 2σ [6]!

B.G. Sidharth, F. Honsell and A. de Angelis. (eds.), Frontiers of Fundamental Physics, 255–261.

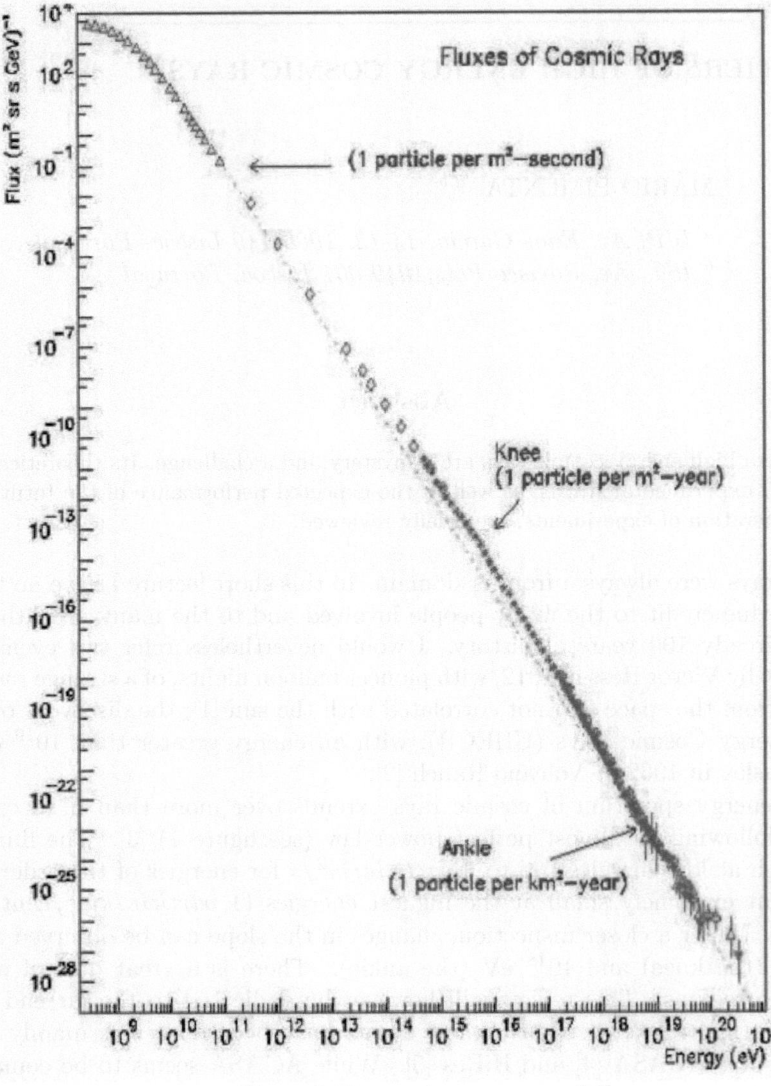

Figure 1. The cosmic ray spectrum [23].

There is however a consensus: UHECR do exist and have been observed. The most energetic event ever registered was collected by the Fly's Eye experiment in 1991 and has an estimated energy of 3.2 10^{20} eV [7]. The geographical distribution of the highest energy data (above 4 10^{19} eV) collected by AGASA is isotropic at large scales but shows some clustering at smaller scales: one triplet and six doublets with $\Delta(\Theta) = 2.5^0$ [8].

There are many papers discussing possible acceleration mechanisms able to produce particles with such energies. The main idea is to have somewhere a shock wave, and particles undergoing successive crossings of the shock front. It is the so called first-order Fermi mechanism [9]. The non-relativistic case is well understood, but the relativistic case is much more difficult. The regions where such events may occur can be characterised by two variables: the size of the region and the intensity of the existing magnetic fields. It is possible to have confinement either in small regions, if there is an intense magnetic field, or, if the magnetic field is weak, in extended regions. Such a feature is usually summarised in the so called Hillas plot (figure 3) [10]. There are several possible sources for 10^{20} eV UHECR but at higher energies it is much more difficult to find good candidates. Furthermore, energy losses due to Bremsstrahlung have to be taken into account whenever the accelerations are very large [11].

In any case, the observed UHECR would have to propagate through very large distances before reaching the Earth. But there is no empty space. The space is, as it is well known, filled with the Cosmic Microwave Background (CMB). The interaction of the UHECR with the CMB γs can be quite dramatic if the centre-of-mass energy is above the first inelastic channel threshold. The net result is that UHECR with energies at the source well above 10^{20}eV would have dropped to about 10^{20}eV by the time they reach Earth. This is the famous GZK effect [12]. GZK is not however a real cut-off. The threshold energy depends on many parameters (the type of the sources and its distribution in space, the existing magnetic fields, the particle type, ...).

But if AGASA is right how to avoid the GZK? One possible solution is to imagine that UHECR are particles with small electromagnetic couplings but able to originate normal air showers. Possible candidates may be neutrinos. In the Z burst model [13], very energetic neutrinos collide with the cosmological neutrino background producing energetic Z^0 bosons near our galaxy. It is an elegant model, but it implies the existence of intense and monochromatic neutrino beams with an energy around 4 10^{21}eV! There are many other proposed solutions [14] such as: enhanced neutrino-nucleon cross-sections, the existence of "new" particles (axions, glueballinos, ...) or even the violation of Lorentz symmetry which could lead to a change on the relevant thresholds. However in this kind of models (the bottom-up scenarios), UHECR would be produced as secondaries of very energetic "ordinary" particles ($E > 10^{22}$eV). There is another way out (the top-down scenarios): UHECR may be decay products of super-heavy particles, which would be produced continuously by cosmological "objects" left over after the phase transitions that the Universe has suffered, or even produced directly in the early Universe. The challenge is to increase the statistics of observed UHECR. If this is achieved, models can be disentangled looking at the energy spectrum power law, the UHECR composition, the geographical distribution, the possible counter-parts in other wavelengths - all a program!

Another major challenge in cosmic ray frontiers is the observation of high energy cosmic neutrinos. In fact astrophysical neutrino beams should exist [15]! A safe prediction is the existence of the cosmological neutrinos produced in the in-

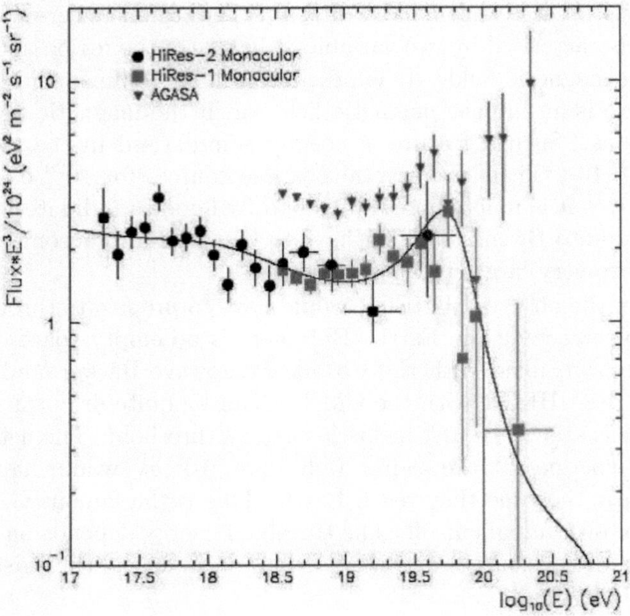

Figure 2. The far end of the cosmic ray spectrum [24].

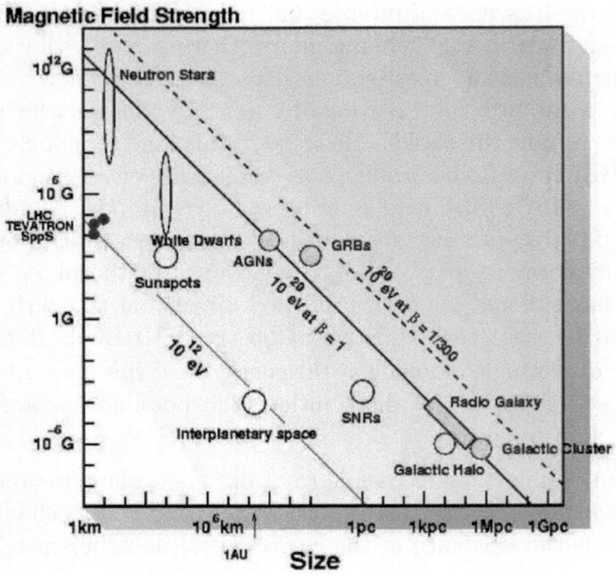

Figure 3. The Hillas plot [10].

teraction of the primordial UHECR with the γ_{CMB} or with clouds of intergalactic gas. These interactions produce collimated pions and kaons, which will originate energetic neutrino beams. But there are many other models: from the Z burst mechanism, to the possible Zevatron sources (AGNs, GRBs, supernovae, ...) and to the top-down scenarios discussed just above (see figure 4). These neutrinos can propagate to Earth almost unscattered and so their detection would open an important channel to observe the Universe. However, so far, there is no experimental measurement above 10^6 GeV. The existing limits, as well as the expected sensitivities of future experiments, are shown in figure 5.

UHECR are the most energetic beams ever accessible. The centre-of-mass energy of a $10^{19} - 10^{20}$ eV UHECR collision in the atmosphere is around 100-400 TeV, which is well above the present and future man made accelerators. But UHECR fluxes are small, the accessible kinematic region is very narrow (very forward region) and the detection capability is much poorer than in conventional Particle Physics experiments. The solution for these low fluxes is to have huge detection areas. The surface of the Pierre Auger observatory is around 3000 Km^2 [16], while the planned space based experiments like EUSO [17] will cover a surface of around 200000 Km^2. The limit is the Earth surface, which is still a factor 1000 larger. Nevertheless, important particle physics results were already obtained or are expected in the near future. For instances: pp cross-sections measurements were extended to higher centre-of-mass energies (figure 6) [18]; heavy quark production is very abundant at these energies and may represent an important channel for QCD studies; new particles (Higgs, SUSY, excited fermions, leptoquarks...) with masses above the TeV can be produced; Lorentz symmetry can be tested at very high boost factors; the existence of extra-dimensions may imply important enhancements in the neutrino-air cross-section [19]. The search for new physics is, therefore, a relevant issue in the next generation of UHECR experiments and an effort for finding new signatures must be pursued. An example of such possible signatures is the so-called double-bang signature. An example would be the observation, in an experiment like EUSO, of the production and the instantaneous decay of a microscopic black hole (first bang), followed, at a measurable distance, by the decay of an energetic tau lepton (second bang) [20].

In the next decade a spectacular increase in the number of observed UHECR is expected (one to two orders of magnitude). Auger is right around the corner. It combines the technique of sampling the shower particles reaching the Earth surface with the technique of detecting the fluorescence light produced by the excitation of Nitrogen molecules by the low energy electrons of the shower. At present, 25% of the 1600 sampling detectors (water tanks) and two of the three sites for fluorescence light detection are already fully working! EUSO, an ESA mission planned for the first years of the next decade, is basically a telescope looking downwards to the Earth in order to collect a fraction of the fluorescence light produced in the interaction of the UHECR with the atmosphere. And there are already plans to go further, OWL [21] is a NASA mission under study which will use two satellites, reaching a field of view about 5 times larger than EUSO.

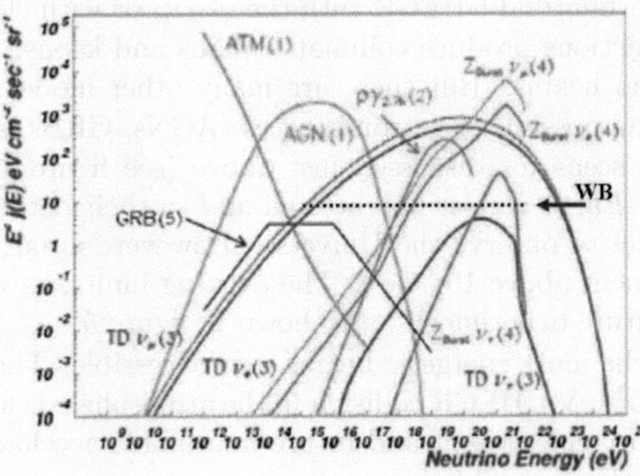

Figure 4. Expected neutrino fluxes [15].

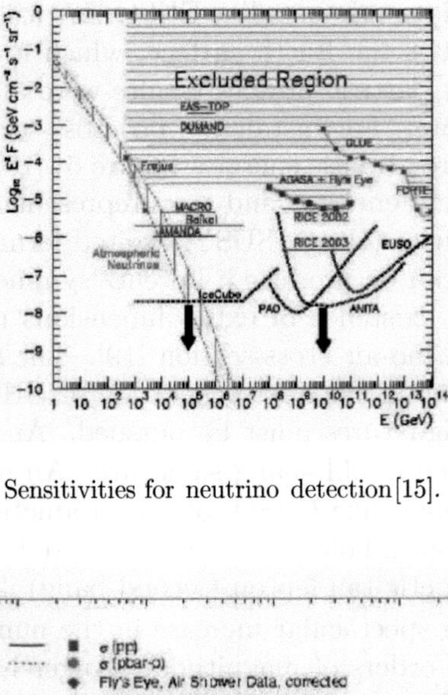

Figure 5. Sensitivities for neutrino detection [15].

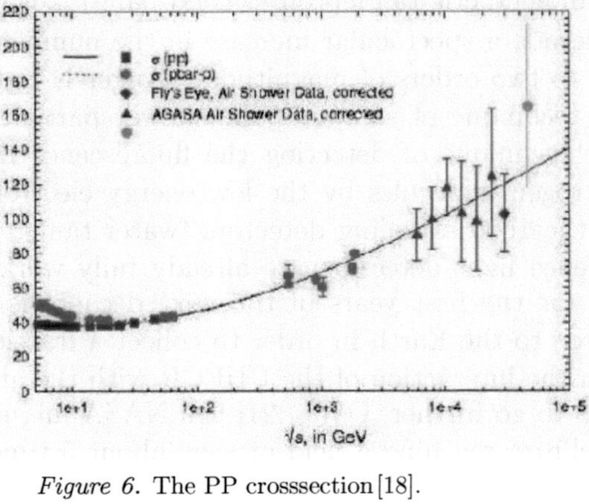

Figure 6. The PP crosssection [18].

Finally, there are several experiments on the way to the detect the coherent radio emission by very energetic showers [22].

New frontiers in Particle Physics, Astrophysics and Cosmology are right ahead of us!

Acknowledgments

I would like to thank Catarina Espírito Santo and Pedro Abreu for reading the manuscript and help in preparing these proceedings, and Alessandro De Angelis for the nice atmosphere during all the conference.

References

[1] //nobelprize.org/physics/laureates/1936/hess-bio.html

[2] J.Linsley, Phys.Rev.Lett., 10,(1963)146.

[3] For a general reference see for example: A.Watson and M.Nagano, Rev.Mod.Phys.,72,689 (2000).

[4] M.Takeda et al., Astro.Part.Phys.,19,(2003)499

[5] D.Bird et al., Astrophys.J.,511,(1999),739

[6] D.De Marco,P.Blasi,A.V.Olinto, Astropart.Phys.20(2003)53

[7] http://www.cosmic-ray.org/reading/flyseye.html

[8] N.Hayashida et al., astro-ph/0008102 (2000)

[9] J.Niemiec, astro-ph/0407531

[10] A.M.Hillas, Ann. Rev. Astron. Astrophys. 22 (1984) 425.

[11] M.V.Medvedv, Physical Review E 67, 045401(R) (2003)

[12] K.Greisen. Phys.Rev.Lett., 16, (1996),748; G.T.Zatsepin and V.A.Kuzmin, Sov.Phys.JETP Lett. (engl.transl.),4,(1996),78

[13] T.J.Weiler, Phys.Rev.Lett., 40 (1982) 234; D.Fargion et al.,Astrophys.J.517(1999)725

[14] G.Sigl, astro-ph/0210049

[15] D.V.Semikoz, Theoretical Predictions of Ultra-High Energy Neutrino Fluxes 28th ICRC Tsukuba, 1439 (2003)

[16] http://www.auger.org/

[17] http://www.euso-mission.org/cgi-bin/pub.cgi?href=welcome.html

[18] M.M.Bock et al., astro-ph/0004232 (2000)

[19] M.Cavagla, Int.J.Mod.Phys. A18,1843 (2003)

[20] V.Cardoso,hep-ph/0405056

[21] http://owl.gsfc.nasa.gov/

[22] H.Falcke, astro-ph/0409229

[23] from S.Swordy

[24] T.Abu-Zayyad et al., astro-ph/0208243

OUTLOOKS ON GAMMA RAY ASTROPHYSICS

RICCARDO GIANNITRAPANI [a]

[a] *Dipartimento di Fisica dell'Universià di Udine and INFN, Italy*

Abstract

Gamma Ray Astrophysics, extending from MeV to PeV, covers a wide range of phenomena in the domain of both particle physics and astrophysics, providing one of the most active field of modern astroparticle physics. Moreover, the study of gamma rays from the universe can unveil some profound implications for the foundation of physics itself (from Quantum Gravity to Cosmology) and provide a challenging domain for technology and computing too, bringing together researchers of different domains in an growing interesting crucible of ideas and experiences.

B.G. Sidharth, F. Honsell and A. de Angelis. (eds.), Frontiers of Fundamental Physics, 263–263.

WILL ANTIHYDROGEN LIGHT SHINE?

ALESSANDRO VARIOLA [a]

[a] *Swansea University of Wales*

Abstract

In 2002 the ATHENA experiment announced the first production of antihydrogen atoms at cryogenic temperature. This achievement is an important step towards antimatter atoms fine spectroscopy. Antihydrogen light is expected to make it possible to carry out very accurate CPT invariance measurements as well as, for the first time ever, tests on the gravitational interaction between matter and antimatter with important consequences for cosmology. Over the last two years significant results have been obtained in the investigation of antihydrogen production mechanisms. They are all the more important given the strong limitations imposed by the neutral atom trapping techniques required for fine spectroscopy measurements. This talk will focus on the main scientific motivations in the study of cold antihydrogen, the results obtained so far and the medium-term plans for the first measurements of the interaction between antiatoms and photons.

B.G. Sidharth, F. Honsell and A. de Angelis. (eds.), Frontiers of Fundamental Physics, 265–265.

PHYSICS POTENTIAL OF THE ATLAS EXPERIMENT

M. COBAL [a], FOR THE ATLAS COLLABORATION

[a] *Dipartimento di Fisica Universitá di Udine, via delle Scienze 208, 33100 Udine, Italy*

Abstract

The ATLAS experiment at the CERN Large Hadron Collider (LHC) will present the most extraordinary challenges that particle physics has ever faced. By studying the collisions of high intensity proton beams at a center-of-mass energy of 14 TeV, ATLAS will explore for the first time the TeV scale region. This paper reviews the experiment's potential for discovery of physics beyond the Standard Model (SM) and for the investigation of the nature of electroweak symmetry breaking.

1 Introduction

The SM has been verified with an impressive accuracy, i.e. to 0.1% or better in most cases [1]. However, there are numerous indications that this is not the ultimate theory of elementary particles. Among them, the evidence for atmospheric [2] and solar [3] neutrino oscillations, and the incapacity of the SM to give satisfactory answers to fundamental questions [4] such as the mass and flavour problems, baryogenesis and the matter-antimatter asymmetry in the universe, the origin of dark matter, the size of the cosmological constant, and the unification of gravity with the other interactions.

One of the most urgent issues is to explain the origin of the particle masses. In the SM particles acquire a mass through their coupling with the Higgs boson. The Higgs mass is not specified by the theory, which provides only an upper bound of ~1 TeV. Direct searches performed at LEP have set a lower limit of

B.G. Sidharth, F. Honsell and A. de Angelis. (eds.), Frontiers of Fundamental Physics, 267–272.

$m_H >$ 114.4 GeV [5], and a fit of the SM to the data collected by various machines (LEP, Tevatron, SLC) gives a 95% C.L. upper bound on m_H of about 250 GeV [1]: experimental data therefore favour a light Higgs boson.

The introduction of the Higgs mechanism brings however some problems: the Higgs mass increases with the energy scale Λ up to which the SM is valid and therefore it requires a large amount of "fine tuning" to be stabilized at the electroweak scale. In addition, the generation of fermion masses spoils the simplicity of the SM with a proliferation of unknown parameters. There are several candidate scenarios for physics beyond the Standard Model, including Supersymmetry (SUSY), Technicolour and theories with Extra-dimensions. All of them predict new particles in the TeV region, as needed to stabilize the Higgs mass.

ATLAS [6] is a general purpose detector which will operate at the LHC at centre-of-mass-energies of 14 TeV from 2007 onwards, in search of the Higgs and of new and even unexpected physics. The design of the detector will allow measurement and identification of leptons, photons up to rapidities $|\eta| < 2.5$ and jets up to rapidities $|\eta| < 5$, thus allowing the exploitation of a wide range of physics signatures. Given the high-energy and high-luminosity which will characterize the LHC, the rate of events produced in the collisions will be huge and LHC will be a factory of all particles with masses up to a few TeV which have reasonable couplings to SM particles. For example, when the machine luminosity will be about a factor of ten lower than the design luminosity $L = 10^{34}$ cm^{-2} s^{-1}, million of events should be collected for many SM channels over only one year of data taking (i.e.: 10^7 $t\bar{t}$, 10^8 W $\rightarrow e\nu$) and $\simeq 130$ $H \rightarrow \gamma\gamma$ with $m_H = 120$ GeV). This will allow not only to explore new territories of discovery, but also to perform many precision measurements (like the determination of the top and W mass, or the study of the Triple Gauge Couplings) with high accuracy. In the following, we will concentrate on the Higgs and SUSY searches.

2 Searches for the SM Higgs boson

The SM predicts that the Higgs couples to fermions and bosons with strength proportional to their masses. Therefore, for $m_H < 120$ GeV the Higgs boson should decay mainly into $b\bar{b}$, whereas for larger masses decays into W pairs and Z pairs should dominate. Gluon-gluon fusion through a top-quark loop is the dominant production channel for all masses. Vector boson (WW, ZZ) fusion contributes about 20% of the cross-section for $m_H \sim$ 120 GeV and becomes more and more important with increasing mass. This process leads to the very distinctive topology of a Higgs boson accompanied by two jets emitted in the forward regions of the detector and very little activity in the central region (since no colour lines are exchanged between the two interacting bosons). Higgs production with a $t\bar{t}$ pair or a W/Z boson has a smaller cross-section; however it allows detection of the purely hadronic $H \rightarrow b\bar{b}$ decay mode, because the reconstruction of the particles produced in association with the Higgs provides additional handles against the large QCD backgrounds. A SM Higgs boson can be discovered over

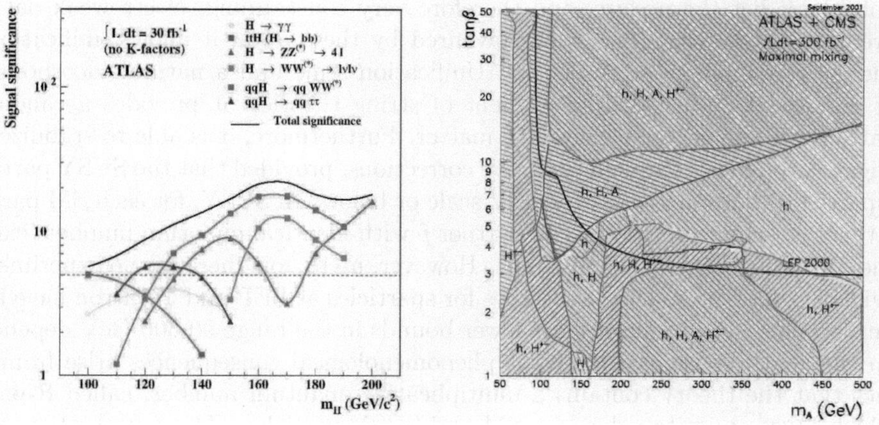

Figure 1. a) The expected SM Higgs signal significance in ATLAS in the low-mass region for 30 fb^{-1}. The total significance (full line) and the contributions of the individual channels (symbols) are shown. b) The regions of the (constrained) MSSM plane $\{m_A\text{-}\tan\beta\}$ where the various Higgs bosons can be discovered at the LHC through their decays into SM particles. The region below the thick curve has been excluded by LEP.

the full allowed mass range (114.4 GeV to 1 TeV) with only 10 fb^{-1} of integrated luminosity, which corresponds (in principle) to only one year of LHC operation at low luminosity. The Higgs boson discovery should be easier and faster for masses above 200 GeV, thanks to the "gold-plated" $H \to ZZ \to 4\ell$ (where $\ell = e, \mu$) channel, which is essentially background-free. In contrast, the most difficult region is the low-mass region close to the LEP limit and at the overlap with the Tevatron reach. The total significance of about 4σ per experiment ($4^{+2.2}_{-1.3}\ \sigma$ including the expected systematic uncertainties) is more or less equally shared among three channels (see Fig. 1 a): $H \to \gamma\gamma$, $t\bar{t}H$ production with $H \to b\bar{b}$, and Higgs production in vector-boson fusion followed by $H \to \tau\tau$. If a Higgs boson were to be discovered at the LHC, ATLAS should be able to perform several precise measurements of its properties. For example, with the ultimate integrated luminosity of 300 fb^{-1} the Higgs mass should be measured with the remarkable experimental precision of 0.1% over the mass region up to $\sim$ 400 GeV. This direct measurement can then be compared to the indirect determination of m_H obtained from the measurements of the W and top masses. The expected precisions at the LHC are $\sim$15 MeV on m_W and $\sim$1 GeV on m_{top} [8], leading to a 25% (indirect) accuracy on the mass of a light Higgs boson.

3 Searches for Supersymmetry and beyond

SUSY [9], a symmetry relating fermions and bosons, that is matter fields and force fields, is one of the relevant scenario today for physics beyond the SM [4]. It does not contradict the precise, and therefore very constraining, electroweak data, it predicts a light Higgs boson, as favoured by these data, it allows unification of the gauge couplings at the Grand Unification scale and a natural incorporation of gravity, it is an essential element of string theories, it provides a candidate particle for the universe cold dark matter. Furthermore, it is able to stabilize the Higgs boson mass, through radiative corrections, provided that the SUSY particles (sparticles) have masses at the TeV scale or below. In SUSY, for each SM particle p there exists a supersymmetric partner $\tilde{p}$ with identical quantum numbers except the spin which differs by half a unit. However, up to now there are no experimental evidences for SUSY. Direct searches for sparticles at LEP and Tevatron have been unsuccessful, and have set mass lower bounds in the range 90-300 GeV depending on the sparticle type. Important phenomenological consequences arise from the fact that the theory contains a multiplicative quantum number, called R-parity, which takes opposite values for SM and SUSY particles. The conservation of R-parity, motivated by cosmological arguments, is assumed here. This implies that sparticles are produced in pairs and that the Lightest Superymmetric particle (LSP), to which all sparticles eventually decay, must be stable. In most models the LSP is the lightest neutralino χ_1^0, which is a stable, massive and weakly-interacting particle, and therefore an excellent candidate for the universe cold dark matter. At the LHC, the dominant SUSY process is expected to be the production of pairs of squarks or gluinos, because these are strongly-interacting particles with QCD-type cross-sections. For instance, a sample of about 10^4 $\tilde{q}\tilde{q}$, $\tilde{g}\tilde{g}$ and $\tilde{q}\tilde{g}$ events should be produced over only one year of data taking at $L = 10^{33}$ cm^{-2} s^{-1} if squarks and gluinos have masses of ~1 TeV. Because these sparticles weigh at least 200-300 GeV, given the present Tevatron limits, they are expected to decay through long chains with several intermediate steps, and hence should give rise to very busy final states containing in general several jets, leptons and missing transverse energy. Squark and gluino masses of 1 TeV are accessible after only one month of data taking at $L = 10^{33}$ cm^{-2} s^{-1}, once the backgrounds (e.g. $t\bar{t}$ production, mismeasured QCD multijet events) and the detector performance (in particular tails in the calorimeter response to jets) will have been well understood. The ultimate mass reach is up to ~3 TeV for squarks and gluinos. As a consequence, SUSY discovery at the LHC could be relatively easy and fast, and if nothing is found at the LHC, TeV-scale Supersymmetry will most likely be ruled out, because of the arguments related to stabilizing the Higgs mass mentioned above. On the other hand if SUSY is there, ATLAS should go beyond the mere discovery phase, being able to perform several precise measurements of the sparticle masses, and thus determining the fundamental parameters of the theory with a precision of ~10% or better in many cases, at least in minimal models like mSUGRA [10]. The correct identification of the underlying theory and the measurements of its fundamental parameters will not be easy for general models with many unknown parameters. It is however hoped that, by exploiting the expected richness of the data with a lot of different measurements (masses, cross-sections, decay modes, etc.) and observations (e.g. excess of events with b-quarks or taus), and with some guidance from theory, it will eventually be possible to narrow the *a priori* large spectrum of models and pin down the correct framework. A rich phenomenology is also expected from the SUSY Higgs sector, which consists of five bosons, three neutral (h, H, A) and two charged $(H^{\pm})$. The mass of the lightest one, h, is predicted to be below 135 GeV, whereas the others are expected to be heavier and essentially mass-degenerate over most of the parameter space. The Higgs sector of the Minimal Supersimmetric Standard Model (MSSM) can be described in terms of the mass of the A boson m_A and of the parameter tanβ (the ratio of the vacuum expectation values of the two Higgs doublets which give rise to the five physical states). Figure 1 b shows the regions of this parameter space where the various Higgs bosons can be discovered at the LHC through their decays into SM particles. It can be seen that over a large fraction of the parameter space two or more Higgs bosons should be observed. The only exception is the region at large m_A and moderate tanβ, where only h can be discovered at the LHC, unless the heavier Higgs bosons have observable decays into SUSY particles. The LHC may therefore miss part of the SUSY Higgs spectrum. Many other examples of physics beyond the SM have been studied by ATLAS: theories with Extra-dimensions, Little Higgs models, Technicolour, Compositeness, etc, to exploit all the relevant expected topology (in particular at the trigger level) and to address as many signatures as possible. The LHC discovery potential for some scenarios beyond the SM is illustrated in Table 1. In most cases the direct reach extends well beyond the 1 TeV "threshold". More details can be found in Refs. [6, 11].

Table 1. LHC discovery potential for some scenarios beyond the Standard Model.

Scenario/channel	Reach
$Z' \to \ell\ell$	m$\sim$5 TeV
$W' \to \ell\nu$	m$\sim$6 TeV
Leptoquarks	m$\sim$1.5 TeV
Compositeness	Compositeness scale $\Lambda \sim 40$ TeV
Excited quarks	m$\sim$6.5 TeV
Extra-dimensions	Gravity scale $M \sim 9$ TeV for 2 extra-dimensions
Monopoles	m$\sim$20 TeV

4 Conclusions

In about three years from now the LHC will start operation, and CERN and experimental particle physics will enter a new epoch. Given the compelling motivations for new physics at the TeV scale, one can anticipate a profusion of exciting results from a machine able to explore this scale in detail, with a direct discovery potential up to particle masses of $\sim$5-6 TeV. As a consequence, the LHC should provide definitive answers about the SM Higgs mechanism, Supersymmetry, and several other TeV-scale predictions that have resisted experimental verification for decades.

References

[1] G. Altarelli and M. Grunewald, *Precision electroweak studies at LEP*, this issue.

[2] Y. Fukuda et al., Phys. Rev. Lett. B81 (1998) 1562.

[3] Q.R. et al., Phys. Rev. Lett. 87 (2001) 071301.

[4] J. Ellis and M. Jacob, *Phenomenology of today and tomorrow*, this issue.

[5] D. Treille, *Searches for new particles at LEP*, this issue.

[6] ATLAS Collaboration, Detector and Physics Performance Technical Design Report, CERN/LHCC/99-15, 1999.

[7] *Large Hadron Collider in the LEP tunnel*, Proceedings of the ECFA-CERN Workshop, ECFA 84/85, CERN 84-10.

[8] *Proceedings of the Workshop on Standard Model Physics (and more) at the LHC*, edited by G. Altarelli and M. Mangano, CERN 2000-004.

[9] For a phenomenological review see for instance P. Fayet and S. Ferrara, Phys. Rep. C32 (1977) 249; H.P. Nilles, Phys. Rep. C110 (1984) 1; H.E. Haber and G.L. Kane, Phys. Rep. C117 (1985) 75.

[10] A.H. Chamseddine, R. Arnowitt and P. Nath, Phys. Rev. Lett. 49 (1982) 970.

[11] B.C. Allanach et al., *Les Houches Physics at TeV Colliders 2003: Beyond the Standard Model Working Group Summary Report*, hep-ph/0402295.

A GLOBAL OPTIMIZATION ALGORITHM FOR FINITE DENSITY QUARK MATTER

RAFFAELE BUFFA [a]

[a] *Dipartimento di Fisica, Università di Udine, via delle Scienze 208, 33100 Udine, Italy*

Abstract

Using a parallel genetic algorithm we study a $SU(3)_f \times SU(3)_c$ Nambu Jona-Lasinio model at finite density and zero temperature within the context of normal and superconducting quark matter. We study the most stable charged configuration in normal quark matter with the use of the Powell constraint techniques and we obtain the phase diagram of superconducting quark matter in the μ, μ_Q plane for different values of the t'Hooft interaction. At the superconducting phase transitions we embed the genetic algorithm in a branch & bound algorithm.

1 Introduction

Stability of compact astrophysical objects and heavy ion collisions are controlled by the behavior of finite density and low temperature hadronic matter. At these densities QCD must be treated in a nonperturbative way. We choose to carry on our analysis by mean field approximation applied to a Nambu Jona-Lasinio (NJL) model. The mean ingredient in the determination of the ground state of a zero temperature field theory is the detection of the global minimum of the effective potential or of the energy. To this aim we adopt a parallel elitist genetic algorithm (for a review about genetic algorithms we send to [1]). We have tried different kinds of parallelization and we have found it convenient to adopt the evolution of parallel populations interacting by the crossing of parents of different populations. This new crossing is able to give richer chromosome variety to the individuals.

B.G. Sidharth, F. Honsell and A. de Angelis. (eds.), Frontiers of Fundamental Physics, 273–277.

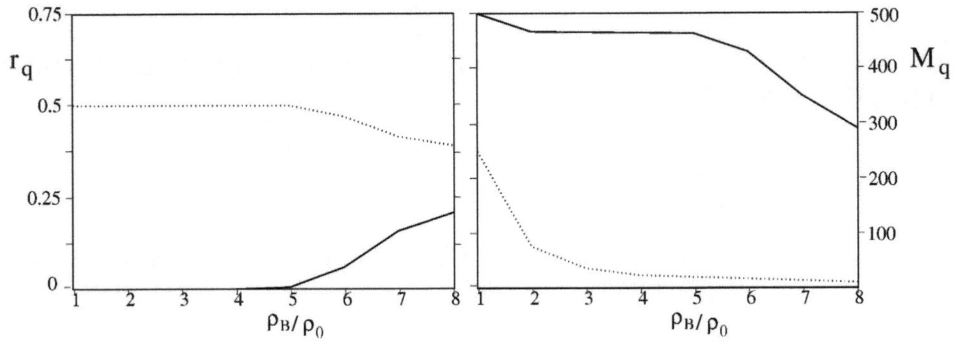

Figure 1. Normal quark matter; fig.1.a left: the fractions $r_s = \rho_s/\rho_B$ (solid line) and $r_u = \rho_u/\rho_B = r_d = \rho_d/\rho_B$ (dotted line); fig.1.b right: the constituent masses (Mev) M_s (solid line) and $M_u = M_d$ (dotted line)

A particular attention must be payed to the proximity of superconducting phase transitions. The genetic algorithm may lose the true minimum of the effective potential. The trivial solution is to increase the number of individuals of the populations and/or the number of generations. We have found it more convenient to embed the genetic algorithm in a simple branch & bound algorithm. The relaxation and the value of the objective function are given by the genetic algorithm applied to the effective potential. Concerning the branching we have divided the variables' domain in distinct regions corresponding to different phases. We have tested the efficiency of our approach with respect to results taken from the literature [2], [3] and we have found good agreement except for what concerns fig.5.6 of [2] at zero temperature and for the parameters set I. In the second section we address the problem of finding the most stable charged solution for $SU(3)_f \times SU(3)_c$ normal quark matter at zero temperature and finite density. The third section is devoted to the determination of the phase diagram in the μ, μ_Q plane at different values of the t'Hooft interaction for $SU(3)_f \times SU(3)_c$ superconducting quark matter at zero temperature (μ is the chemical potential relative to the total quark number density and μ_Q is the chemical potential relative to the electric charge density). The two problems are non linear and non convex. In the last section we report the conclusions. The analysis has been carried on on a cluster composed of three nodes.

2 Normal quark matter

In this section we consider a NJL Lagrangian without superconducting interactions:

$$L = \bar{\Psi}(i\partial\!\!\!/ - m_{bare})\Psi + G_S\left((\bar{\Psi}\lambda_a\Psi)^2 + (\bar{\Psi}i\gamma_5\lambda_a\Psi)^2\right) \\ - K\ \left(Det_f(\bar{\Psi}(1+\gamma_5)\Psi) + Det_f(\bar{\Psi}(1-\gamma_5)\Psi)\right) \qquad (1)$$

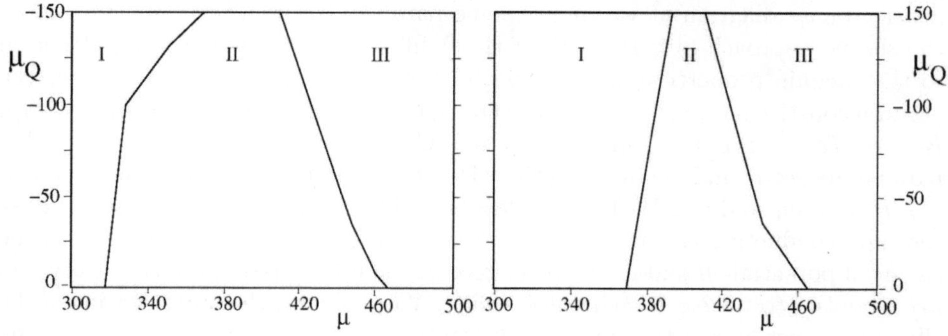

Figure 2. Superconducting quark matter; fig.2.a left: the phase diagram in μ (Mev), μ_Q (Mev) plane for $K = 0\ Mev^{-5}$; fig2.b right: the phase diagram in μ (Mev), μ_Q (Mev) plane for $K = 7.795\ 10^{-14}\ Mev^{-5}$. Region I corresponds to $D_2 = D_5 = D_7 = 0$, region II corresponds to $D_2 \neq 0, D_5 = D_7 = 0$, region III corresponds to $D_2 \simeq D_5 \simeq D_7 \neq 0$

where summation over repeated indices is assumed, λ_a are the SU(3) flavor matrices and Det_f is the determinant in flavor space. Our aim is to determine the zero temperature ground state of the theory in the bulk without leptons and without imposing constraints over the charge and the β-equilibrium. The freedom in charge and β-equilibrium makes necessary to minimize the energy with respect to the chiral condensates ϕ_u, ϕ_d, ϕ_s and the densities ρ_u, ρ_d, ρ_s of up, down and strange quarks maintaining fixed the total density at the desired value. To fix the total density we use a simple repair algorithm in the generation of the initial populations and a Powell [1] constraint method during the evolution of the populations. We have chosen the following set of parameters: $\Lambda = 602.3$, $G_S = 5.0575\ 10^{-06}\ Mev^{-2}$, $K = 1.5590\ 10^{-13}\ Mev^{-5}$, $m_u = 5.5\ Mev$, $m_d = 5.5\ Mev$, $m_s = 140.7\ Mev$ where Λ is a sharp cut-off in momentum space. We send to [4] for the derivation of the energy of the system in mean-field approximation. We report in fig.1.a and fig.1.b respectively the fractions ρ_q/ρ_B of the densities and the constituent masses M_q of up, down and strange quarks with respect to ρ_B/ρ_0 (ρ_B is the total density and ρ_0 is the nuclear density). The most stable solution is positively charged.

3 Superconducting quark matter

In this section we consider a NJL model with superconducting interactions. The total Lagrangian is given by adding L_{qq} to eq.1:

$$L_{qq} = H(\bar{\Psi} i\gamma_5 \lambda_a \lambda_A C \bar{\Psi}^T)(\Psi^T C i\gamma_5 \lambda_a \lambda_A \Psi)$$

where summation over repeated indices is assumed, λ_A are the SU(3) color matrices and C is the charge conjugation operator. NJL is a low energy effective theory of QCD. There is no reason to assume that the parameters of the NJL at finite density must reproduce the vacuum quark masses. We would aspect a density dependence of the parameters. Therefore, it is of some interest to study the finite density phase diagram at varying of some parameters. In this section we choose the simple approach of varying the t'Hooft interaction without paying attention to the vacuum properties: we start with a set of parameters that reproduce the vacuum constituent quark masses $\Lambda = 602.3$, $G_S = 6.3926\, 10^{-06}\, Mev^{-2}$, $H = G_S$, $K = 0\ Mev^{-5}$, $m_u = 5.5\ Mev$, $m_d = 5.5\ Mev$, $m_s = 112\ Mev$ (taken from [2], parameters set I) and we increase the t'Hooft interaction K maintaining fixed Λ, G_S, H, m_u, m_d and m_s. We fix the color chemical potentials to zero and we analyze the superconducting zero temperature phase diagram in the bulk at varying of the chemical potentials μ and μ_Q (μ is relative to the total quark number density and μ_Q is relative to the electric charge density). We deal with the minimization of the effective potential over the chiral condensates ϕ_u, ϕ_d, ϕ_s and the superconducting condensates D_2, D_5, D_7. The evaluation of the effective potential has been carried out without approximations other than the mean field approximation in the same spirit of [5]. We report in fig.2.a and fig.2.b respectively the phase diagrams for $K = 0\ Mev^{-5}$ and for $K = 7.795\ 10^{-14}\ Mev^{-5}$. We can see: i) the region of superconductivity is restricted at increasing of the t'Hooft interaction; ii) the line separating the two phases relative to 2SC ($D_2 \neq 0,\ D_5 = D_7 = 0$) and to CFL ($D_2 \simeq D_5 \simeq D_7 \neq 0$) is practically unaffected by the change of the t'Hooft interaction.

4 Conclusions

We have applied a parallel genetic algorithm to two non convex problems: i) the determination of the most stable charged solution of zero temperature normal quark matter in the bulk; ii) the determination of the phase diagram in the μ, μ_Q plane of zero temperature superconducting quark matter in the bulk for different values of the t'Hooft interaction. Concerning the first application we have obtained a positively charged solution. The result must be taken with care because we have neglected leptons degrees of freedom and β-equilibrium. Our result could be useful for mixed phase studies [6]. The second application is much more time consuming and the parallelization becomes really useful. With our choices of the parameters we have found superconducting gaps of the order of 200 Mev. At the superconducting phase transitions we have improved the genetic algorithm by a branch & bound algorithm.

References

[1] Z. Michalewicz, "Genetic Algorithms + Data Structures = Evolution Programs", Springer-Verlag, 1996; D. Powell, M.M. Skolnick, Proceedings of the Fifth International Conference on Genetic Algorithms, Morgan Kaufmann, San Matteo, CA, (1993), 424

[2] M. Buballa, hep-ph/0402234

[3] M. Kitazawa, T. Koide, T. Kunihiro, Prog. Theor. Phys. 108, (2002), 929

[4] I.N. Mishustin, L.M. Satarov, H. Stocker, W. Greiner, Phys. Rev. C59, (1999), 3343

[5] M. Buballa, M. Oertel, Nucl. Phys. A703, (2002), 770; A.W. Steiner, S. Reddy, M. Prakash, Phys. Rev. D66,(2002), 094007

[6] N.K. Glendenning, Phys. Rev. D46, (1992), 1274

MULTIPHOTON APPROACH ON PAIR PRODUCTION UNDER THE LIGHT OF RECENT EXPERIMENTAL AND THEORETICAL INVESTIGATIONS

C. KABERIDIS,[a] I. TSOHANTJIS [a] AND S. MOUSTAIZIS [a]

[a] *Technical University of Crete, Institute of Matter Structure and Laser Physics, Chania 73100, Crete, Greece*

Abstract

We explore the range of applicability of the imaginary time approach and the two level on resonance approach of e^+e^- pair production from vacuum both theoretically and on the E-144 experiment. We show that the resonance approximation is more efficient by many orders of magnitude.

1 Introduction

One of the most intriguing non-linear phenomenon in QED is particle-antiparticle pair production from vacuum in the presence of strong elm fields[1]. Schwinger [2] implementing the proper time method obtained conditions under which pair production is possible: $\mathcal{F} = \frac{1}{4}F_{\mu\nu}F^{\mu\nu}$, $\mathcal{G} = \frac{1}{4}F_{\mu\nu}\tilde{F}^{\mu\nu}$, where $F_{\mu\nu}$ and $\tilde{F}_{\mu\nu}$ are the elm field tensor and its dual respectively, must be such that such that neither $\mathcal{F} = 0$, $\mathcal{G} = 0$ nor $\mathcal{F} > 0$, $\mathcal{G} = 0$. For the case of a uniform, constant electric field he obtained a probability density $w \sim \frac{1}{n^2}e^{-\frac{m^2\pi n}{eE}}$. Brezin and Itzykson [3] examined the case of pair creation in the presence of a pure oscillating electric field E by applying a version of WKB approximation and treating the problem in an analogous way as in the ionization of atoms(where the three basic mechanisms multiphoton, tunnelling and over the barrier ionization are present), considering the pairs as bound in vacuum with binding energy $2mc^2$. With the *critical value* of electric field strength E_c being $E_c = \frac{mc^2}{e\lambda_c} \simeq 1.3 \times 10^{18} V/m$ and under the

B.G. Sidharth, F. Honsell and A. de Angelis. (eds.), Frontiers of Fundamental Physics, 279–283.

conditions $E \ll E_c$, $\hbar\omega \ll mc^2$, they obtained a probability per 4-Compton volume of e^+e^- pair creation given by $w_{BI} = \frac{e^2E^2}{\pi\hbar c}\frac{1}{g(\gamma)+\frac{\gamma g'(\gamma)}{2}}\exp\left(-\frac{\pi m^2}{eE}g(\gamma)\right)$, where $g(\gamma) = \frac{4}{\pi}\int_0^1 \left(\frac{1-y^2}{1+\gamma^2y^2}\right)^{\frac{1}{2}} dy$. The parameter $\gamma = \frac{mc\omega}{eE}$ is the equivalent of the Keldysh parameter in the ionization of atoms and the above formula interpolates between the regimes $\gamma \ll 1$ (adiabatic non-perturbative tunneling mechanism regime) and $\gamma \gg 1$ (multiphoton mechanism regime) . Popov [4] using the imaginary time method obtained more accurate formulas for pair production taking also in to account interference effects. The condition $\mathcal{F}$ <0 is achieved by using two oppositely propagating laser beams. The probability over a 4-Compton volume w is obtained as $w = \sum_{n>n_0} w_n$ where $w = \frac{m^4}{2^{\frac{3}{2}}\pi^4}\left(\frac{E}{E_c}\right)^{\frac{5}{2}} e^{-\pi\frac{E_c}{E}\ 1-\frac{\gamma^2}{8}+O(\gamma^4)}$ for $\gamma \ll 1$. For $\gamma \gg 1$

$$w_n = \frac{2}{\pi^3}m^4\left(n_0\right)^{-\frac{5}{2}}\left(\frac{e}{4\gamma}\right)^{2n} q\left(n-n_0\right), \text{ and } w \approx \frac{2m^4}{\pi^3}\left(n_0\right)^{-\frac{5}{2}}\left(\frac{e}{4\gamma}\right)^{2n_0}, \quad (1)$$

where $q\left(n-n_0\right) = \frac{1}{2}e^{-2(n-n_0)}\int_0^{2(n-n_0)} e^t t^{-\frac{1}{2}}dt$ and $n_0 = 2mc^2/\hbar\omega$ is the threshold number of photons. The number of e^-e^+pairs created can be obtained by multiplying the above relations by the 4-volume $\lambda^3\tau = \lambda^4/c$ where τ is the pulse duration and λ is the elm wavelength. Specially for the $\gamma \gg 1$ regime

$$N(\tau) \approx 2\pi n_0^{\frac{3}{2}}\left(\frac{\omega}{m}\right)\left(\frac{4\gamma}{e}\right)^{-2n_0} N', \quad (2)$$

where $N' = \omega\tau/2\pi$ is the number of oscillations of the electric field in a laser pulse.

In a recent paper Avetissian et al [5] treated the problem of e^+e^- production in a standing wave of oppositely directed laser beams using a two level multiphoton on resonance approximation. Pairs are produced close to antinodes and in spacial dimensions $l \ll \lambda$. Transitions can be thought to occur between two energy levels from $-\mathcal{E}$ to $\mathcal{E}$ by the absorption of n photons and the multiphoton probabilities will have maximum values for resonant transitions $2\mathcal{E} = n\omega$. The probability of n-photon e^+e^- production process with certain energy $\mathcal{E}$, summing over the spin states, obtained as $W_n = \frac{2f_n^2}{\Omega_n^2}\sin^2\left(\Omega_n^2\tau\right)$, where $f_n = \frac{\mathcal{E}}{2p_y}\left(1-\frac{p^2\cos^2\theta}{\mathcal{E}^2}\right)^{\frac{1}{2}} n\omega J_n(4\xi\frac{mp_y}{\mathcal{E}\omega})$, θ is the angle between the momentum of produced $e^-(e^+)$ and the amplitude of the electric field, the parameter $\xi = \frac{1}{\gamma}$, $\Omega_n = \sqrt{f_n^2+\frac{\Delta_n^2}{4}}$ is the 'Rabi frequency' , τ is the interaction time and $\Delta_n = 2\mathcal{E} - n\omega$ is the detuning of resonance. With the usual condition $\Omega_n \ll \omega$, field intensities are such that $\xi \lesssim 1$ (i.e. $\gamma > 1$). The function f_n (Rabi frequency on exact resonance) and correspondingly the probabilities obtained maximize when $n = N = 2\sqrt{2}m/\omega$. Then the angular distribution of a n-photon probability density on exact resonance is given by $\frac{dw_n}{do} = \frac{n^3\omega^3}{64\pi^2}\frac{n^2\omega^2\sin^2\theta+4m^2\cos^2\theta}{\left(n^2\omega^2-4m^2\right)^{\frac{1}{2}}\cos^2\theta}J_n^2\left(4\xi\frac{m\left(n^2\omega^2-4m^2\right)^{\frac{1}{2}}\cos\theta}{n\omega^2}\right)$ where

$do = \sin\theta d\theta d\varphi$. $\frac{dw}{do} = \sum_{n=n_0} \frac{dw_n}{do}$ and integrating over the solid angle, the total probability density can be found in [5]. We shall use only $\frac{dw_n}{do}$ at angle $\theta = 0$ and for $n = N$ as this is quite adequate to give an estimate of the power of this approach compared to others.

2 Comparison of the imaginary time and two level resonance approximation

The relation for $\frac{dw_n}{do}$ at angle $\theta = 0$ and for $n = N$ becomes

$$\frac{dw_N}{do}\mid_{\theta=0} = \frac{1}{\sqrt{2}}\frac{m^4}{\pi^2} J_N^2(\xi N) \tag{3}$$

Numerical simulations can be performed for the common Nd laser [5] with $\omega = 1.7eV$, $\lambda = 1.074\mu m$, $I = 1.35W/m^2$ and $\xi = 0.9995$. As $N \sim 10^6$, $J_N(\xi N) \sim \frac{\exp(N \tanh a - Na)}{\sqrt{2\pi N \in \tanh a}}$, where $a = \text{sech}^{-1}\xi$, giving $\frac{dw_N}{do}\mid_{\theta=0} = 1.9 \times 10^{40} m^{-3} s^{-1}$. To meet the conditions for two level resonance approximations a 4-volume $V = d^2 l\tau = (10\lambda)^2(0.1\lambda)\, 10fs$ can be chosen (i.e maximum laser intensity close to the diffraction limit, where the radius of focus is $d \sim 10\lambda$ and duration of the pulses $\tau = 10fs$), leading to $10^8 - 10^9$ number of pairs per laser pulse. On the other hand if we use relation (2) for the same laser characteristics as above we obtain that in the above four volume the number of pairs $N(\tau)$ predicted is approximately $10^{-216546}$ per laser shot, which suggests that the multiphoton on resonance approximation is by far more efficient. Moreover requiring the same number of pairs, for $\omega = 1.7eV$, the imaginary time method would require very high E of the order of $10^{17} V/m$ while being in the regime $\gamma < 1$. The above analysis shows that implementing the resonance approximation, one does not need to use laser such as XFEL [6] or the laser of NIF (National Ignition Facility) [10] in order to observe e^+e^- pair creation as present lasers technology is adequate.

3 Application of the imaginary time and the multiphoton resonance approximations on E-144 experiment.

The first experimental verification of e^-, e^+ pair production took place in SLAC and it is known as E-144 experiment [7]. In the first stage non linear Compton scattering [9], of highly energetic electrons (maximum 46.6GeV) with n laser photons (527nm, $\omega = 2.35eV$ and laser pulses of nominal energy 500mJ) , $e + n\omega \to e' + \omega'$, produces backscattered photons ω'(27-30Gev) which in the second stage interact via the multiphoton Breit -Wheeler [8] mechanism with the laser photons, $\omega' + n\omega \to e^- \, e^+$,to produce e^-, e^+ pairs . The scattering occurs at angle $a = 17°$ between the initial electron and the laser beam (and thus for pair creation $\mathcal{F} < 0$ holds). The minimum number of photons required for pair production is given by $n_0 = \frac{2m^2(1+\eta^2)}{\omega\omega'(1+\beta\cos a)}$ where $\eta = \xi = \gamma^{-1}$. The number of positrons measured in

21962 laser pulses was 175 ± 13 for a $n = 5.1 \pm 0.2(statistical)^{+0.5}_{-0.8}(systematic)$ multiphoton order process, in very good agreement with the theory. We shall examine whether the result of E-144 experiment can be in agreement with the two methods mentioned above. In E-144 experiment $\omega' \sim 30GeV$ in the laboratory frame. Changing to the electron's frame in the beam, both the laser photons and the backscattered ones have the same energy $\omega = 2.35 \times \gamma_L eV$ where $\gamma_L = \frac{46GeV}{0.511MeV} \sim 10^5$ is the Lorenz factor, $\lambda = 5.78 \times 10^{-12}m$ and period of the field $T = 1.9 \times 10^{-20}s$. Similarly the electric field $E_L = \gamma_L\sqrt{377 I_{laser}}$ $\sim 3 \times 10^{17}V/m$ (with maximum laser intensity at focus $I_{laser} \sim 1.35 \times 10^{22}W/m^2$), very close to E_c for pair production. Thus the E-144 experiment, in the electrons' frame, can be viewed as an e^+e^- production from vacum in the presence of strong field. Applying the imaginary time approach with relativistic invariant $\gamma = 2.7$, the probability density w_n of (1) (with $n_0 = 4.77$) has a pick when $q(n - n_0)$ maximizes and this happens for $n \simeq 5$, thus verifying that the multiphoton process is a 5th order one . From (1) $w_5 \sim 9 \times 10^{48}m^{-3}s^{-1}$, which for pulse duration $\tau = 1.9 \times 10^{-18}s$ in the 4-volume $V\tau = \lambda^3\tau$ and for 21962 shots, we obtain ~ 73 pairs. Moreover from $N(\tau)$ of (2) (where the choice $\tau \sim 10^{-18}$ corresponds to $N' \sim 10^2$) we can obtain an estimate for $N(\tau)$ to be 155 to 182 pairs, in very good agreement with E-144 experiment .

Applying the resonance approximation on the E-144 experiment (electron's frame) and using (3) we first note that $N = \frac{2\sqrt{2}m}{\omega}$ must be taken to be the closest integer to the ratio $\frac{2\sqrt{2}m}{\omega}$. For $\omega = 235keV$, $N \sim 6$ which suggests that with a fine tuning of ω i.e fine tuning of the electrons' energy in the electron beam, we can achieve exact resonance with $N = 6$. Thus using $\frac{2\sqrt{2}m}{\omega} = 6$, we find that the backscattered photons on the electrons frame will have energy $240.88keV$ ($\lambda = 5.15 \times 10^{-12}$ and period $T = 1.7 \times 10^{-20}s$). Then with $\xi = \frac{1}{2.7} = 0.37$, $n = 6$ we find from (3) $\frac{dw_n}{do}|_{\theta-0} = 4.56 \times 10^{51} pairs/m^3s^1$ and in the volume $V\tau = (10\lambda)^2(0.1\lambda)10^{-18}\ m^3s^1$and for 21962 shots we would have observed on exact resonance, 1.37×10^5 pairs. Thus had we the opportunity to calibrate the energy of the electrons in the electron beam of SLAC experiment in order to be on exact resonance for a 6th order multiphoton process, we should have observed many orders of magnitude more pairs per laser shot than it was observed.

References

[1] W. Greiner, B. Muller, J. Rafelski, 'Quantum Electrodynamics of Strong Fields', Springer Verlag, Berlin; E. S. Fradkin, D. M. Gitman and Sh. M. Shvartsman, 'Quantum Electrodynamics with unstable vacuum' Springer-Verlag, Berlin, 1991;

[2] J. W. Schwinger, Phys. Rev., 82, 664 (1951);

[3] E. Brezin and C. Itzykson, Phys. Rev. D, 2, 1191 (1970);

[4] V.S. Popov, JETP Lett. 13 (1971) 185; Sov. Phys. JETP 34 (1972) 709;Sov. Phys. JETP 35 (1972) 659; V.S. Popov and M. S. Marinov, Sov.J. Nucl. Phys.16 (1973) 449; JETP Lett. 18 (1974) 255;Sov. J. Nucl. Phys.,19, (1974) 584;V.S. Popov, Phys. Let. A, 298, 83 (2002); Laser Phys. 10(2000) 1033;

[5] H. K. Avetissian et. al., Phys. Rev. E, 66, 016 502 (2002);

[6] A. Ringwald, Phys. Let. B, 510, 107 (2001);

[7] D.L. Burke et. al., Phys. Rev. Let., 79, 1626 (1997);

[8] G. Breit and J.A. Wheeler, Phys. Rev. 46,1087 (1934);

[9] A.I. Nikishov and V.I. Ritus, Sov. Phys. JETP, 19, 529 (1964) ; 19, 1191 (1964) ; 20, 757 (1965) ; 25, 1135 (1967);N.B. Narozhny et al., Sov. Phys. JETP 20, 622 (1965)

[10] Science of High Energy Lasers : 'From Today to the NIF', R. Lee, R. Petrasso, R. Falcone

NIELSEN IDENTITY, WILSON LINE AND CONSTRAINED EFFECTIVE ACTION: THE HIGH TEMPERATURE STANDARD MODEL

RAFFAELE BUFFA [a]

[a] *Dipartimento di Fisica, Università di Udine, via delle Scienze 208, 33100 Udine, Italy*

Abstract

Using Nielsen Identities we study the high temperature Electroweak Standard Model. We show the gauge fixing parameter independence at the first non trivial order of the Fukuda effective action constrained over the phase of the Polyakov loop.

1 Introduction

Nielsen identities originate from enlarged BRST symmetry [1]. They provide a good framework to study problems related to gauge dependence. Moreover they don't depend on the topology of space time. For this reason they are suitable for studying finite temperature field theory. Thermal gluodynamics break spontaneously the centergroup symmetry developing a A_0 condensation, primarily due to the breaking of the Lorentz invariance. The lack of gauge invariance for the effective potential at 2 loops order and for the effective action for a slowly varying background at 1 loop is well known [2],[3]. Moreover the centergroup symmetry is not explicit. This problem has been solved by Belyev [2] by the so called renormalization of the Polyakov loop, the trace of the thermal Wilson line: $\Omega(x) = tr\ P\ exp\left(i \int_0^{1/T} d\tau\ A_0(x,\tau)\right)$, where P stands for the path ordering. It is clearly a gauge invariant operator. Adding dynamical fermions and scalars in the fundamental representation lifts the centergroup degeneracy, the Z(N) non-trivial

B.G. Sidharth, F. Honsell and A. de Angelis. (eds.), Frontiers of Fundamental Physics, 285–289.

vacua become metastable with interesting phenomenological consequences [4]. In this presentation we show the link between Nielsen identity and the Fukuda effective action [5] constrained over the phase of the Polyakov loop. As a concrete example we treat the Electroweak Standard Model at temperatures well above the phase transition where we can ignore the bare masses of fermions and the electroweak symmetry is restored. We show for the Standard Model case how the gauge invariance of the off shell Fukuda effective action comes from the perturbative arrangement of new terms coming from the non linearity of the constraint. Moreover we show a very convenient way to extract the gauge dependence at 1 loop for a slowing varying background without the occurrence of the steepest descendant. We work with the imaginary time formalism and zeta function regularization.

2 Nielsen identity and 1PI effective action

In this section we are interested in the gauge dependence of the perturbative 1PI effective action. Within the context of high temperature SM A_0 and B_0 condense [4]. Thus we split the fields into background plus small quantum fluctuations:$A_\mu = \delta_{\mu 0}\tilde{a} + gA_q$, $B_\mu = \delta_{\mu 0}\tilde{b} + g'B_q$ where A_μ and B_μ are respectively the SU(2) and U(1) gauge fields. Other fields have only the quantum part. For the sake of simplicity we choose $\tilde{a}$ a time independent diagonal matrix in the Lie algebra of SU(2). We need to fix the gauge, we have chosen to use background gauge fixing: $S_{g.f.} = \frac{g^2}{2\xi}\, tr\left(D^{\tilde{a}}_\mu A^\mu_q\right)^2 + \frac{g'^2}{\xi'}\left(\partial_\mu B^\mu_q\right)^2$ where $D^{\tilde{a}}_\mu = \partial_\mu + \delta_{\mu 0}[\tilde{a}, .]$

In this context the Nielsen identity takes the form [1]:

$$\int_x \int_{x'} \left(\frac{\delta \Gamma}{\delta A_0(x')} \frac{\delta \Gamma_{O^{\tilde{a}}(x)}}{\delta k_{A_0}(x')} + \frac{\delta \Gamma}{\delta B_0(x')} \frac{\delta \Gamma_{O^{\tilde{b}}(x)}}{\delta k_{B_0}(x')} \right) = -\xi \partial_\xi \Gamma \tag{1}$$

where Γ is the 1PI effective action, Γ_O is the 1PI effective action with a one O insertion ($O^{\tilde{a}} = \frac{1}{2}\bar{\eta}_A D^{\tilde{a}}_\mu A^\mu_q$, $O^{\tilde{b}} = \frac{1}{2}\bar{\eta}_B \partial_\mu B^\mu_q$), k_{A_0} is the source of sA_{q0} (which is the BRST transformation of A_{q0}) and k_{B_0} is the source of the BRST transformation of B_{q0}. We consider $\tilde{b}$ constant in time, so the second term on the left hand side is zero. By the use of identity (1) we see that the gauge dependence at the order of 1 loop is factorized in a classic term plus a 1 loop term. For a constant background field the classical action does not depend on $\tilde{a}$. On the other side $< O^{\tilde{a}} sA_{q0} >$ at the classical level is zero. Thus the 1 loop effective action is independent from gauge fixing parameter for constant background. At 1 loop order the effective potential depends on the background and the derivative is no longer zero. Again using eq.(1) we obtain:

$$\xi \partial_\xi V^{2loop} = g^2 \frac{\delta V^{1loop}}{\delta A_0} < \bar{\eta} D^{\tilde{a}}_\mu A^\mu_q D_0 \eta >^{1loop} \tag{2}$$

Other terms are zero following the same arguments followed in the 1 loop effective potential gauge fixing independence.

For a slowly varying $\tilde{a}$ we consider the spatial derivative of the background as a new scale. The effective action already has a non trivial gauge dependence at 1 loop order in the coupling constant and at second order in $\partial_i \tilde{a}$. For a slowing varying background the classical action has a kinetic term different from zero of the order $\partial_i^2 \tilde{a}$ so that we can use the powerful technique of Nielsen identity to extract $\partial_i^2 \tilde{a}$ and treat the rest of the calculation for a constant background. The Nielsen identity takes the form:

$$\xi \partial_\xi \Gamma^{1loop} = \frac{\delta S_{cl}}{\delta \tilde{a}} < \bar{\eta} D_\mu^{\tilde{a}} A_q^\mu D_0 \eta >^{1loop} \tag{3}$$

3 Constrained effective action

The constrained effective action first introduced by Fukuda et al. [5] has been applied to finite temperature SU(N) gluodynamics with N_f fermions for a constant background field, to a pure gauge Z(N) interface and to a pure gauge Z(N) dimensionally reduced interface [3]. In this section we consider the Polyakov loop to be constant and slowly varying in the z direction and we show how the non-linearity of the constraint takes care of the gauge dependence of the 1PI effective action. The effective action constrained over the phases of the Polyakov loop is defined in the following way:

$$\begin{aligned} exp - V\Gamma(a,b) = \int DA\ DB\ D\bar{\psi}\ D\psi\ D\bar{\eta}\ D\eta\ D\Phi\ \delta\left(a - \bar{p}(\tilde{a} + gA_{q0})\right) \\ \delta\left(b - \bar{p}(\tilde{b} + g' B_{q0})\right)\ exp\left(-\frac{1}{g^2}(S_{cl} + S_{g.f.} + S_{ghosts})\right) \end{aligned} \tag{4}$$

where ψ are left and right fermions, η are the ghosts, Φ is the Higgs, $p(.)$ denotes the phase of the Polyakov loop and $\bar{.}$ means the spatial average for a constant Polyakov loop and the transverse spatial average for a Polyakov loop varying in the z direction. The second constraint is linear and does not introduce new terms to the effective action. Its treatment is standard [5], in the following we ignore it.

As in [3] we Fourier transform the δ function: $\delta(t-\tilde{t}) = \int D\Lambda\ exp\left(i\frac{\Lambda}{g^2}(t-\tilde{t})\right)$. We insert this representation of the constraint in eq.(4) and we expand the new field into background plus quantum fluctuation: $\Lambda = \lambda + g\Lambda_q$.

In the constrained effective action formulation the saddle point solution at the thermodynamical limit takes care of the Legendre transformation. We obtain 3 saddle point equations:

$$\begin{aligned} a - \tilde{a} &= 0 \\ i\lambda \bar{p}^{(1)}(\tilde{a}) - gS_{cl}^{\prime(1)}(\tilde{a}) &= 0 \\ S_{cl}^{(1)}(\tilde{a},\tilde{b}) &= 0 \end{aligned}$$

where $\bar{p}^{(1)}(\tilde{a}) = -g$ is the development of $\bar{p}(.)$ at first order with respect to A_{q0}, $S_{cl}^{\prime(1)}(\tilde{a})$ is the linear development of the action with respect to the same quantum degrees of freedom that are linked to $\bar{p}^{(1)}(\tilde{a})$, $S_{cl}^{(1)}(\tilde{a},\tilde{b})$ is what remains of the linear development of the action with respect to quantum fluctuations. The classical action for a constant Polyakov loop does not depend on $\tilde{a}$ and the saddle point equations impose $\lambda = 0$. At one loop order the constraint has no effect other than the Legendre transformation. At 2 loops there is a new vertex coming from the dynamical propagation of Λ_{1q}. It couples the derivative of the 1 loop effective potential to the renormalization of the Polyakov loop (see [3] for more details):

$$\delta V_{constraint}^{2loops} = g^2 \frac{\delta V^{1loop}}{\delta \tilde{a}} \frac{\bar{p}^{(2)}(\tilde{a}) * < A_{q0} A_{q0} >}{(p^{(1)}(\tilde{a}))^2} \tag{5}$$

where $\bar{p}^{(2)}(\tilde{a}) * A_{q0}A_{q0}$ is the development of $\bar{p}(\tilde{a} + gA_{q0})$ at second order in A_{q0}. For a background varying in the z direction the classical action is no longer zero and the saddle point equations give: $\lambda = -i\frac{g}{p^{(1)}}\frac{\delta S_{cl}}{\delta \tilde{a}}$. At 1 loop order λ couples to $p^{(2)}(\tilde{a}) * < A_{q0}A_{q0} >$ and we arrive at the following expression for the constraint contribution to the effective action:

$$\delta\Gamma_{constraint}^{1loop} = -\frac{\delta S_{cl}}{\delta \tilde{a}} \frac{\bar{p}^{(2)}(\tilde{a}) * < A_{q0} A_{q0} >}{g\, p^{(1)}(\tilde{a})} \tag{6}$$

What remains to be done is the explicit calculation of $\bar{p}^{(2)}(\tilde{a}) * < A_{q0}A_{q0} >$ and $< \bar{\eta} D_{\mu}^{\tilde{a}} A_q^{\mu} D_0 \eta >$. The first gives $(3-\xi)(\tilde{a}/2\pi T - 1/2)$ and second $\xi(\tilde{a}/2\pi T - 1/2)$. We have shown the gauge invariance of the constrained effective action.

4 Conclusions

We have shown the connection between the Nielsen identity and the effective action constrained over the phase of the Polyakov loop.

We have shown the gauge invariance of the off-shell constrained effective action in the high temperature Standard Model for both constant and slowly varying background fields.

The constrained path integral has also found application in the study of the Casimir effect [6]. We believe that the work undertaken in this presentation can be applied in this context.

4.1 ACKNOWLEDGMENTS

I would like to thank C.P.Korthals Altes and S.Bronoff for useful discussions about the constrained effective action.

References

[1] N.K. Nielsen, Nucl.Phys. **B101** (1975), 173; I.J.R. Aitchison, C.M. Fraser, Annals of Phys. **156**, (1984), 1; O.M. Del Cima, D.H.T. Franco, O. Piguet, Nucl.Phys. **B551** (1999), 813

[2] V.M. Belyaev, Phys.Lett. **B254** (1991), 153

[3] C.P. Korthals Altes, Nucl.Phys. **420** (1994), 637; T. Bhattacharya, A. Gocksch, C.P. Korthals Altes, R.D. Pisarski, Nucl.Phys. **B383** (1992), 497; S. Bronoff, R. Buffa, C.P. Korthals Altes, hep-ph/9801333

[4] C.P. Korthals Altes, K.M. Lee, R.D. Pisarski, Phys.Rev.Lett. **73** (1994), 1754

[5] R. Fukuda, E. Kyriakopoulos, Nucl.Phys. **B85** (1975), 354

[6] M. Bordag, J. Lindig, Phys.Rev. **D58** (1998), 045003

THE MAGIC EXPERIMENT AND ITS FIRST RESULTS

D. BASTIERI [a], R. BAVIKADI [b], C. BIGONGIARI [a], E. BISESI [b], P. BOINEE [b], A. DE ANGELIS [b], B. DE LOTTO [b], A. FORTI [b], T. LENISA [b], F. LONGO [c], O. MANSUTTI [b], M. MARIOTTI [a], A. MORALEJO [a], D. PASCOLI [a], L. PERUZZO [a], A. SAGGION [a], P. SARTORI [a], V. SCALZOTTO [a] AND THE MAGIC COLLABORATION

[a] *Dipartimento di Fisica dell'Università di Padova and INFN, Italy*
[b] *Dipartimento di Fisica dell'Università di Udine and INFN, Italy*
[c] *Dipartimento di Fisica dell'Università di Trieste and INFN, Italy*

Abstract

With its diameter of 17m, the MAGIC telescope is the largest Cherenkov detector for gamma ray astrophysics. It is sensitive to photons above an energy of 30 GeV. MAGIC started operations in October 2003 and is currently taking data. This report summarizes its main characteristics, its first results and its potential for physics.

1 Introduction

The MAGIC (Major Atmospheric Gamma Imaging Cherenkov) telescope was designed in 1998 [1] with the main goal of being the Imaging Atmospheric Cherenkov Telescope (IACT) with the lowest gamma energy threshold. It is based on the experience acquired with the first generation of Cherenkov telescopes, and includes a large number of technological improvements [2]. With a reflector diameter of 17 m, it is the largest Imaging Cherenkov Telescope in the world.

B.G. Sidharth, F. Honsell and A. de Angelis. (eds.), Frontiers of Fundamental Physics, 291–296.

Figure 1. The MAGIC detector.

The study of γ rays is fundamental for our understanding of the universe [3]: γ rays probe the most energetic phenomena occurring in nature, and several signatures of new physics are associated with the emission of γ rays. Photons can travel essentially undeflected and unabsorbed in space, and thus they point with excellent approximation to the source of their emission.

Together with HESS [4] in Namibia, VERITAS [5] in the USA, and CANGAROO [6] in Australia, MAGIC is one of the "big four" ground-based gamma experiments; among them, it is the only one consisting for the moment of a single telescope.

MAGIC comes after the scientific success of the Energetic Gamma Ray Experiment Telescope (EGRET) instrument on the Compton Gamma Ray Observatory [7]. Launched in 1991, EGRET made the first complete survey of the sky above 30 MeV. EGRET increased the number of identified γ sources producing a catalog which is a reference. Still a large fraction of EGRET sources are unidentified and, besides other fundamental physics goals, MAGIC aims at their identification.

MAGIC is located at the Roque de los Muchachos Observatory (ORM) at 2200 m asl (28.8^o north, 17.9^o west) on the Canary island of La Palma.

MAGIC has an effective area of about $4{\cdot}10^4$ m^2, angular resolution of about 0.2 degrees, relative energy resolution of the order of 20% and can well separate gammas from background (mainly due to hadrons).

The MAGIC detector is shown in Figure 1.

The sensitivity of MAGIC as calculated from Monte Carlo is shown together with the expected sensitivity from other gamma-ray detectors in the GeV-TeV range in Figure 2.

MAGIC has already observed a source, 1ES1959, 10 times fainter than Crab.

By observing gamma rays in the energy range from a few tenth of GeV upwards, in overlap with satellite observations and with a substantial improvement in sensitivity, energy and angular resolution, results in several fields of science can

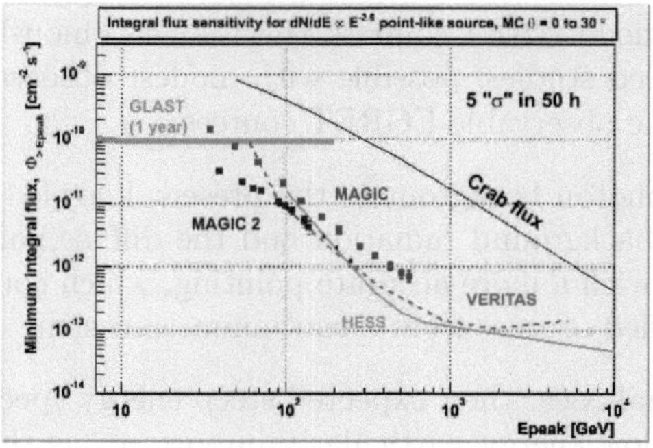

Figure 2. Sensitivities for some operating and proposed gamma detectors.

be achieved:

- Fundamental physics.
  - Gamma Ray Horizon, with a larger sensitivity for tests of light propagation effects.
  - Dark matter, extending the sensitivity to many theoretical models.
  - Quantum gravity: the search for effects using time delays in the arrival of gammas dependent on the energy, will profit of an improved energy resolution.
- Astrophysics.
  - Gamma Ray Bursts (GRBs): several tests of fundamental physics can be performed based on the characteristics of the transient emission. GRBs are among the most distant and powerful sources in the Universe and key informations on space-time structure and cosmology can be gained. In addition a more accurate localization of such sources can help in correlating GRBs with known astrophysical objects.
  - Supernova remnants and plerions: improved energy resolution will help in discriminating between the various acceleration mechanisms assumed to be at the origin of VHE gammas.
  - Active Galactic Nuclei: the precise gamma-ray observation of more AGNs at different redshifts will much contribute to answer one of the major open questions in extragalactic astronomy, the formation and evolution of galaxies in the early universe.
  - Pulsars: the known pulsars have cutoff energies of their pulsed emission in the few GeV range, hence their detection will become possible by lowering the energy threshold.

- Unidentified EGRET sources: this is an enormously rich field of activity for detailed studies, possible with modest observation times on nearly half of the observable EGRET sources.
- Diffuse photon background: the present knowledge of both the extra-galactic background radiation and the diffuse galactic emission would improve with a more accurate pointing, which could help in separating unidentified sources from a continuum emission.
- Nearby galaxies: their expected steep energy spectrum makes observations at low energy particularly important, as they allow enough flux to be detected in the gamma ray domain.
- Studies on the galactic center, where a precise pointing might help in reducing the background on nearby sources.

2 First results

All the results presented here are preliminary since the analysis software is still in its development phase.

In the first months of data taking, we have mainly concentrated on low zenith angle ($<40^\circ$) observations of TeV sources like Crab Nebula, Mrk 421, Mrk 501, 1ES 1426 and 1ES 1959. A roughly equivalent amount of OFF source data have been recorded under the same conditions as the ON data for background substraction.

We have applied standard analysis based on the angle α between the major axis of the ellipsis which gives the best approximation of the energy deposition in the detector and the line joining the tenter of such ellipsis to the center of the camera [9].

We give in what follows a flavor of the results collected during the first semester of 1004 on two sources: Crab Nebula and Markarian 421.

2.1 CRAB NEBULA

The Crab nebula is one of the most studied sources in the sky. It radiates a wide range of electromagnetic radiation from radio to 100 TeV γ rays. The spectrum of this source has been measured in the GeV range by EGRET and at energies above 300 GeV by a number of Cherenkov telescopes. MAGIC can determine for the first time the peak of the emission by inverse Compton, where no experimental data have been published, yet.

A significance of around $\sim 20\sigma$ per hour of data taking is presently reached by MAGIC. The data can be subdivided in ranges of energy to obtain an energy scan in the region of interest. Presently we are sensitive to energies above $\simeq 50$ GeV and below 2 TeV (Figure 3).

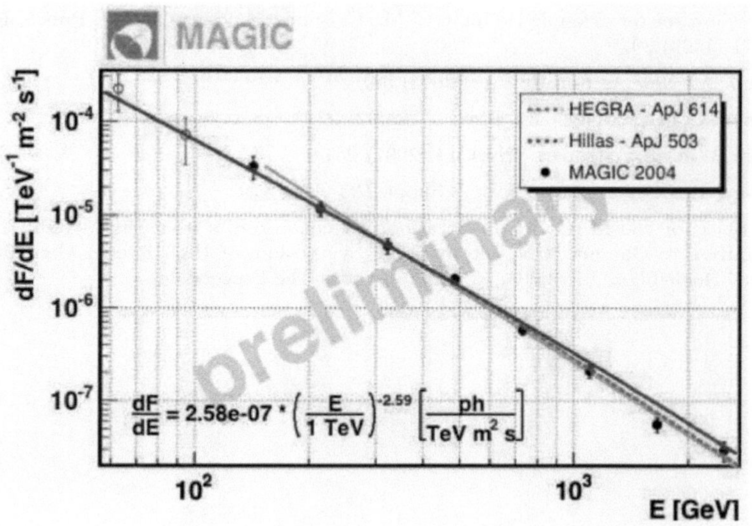

Figure 3. Differential energy spectrum from Crab; data are preliminary. The study of the efficiency on the first two bins is still under study.

2.2 MRK 421

Mrk 421 undergoes the fastest flares that have been observed at the TeV energies, with durations as short as 20 minutes. During 2004 the source has undergone an episode of intense flaring.

The preliminary MAGIC data show a significant signal also at energies as low as 50-60 GeV.

3 Conclusions

MAGIC's preliminary results confirm its instrumental capabilities to scan gamma emissions from the Universe above 30 GeV. MAGIC has demonstrated the observability of sources as faint as 10% of Crab.

The construction of a second detector (MAGIC II) to be put at 80 m from MAGIC has started, and completion is expected in 2006.

References

[1] Barrio J. et al., "The MAGIC Telescope Design report", MPI Institute Report MPI-PhE/98-5 (March 1998)

[2] Mirzoyan R. et al., "Technical Innovations for the MAGIC project", Proc. International Cosmic Ray Conference, Tokyo, 2003

[3] For a review, see for example Hoffman C. M., C. Sinnis, P. Fleury and M. Punch, Rev. Mod. Phys. 71 (1999) 897

[4] Hinton J. A., New Astron. Rev. 48 (2004) 331

[5] Weekes T. C. et al., Astropart. Phys. 17 (2002) 221

[6] Kawachi A. et al., Astropart. Phys. 14 (2001) 261

[7] Hartmann R. C. et al., Astrophys. J. Suppl. 123 (1999) 79

[8] R. K. Bock for the MAGIC Collaboration, "Recent Results from the MAGIC Telescope" Contribution to Gamma 2004, International Symposium of High Energy Gamma-Ray Astronomy, Heidelberg, July 2004, to be published in the Proceedings

[9] Weekes T. C. et al., Astrophys. J. 342 (1989) 379

NEURAL NETWORKS FOR GAMMA-HADRON SEPARATION IN MAGIC

P. BOINEE [a,b], F. BARBARINO [a,b], A. DE ANGELIS [a,b], A. SAGGION [c], M. ZACCHELLO [c]

[a] *Dipartimento di Fisica, Università di Udine, via delle Scienze 208, 33100 Udine, Italy*
[b] *INFN, Sezione di Trieste, Gruppo di Udine, via delle Scienze 208, 33100 Udine, Italy*
[c] *Dipartimento di Fisica, Università di Padova, via Marzolo 8, 35131 Padova, Italy*

Abstract

Neural networks have proved to be versatile and robust for particle separation in many experiments related to particle astrophysics. We apply these techniques to separate gamma rays from hadrons for the MAGIC Čerenkov Telescope. Two types of neural network architectures have been used for the classification task: one is the MultiLayer Perceptron (MLP) based on supervised learning, and the other is the Self-Organising Tree Algorithm (SOTA), which is based on unsupervised learning. We propose a new architecture by combining these two neural networks types to yield better and faster classification results for our classification problem.

1 Introduction

Many gamma ray experiments have to deal with the problem of separating gammas from hadrons. The experiments usually generate large data sets with many attributes in them. This multi-dimensional data classification problem offers a daunting challenge of extracting small number of interesting events (gammas) from an overwhelming sea of background (hadrons). Many techniques are in active re-

B.G. Sidharth, F. Honsell and A. de Angelis. (eds.), Frontiers of Fundamental Physics, 297–302.

search for addressing this problem. The list includes classical statistical multivariate techniques to more sophisticated techniques like neural networks, classification trees and kernel functions.

The class of neural networks provides an automated technique for the classification of the data set into given number of classes [3]. It is in active research in both artificial intelligence and machine learning communities. Several neural network models have been developed to address the classification problem. Usually, one makes the distinction between supervised and unsupervised classifiers: the former are trained with data for which the classification is known and then used to classify raw data, while the latter attempt to find the best-fitting class structure in the input data by using some measure of merit (usually an euclidean metric is used [6]). From a mathematical perspective, a neural network is simply a mapping from $R^n \to R^m$, where R^n is the input data set dimension and R^m is the output dimension of the neural network. The network is typically divided into various layers; each layer has a set of neurons also called nodes or information units, connected together by the links. The artificial neural networks are able to classify data by learning how to discriminate patterns in features (or parameters) associated with the data. The neural network learns from the data set when each data vector from the input set is subjected to it. The learning or information gain is stored in the links associated with the neurons. The output structure of the network is dependent on both the problem and the network type. For a gamma/hadron separation problem the network maps each input vector onto the [0,1] interval in supervised networks, whereas in unsupervised networks the nodes are adapted to the input vector in such a way that the output of the network represents the natural groups that exist in the data set. The output of the unsupervised network is generally stored in an ASCII file. A visualization technique is then used to view the groups by processing the output file generated by the network.

Section 2 describes the data sets used for the classification. Section 3 deals with the MultiLayer Perceptron network and its classification results. Section 4 deals with the Self-Organizing Tree Algorithm and its variant along with their classification results. Conclusions and future perspectives are discussed in the section 5.

2 Data set description

The data sets are generated by a MonteCarlo simulation program, CORSIKA [2]. They contain 12332 gammas, 7356 ON events (mixture of gammas and hadrons), and 6688 hadron or OFF events. These events are stored in different files. The files contain event parameters in ASCII format, each line of 10 numbers being one event [5] with the parameters defined below.

1. *Length*: major axis of ellipse [mm]

2. *Width*: minor axis of ellipse [mm]

3. *Size*: 10-log of sum of content of all pixels
4. *Conc*: ratio of sum of two highest pixels over fSize [ratio]
5. *Conc1*: ratio of highest pixel over fSize [ratio]
6. *Asym*: distance from highest pixel to centre, projected onto major axis [mm]
7. *M3Long*: 3rd root of third moment along major axis [mm]
8. *M3Trans*: 3rd root of third moment along minor axis [mm]
9. *Alpha*: angle of major axis with vector to origin [deg]
10. *Dist*: distance from origin to centre of ellipse [mm]

These Hillas image parameters [1] are derived from pixel analysis and are used for classification.

3 Multi-Layer Perceptron

For this approach we used the ROOT Analysis Package (v4.00/02) and in particular the MultiLayer Perceptron class [4] which implements a generic layered network. Since this is a supervised network we took two thirds of gamma and OFF data to train the network and the remaining data to test it.The code of the ROOT package is very flexible and simple to use. It allowed us to create a network with a 10 nodes input layer, a hidden layer with the same number of nodes and an output layer with just a single neuron which should return "0" if the data represent hadrons or "1" if they are gammas. Weights are put randomly

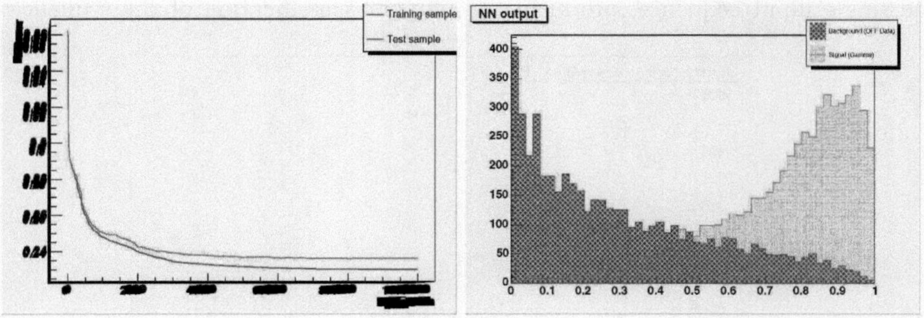

(a) The error functions for training and test data took on 1000 runs

(b) The histogram of distributions for gamma and hadron parameters

Figure 1. MLP classification results using the BFGS default learning method.

at the beginning of the training session and then adjusted from the following runs in order to minimize errors (back-propagation). Errors at cycle i are defined as: $err_i = \frac{1}{2}\, o_i^2$ where o_i is the error of the output node. Data to input and output nodes are transferred linearly, while for hidden layers they use a sigmoid (usually: $\sigma(x) = 1/(1 + \exp(-x)))$.

We have tested the same network using different learning methods proposed by the code authors, as for example the so called "Stochastic minimization", based on the Robbins-Monro stochastic approximation, but the default "Broyden, Fletcher, Goldfarb, Shanno" (BFGS) method has proved to be the quickest and with the best error approximation.

Figures 1.a and 1.b represent a possible output when using the ROOT package on those data. The first one depicts the error function for each run of the network, comparing the training and the test data. Note that the greater is the number of runs, the better the network behaves. The second one shows the distributions of output nodes, that is how many times the network decides to give a value near to "0" or to "1".

4 Self-Organizing Tree Algorithm (SOTA)

The Self-Organizing Tree Algorithm [8] is an unsupervised neural network which implements a growing hierarchical clustering and is based on the self organising map network [7]. It hierarchically clusters the data into a binary tree of natural groups that exist in the data set. Initially the tree consists of one root node linked to 2 child cells. All the input events are randomly distributed between the 2 initial child nodes. The tree grows by expanding the child node having the most heterogeneous population of associated inputs. Two new descendants are generated from this heterogeneous cell that changes its state from cell to node. The tree then grows by descending the cells into child nodes until each cell has one single input sequence, producing a complete classification of the sequences.

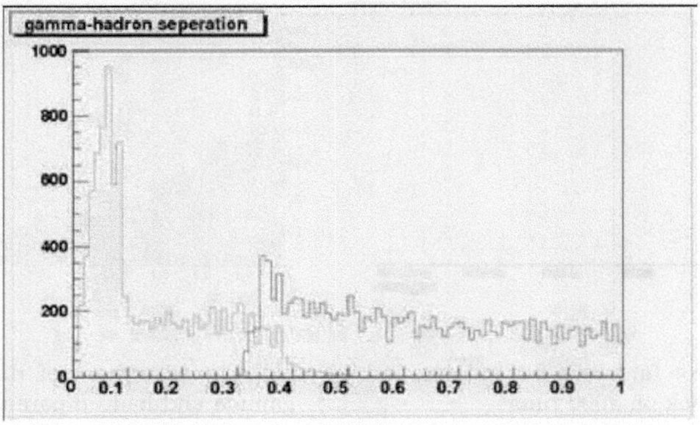

Figure 2. SOTA classification results.

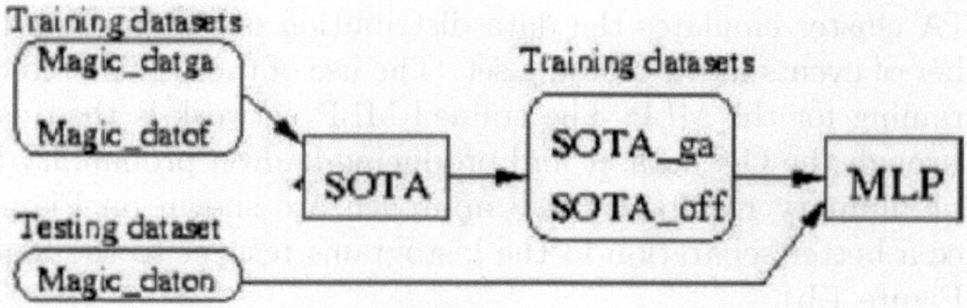

Figure 3. Data flow for a SOTA-MLP network using MonteCarlo datasets.

Alternatively, the expansion can be stopped at the desired level of heterogeneity in the cells, producing in this way a classification of sequences at a higher taxonomic level.

This kind of classification could be useful for astrophysics when a multi-event separation is needed on the same dataset, that is when multiple particles have been detected simultaneously and the analysis software should assign them a label (as "proton", "muon", "gamma", etc.). They are also used as a data mining tool to explore the natural groups that exists in data sets. Using the MAGIC datasets the tree has grown up to 10 levels, with the training sets taken from the MonteCarlo simulations (Figure 2). This approach gives a hierarchical view of data, is robust for noisy data and is faster than traditional hierarchical clustering.

4.1 SOTA + MLP

Figure 3 shows the combination of SOTA with MLP for the separation task. The SOTA method is applied to the initial MonteCarlo datasets (gamma, ON and OFF) to find the natural clusters that exist in the datasets. The SOTA tree produced two clusters of gamma and hadron which are used to train the MLP.

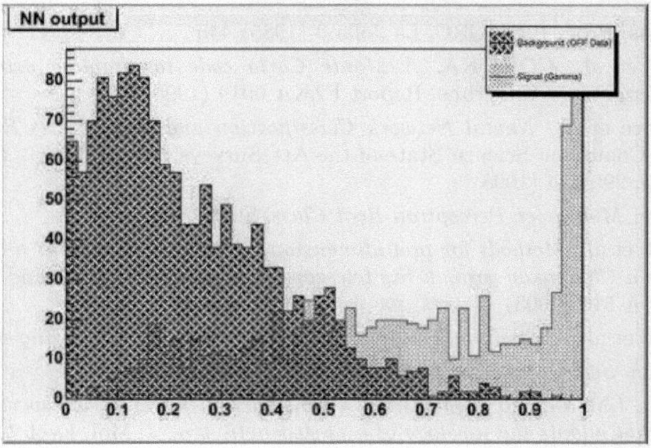

Figure 4. A preliminary result using an MLP feeded with SOTA labelled datasets.

The SOTA cluster emulates the data distribution of the patterns, thus reducing the number of events in the training set. The use of these clustered data can result in fast training for the MLP. The trained MLP network is then used to perform testing through the ON dataset and producing hadron probability for each event.

The preliminary results for this approach are shown on Figure 4 where we can notice a better separation in the histograms respect to the non-treated MLP results (Figure 1.b).

5 Discussion

In this article we classified the gamma ray data using MLP and SOTA. Both MLP and SOTA shown some good classification results. The algorithms used here suggest that a complex problem could not be solved using standalone methods even if they are suitable for a large part of other data analysis problems.

SOTA algorithm clusters the data set into groups thus reducing the number of events in the training set. This can be useful for the MAGIC experiment where there are overwhelming events to be classified. MLP based on supervised technique identifies the group labels, but the training session could be longer. By combining SOTA with MLP we can significantly decrease the training period and yield better classification results.

The work can be further extended by using combination of different models in both self-organizing networks and supervised networks. Future experiments can be done using Growing Cell Structures and Growing Neural Gas models [9] in the unsupervised category. In supervised networks, tests can be performed by using probabilistic networks and MLP trained with fast back-propagation.

References

[1] A.M. Hillas, Proc. 19th ICRC, La Jolla 3 (1985) 445

[2] D. Heck, et al., CORSIKA, *A Monte Carlo code to simulate extensive air showers*, Forschungszentrum Karlsruhe, Report FZKA 6019 (1998)

[3] S. Lawrence et al., *Neural Network Classification and Prior Class Probabilities*, Lecture Notes in Computer Science State-of-the-Art Surveys, edited by G. Orr et al., Springer Verlag, pp. 299-314 (1998)

[4] C. Delaere, *Multilayer Perceptron Root Class*, http://root.cern.ch

[5] R.K. Bock et al., *Methods for multidimensional event classification: a case study using images from a Cherenkov gamma-ray telescope*, Nuclear Instruments and Methods in Physics Research, A 516 (2003)

[6] Rajaniemi et al., *Classifying Gamma-ray bursts using self-organizing maps*

[7] T. Kohonen *Self-Organizing Maps*, 2nd Ed., Springer (1997)

[8] J. Dopazo, J.M. Carazo *Phylogenetic reconstruction using an unsupervised growing neural network that adopts the topology of a phylogenetic tree*, J. Mol. Evol. (1997) 44:226-233

[9] B. Fritzke, *A growing neural gas network learns topologies*, in *Advances in Neural Information Processing Systems 7*, G. Tesauro, D.S. Touretzky and T. K. Leen, editors, MIT Press (1995)

GAMMA-RAY ASTROPHYSICS WITH AGILE

F. LONGO [a,b], A. ARGAN [c], G. BARBIELLINI [a,b], M. BASSET [b], F. BOFFELLI [d], A. BULGARELLI [e], P. CARAVEO [c], P.W. CATTANEO [d], E. CELESTI [e], A. CHEN [c,f], V. COCCO [f,g,h], E. COSTA [i], E. DEL MONTE [i], G. DI COCCO [e], G. DI PERSIO [i], I. DONNARUMMA [i], M. FEROCI [i], M. FIORINI [c], T. FROYSLAND [f,h], M. FRUTTI [i], M. GALLI [j], F. GIANOTTI [e], A. GIULIANI [c,f], C. LABANTI [e], I. LAPSHOV [i], F. LAZZAROTTO [i], F. LIELLO [a,b], P. LIPARI [k], M. MARISALDI [e], M. MASTROPIETRO [i], A. MAURI [e], F. MAURI [d], S. MEREGHETTI [c], E. MORELLI [e], A. MORSELLI [h], L. PACCIANI [i], A. PELLIZZONI [c], F. PEROTTI [c], P. PICOZZA [g,h], C. PITTORI [f,g,h], C. PONTONI [b,f], G. PORROVECCHIO [i], M. PREST [b,l], M. RAPISARDA [m], E. ROSSI [e], A. RUBINI [i], P. SOFFITTA [i], M. TAVANI [i], A. TRACI [e], M. TRIFOGLIO [e], E. VALLAZZA [b], S. VERCELLONE [c], D. ZANELLO [k]

[a] *Università di Trieste*
[b] *INFN, Sezione di Trieste*
[c] *IASF-CNR, Sezione di Milano*
[d] *INFN, Sezione di Pavia*
[e] *IASF-CNR, Sezione di Bologna*
[f] *CIFS, Torino*
[g] *Università di Roma "Tor Vergata"*
[h] *INFN, Sezione di Roma2*
[i] *IASF-CNR, Sezione di Roma*
[j] *ENEA, Sezione di Bologna*
[k] *Università di Roma "La Sapienza, INFN, Sezione di Roma1*
[l] *Università dell'Insubria*
[m] *ENEA, Sezione di Roma*

B.G. Sidharth, F. Honsell and A. de Angelis. (eds.), Frontiers of Fundamental Physics, 303–308.

Abstract

A brief description of the new generation high-energy gamma-ray astrophysics experiment AGILE and its main science objectives are presented.

1 Introduction

The Energetic Gamma-Ray Experiment Telescope (EGRET) aboard CGRO (1991-2000) was very successful in detecting γ-rays (in the energy band 100 MeV - few GeV) from around 70 AGN, 8 pulsars, and 170 sources not yet identified firmly with known objects [1]. EGRET has also measured the spectrum and the spatial distribution of the diffuse galactic γ-ray emission with unprecedented sensitivity and resolution [2].

The italian AGILE satellite (Astro-rivelatore Gamma a Immagini LEggero), to be launched in 2006, will offer a sensitive area similar to that of EGRET and an angular resolution somewhat better than EGRET [3].

2 Detection technique

All GeV gamma-ray experiments use pair production in thin foils of high-Z material to actually detect the γ-rays . Different techniques are used to track the e^+/e^--pairs and to measure their energy, though. In principle the gamma-ray energy threshold is around 10 MeV, but the short range of the pairs and small-angle scattering in the tracker significantly deteriorate the detector performance below 100 MeV. Towards high γ-ray energies self-vetoing and the finite thickness of the calorimeter can reduce the quality of measurement. The main problems with satellite-based γ-ray detectors, however, are the technical constraints which prohibit satellite payloads with an effective area of much more than a squaremeter. The flux of all cosmic γ-ray sources falls off with photon energy and therefore the scientific return of the γ-ray detectors at high photon energies is limited by statistics rather than inapplicability of the technique of measurement.

3 AGILE

The AGILE scientific instrument is based on an innovative design based on three detecting systems: (1) a Silicon Tracker, (2) a Mini-Calorimeter (MC), and (3) an ultralight coded mask system with Si-detectors (Super-AGILE). AGILE is designed to provide: (1) excellent imaging in the energy bands 30 MeV–50 GeV (5–10 arcmin for intense sources) and 15-40 keV (1–3 arcmin); (2) optimal timing capabilities, with independent readout systems and minimal deadtimes for the Silicon tracker, Super-AGILE and Mini-Calorimeter; (3) large fields of view for the gamma-ray imaging detector (GRID) ($\sim$3 sr) and Super-AGILE ($\sim$1 sr) [3].

Figure 1. Schematic view of the AGILE instrument

Despite of its smaller dimensions AGILE will have comparable performances to EGRET on axis and substantially better off axis. The innovative technology will allow AGILE to achieve the smallest deadtime in high-energy astrophysics.

Fig. 1 shows the AGILE instrument configuration of total weight of ~ 120 kg including the Si-Tracker, Super-AGILE, Mini-Calorimeter, the Anticoincidence system and electronics.

- The **Silicon-Tracker**, is a gamma-ray pair-converter and imager made of 12 planes, with two Si-layers per plane providing the X and Y coordinates of interacting charged particles. The fundamental Silicon detector unit is a tile of area 9.5×9.5 cm^2. Each Si-Tracker layer is made of 4 ladders (each composed of 4 Si tiles), for a total geometric area of 38×38 cm^2 and 1,536 readout channels. The first 10 planes are made of three elements: a first layer of Tungsten (0.07 X_0) for gamma-ray conversion, and two Si-layers (views) with microstrips orthogonally positioned. Both digital and analog information (charge deposition in Si-microstrip) is read by front-end electronics (FEE). The GRID has an *on-axis* total radiation length near ~ 0.8 X_0. Special algorithms applied off-line to telemetered data will allow optimal background subtraction and reconstruction of the photon incidence angle. Both digital and analog information are crucial for this task.

- **Super-AGILE** is made of four square Silicon detectors ($19 \times 19\,\mathrm{cm}^2$ each) and associated FEE placed on the first GRID tray plus an ultra-light coded mask system supporting a Tungsten mask placed at a distance of 14 cm from the Silicon detectors. Super-AGILE tasks are: *(i)* photon-by-photon detection and imaging of sources in the energy range 15-40 keV, with a field-of-view (FOV) of ~ 0.8 sr, good angular resolution (1-3 arcmins, depending on source intensity and geometry), and good sensitivity (~ 5 mCrab for 50 ksec integration, and < 1 Crab for a few seconds integration); *(ii)* simultaneous X-ray and gamma-ray spectral studies of high-energy sources; *(iii)* excellent timing ($\sim 4\,\mu$s); *(iv)* burst trigger for the GRID and MCAL; *(v)* GRB alert and quick on-board positioning capability.

- The **Mini-Calorimeter** (MCAL) is made of two planes of Cesium Iodide (CsI) bars, for a total (on-axis) radiation length of 1.5 X_0. The signal from each CsI bar is collected by two photodiodes placed at both ends. The MCAL tasks are: *(i)* obtaining additional information on the energy of particles produced in the Si-Tracker; *(ii)* detecting GRBs and other impulsive events with spectral and intensity information in the energy band $\sim 0.3-100$ MeV. We note that the problem of "particle backsplash" for AGILE is much less severe than in the case of EGRET. AGILE allows a relatively efficient detection of (inclined) photons near 10 GeV and above also because the AC-veto can be disabled for events with more than ~ 100 MeV total energy collected in the MCAL.

Table 1 summarizes the main characteristics of the AGILE gamma-ray instrument and its performance compared to that of EGRET. We assumed a typical 2-week pointing duration and a $\sim 50\%$ exposure efficiency.

	EGRET	AGILE
Mass	1830 kg	120 kg
Gamma-ray energy band	30 MeV–30 GeV	30 MeV–50 GeV
Field of View	~ 0.5 sr	~ 3 sr
PSF	5.5°	4.7° (@ 0.1 GeV)
(68% containment radius)	1.3°	0.6° (@ 1 GeV)
	0.5°	0.2° (@ 10 GeV)
Deadtime for γ-ray detection	$\gtrsim 100$ ms	$\lesssim 100\,\mu$s
Sensitivity	8×10^{-9}	6×10^{-9} (@ 0.1 GeV)
(ph cm^{-2} s^{-1} MeV^{-1})	1×10^{-10}	4×10^{-11} (@ 1 GeV)
	1×10^{-11}	3×10^{-12} (@ 10 GeV)

Table 1. A comparison between EGRET and AGILE

4 Science Objectives

4.1 GALACTIC DIFFUSE EMISSION

The galactic diffuse emission is produced in interactions of cosmic rays with gas and ambient photon fields and thus provides us with an indirect measurement of cosmic rays in various locations in the Galaxy. A significant fraction of the diffuse galactic γ-rays is supposedly produced in decays of neutral pions following inelastic collisions of cosmic ray nucleons. Leptonic emission is particularly important at γ-ray energies below 100 MeV, where bremsstrahlung is presumably the main emission mechanism. Inverse Compton scattering of relativistic electrons on soft ambient photons is expected to provide γ-rays with a hard spectrum, thus eventually dominating over the π^0-decay γ-rays at high energies. A new model for Galactic diffuse emission suited for the AGILE observations has been developed[4].

4.2 GALACTIC SOURCES

The capabilities of AGILE with respect of the astrophysics of Galactic γ-ray sources have been recently discussed elsewhere[5].

- **Supernova remnants**: SNR are considered the most likely sources of galactic cosmic rays. Observational evidence in favor of this scenario has been found only for cosmic ray electrons, not for the nucleons. The signal of π^0 decaying into $\gamma\gamma$ could indicate the dominant role of the acceleration of nuclei.
- **Unidentified EGRET sources**: EGRET has left a legacy of about 170 sources not yet identified firmly with known sources. Various population studies have been performed to search for correlations with classes of galactic objects.
- **Pulsars**: To date eight pulsars have been identified in the EGRET data on account of pulsed emission. AGILE will offer the first possibility of detecting several young and energetic radio pulsars that have been discovered since the end of the CGRO mission[6].

4.3 EXTRAGALACTIC SOURCES

- **Active galactic nuclei** These sources show very intense emission, which in many cases is variable. The variability has been observed on all time scales accessible with the available measurement techniques down to about one hour. Simultaneous monitoring of a large number of AGNs per pointing will be possible with the new generation satellites. Several outstanding issues concerning the mechanism of AGN gamma-ray production and activity can be addressed in the near future (e.g. [7]) including: (1) the study of transient vs. low-level gamma-ray emission and duty-cycles; (2) the relationship between the gamma-ray variability and the radio-optical-X-ray-TeV emission; (3) the correlation between relativistic radio plasmoid ejections and gamma-ray flares; (4) hard X-ray/gamma-ray correlations.
- **Gamma-ray bursts**: About ten GRBs were detected by the EGRET spark chamber during ~ 7 years of operations [8]. This number was limited by

the EGRET FOV and sensitivity and, from what we know today, not by the GRB emission mechanism normally producing gamma-rays above 100 MeV[9]. The small deadtime of the new generation satellites allows a better study of the initial phase of GRB pulses (for which EGRET response was in many cases inadequate). The remarkable discovery of 'delayed' gamma-ray emission up to $\sim$ 20 GeV from GRB 940217 [10] is of great importance to model prompt and afterglow acceleration processes. Measuring the HE component of GRBs may be critical to the understanding of the charged particle acceleration. AGILE will be able to detect most of the bright long bursts and its two detectors (Super-AGILE and MCAL), with their independent trigger algorithms, compensate each other in detecting soft and high energy GRBs[11].

5 Conclusions

The AGILE scientific instrument is innovative in many ways, and is designed to obtain an optimal gamma-ray detection performance despite its relatively small mass and absorbed power. The combination of hard X-ray and gamma-ray imaging capabilities in a single integrated instrument is unique to AGILE. We anticipate a crucial role of Super-AGILE for studies of AGNs, GRBs, and Galactic sources. AGILE will provide an important step forward in γ astronomy. We are confident that the partnership between High Energy Physics and High Energy Astrophysics will be the source of new discoveries over a wide range of subjects.

References

[1] Hartman R.C. et al., ApJS, 123, 79, 1999

[2] Hunter S.D. et al., ApJ, 481, 205, 1997

[3] Tavani M. et al., *Proceedings of Gamma2001 Conference*, AIP Conf. Proceedings, eds. Ritz S., Gehrels N. and Shrader C.R., 2001, Vol. 587, p. 729

[4] Giuliani A., PhD thesis, University of Milano

[5] Tavani M., Pellizzoni A. and Vercellone S. (eds.), *X and Gamma-Ray Astrophysics of Galactic Sources*, Rome 2004

[6] Pellizzoni A. et al., Adv.Sp.Res, 33, 625, 2004

[7] Vercellone S., Soldi S., Chen A.W., Tavani M., MNRAS, 353, 890, 2004

[8] Schneid E.J. et al., in AIP Conf. Proc., Vol. 384, p. 253, 1996

[9] Tavani M., PRL, 76, 3478, 1996

[10] Hurley K. et al., Nature, 372, 652, 1994

[11] Ghirlanda G. et al., AIP Conf. Proc., 727, 704, 2004

SIMULATING THE HIGH ENERGY GAMMA-RAY SKY SEEN BY THE GLAST LARGE AREA TELESCOPE

F. LONGO [a], P. AZZI [b], D. BASTIERI [b], G. BUSETTO [b], Y. LEI [b], R. RANDO [b], O. TIBOLLA [b], L. BALDINI [c], M. KUSS [c], L. LATRONICO [c], N. OMODEI [c], M. RAZZANO [c], G. SPANDRE [c], P. BOINEE [d], A. DE ANGELIS [d], M. FRAILIS [d], M. BRIGIDA [e], F. GARGANO [e], N. GIGLIETTO [e], F. LOPARCO [e], M.N. MAZZIOTTA [e], C. CECCHI [f], P. LUBRANO [f], F. MARCUCCI [f], M. PEPE [f], G. TOSTI [f], A. LIONETTO [g], A. MORSELLI [g], C. PITTORI [g]

[a] *University of Trieste and INFN, Sezione di Trieste*
[b] *University of Padova and INFN, Sezione di Padova*
[c] *University of Pisa and INFN, Sezione di Pisa*
[e] *University of Bari and INFN, Sezione di Bari*
[d] *University of Udine and INFN, Sezione di Trieste*
[f] *University of Perugia and INFN, Sezione di Perugia*
[g] *University of Roma "Tor Vergata" and INFN, Sezione di Roma2*

Abstract

This paper presents the simulation of the GLAST high energy gamma-ray telescope. The simulation package, written in C++, is based on the Geant4 toolkit, and it is integrated into a general framework used to process events. A detailed simulation of the electronic signals inside Silicon detectors has been provided and it is used for the particle tracking, which is handled by a dedicated software. A unique repository for the geometrical description of the detector has been realized using the XML language and a C++ library to access this information has been designed and implemented. A new event display based on the HepRep protocol was implemented. The full simulation

B.G. Sidharth, F. Honsell and A. de Angelis. (eds.), Frontiers of Fundamental Physics, 309–314.

was used to simulate a full week of GLAST high energy gamma-ray observations. This paper outlines the contribution developed by the Italian GLAST software group.

1 Introduction

The Gamma-ray Large Area Space Telescope (GLAST) is an international mission that will study the high-energy phenomena in gamma-rays universe [1]. GLAST is scheduled for launch in mid 2007.

GLAST is instrumented with a hodoscope of Silicon planes with slabs of converter, followed by a calorimeter; the hodoscope is surrounded by an anticoincidence (ACD). This instrument, called the Large Area Telescope LAT, is sensitive to gamma rays in the energy range between 30 MeV and 300 GeV. The energy range, the field of view and the angular resolution of the GLAST LAT are vastly improved in comparison with those of its predecessor EGRET (operating in 1991-2000), so that the LAT will provide a factor of 30 or more advance in sensitivity. This improvement should enable the detection of several thousands of new high-energy sources and allow the study of gamma-ray bursts and other transients, the resolution of the gamma-ray sky and diffuse emission, the search for evidence of dark matter and the detection of AGNs, pulsars and SNRs. A detailed description of the scientific goals of GLAST mission and an introduction to the experiment can be found in [2].

GLAST is a complex system, and detailed computer simulations are required to design the instrument, to construct the response function and to predict the background in the orbit. To accomplish these tasks an object-oriented C++ application called *Gleam* (GLAST LAT Event Analysis Machine) was adopted and implemented by the GLAST LAT collaboration. A brief description of Gleam could be found in [3]. Its structure is described in figure 1. An entire week of the gamma-ray observations by GLAST-LAT was simulated using Gleam in order to develop and test GLAST LAT scientific analysis softaware.

2 Simulation and Reconstruction Software

The GLAST off-line software is based manly on Gaudi, a C++ framework, originally developed at CERN [4]. Gaudi manages the loop of particles to be simulated, then a series of algorithms are applied to each of them to get the result of the complete simulation and reconstruction chain. The Source Generation is the first algorithm called within the particle loop. Its task is to generate particles according to certain characteristcs. This algorithm must store the information on the temporal and spectral behaviour of the source, as well as on the orbital characteristics of GLAST. It provides a user interface to produce additional incoming particles and is responsible for setting the current time, the particle energy, direction, and type. Within this package a series of default sources are implemented.

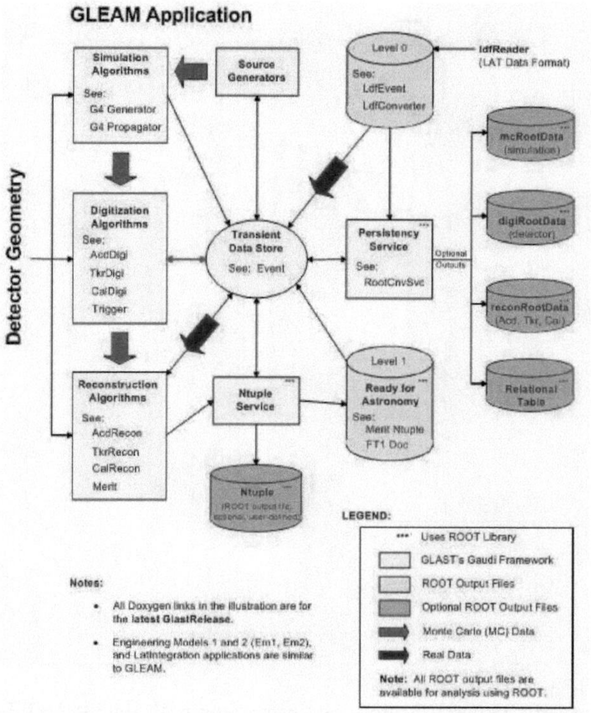

Figure 1. General scheme for simulation and reconstruction within the GLAST off-line software framework

They include source for testing purposes as well as the description of astrophysical spectra and the expected particle and albedo gamma backgrounds. An extension of this framework has been implemented for simulating transient sources such as Gamma-Ray Bursts (GRB). It can be used for studying the capability of GLAST for the observation of rapid transient fluxes in general[5].

The algorithm which is responsible for generating the interactions of particles with the detector is based on the Geant4 MonteCarlo toolkit [6] which is an Object Oriented (OO) simulator of the passage of particles through matter. Its application areas include high energy physics and nuclear experiments, medical science, accelerator and space physics. Within the Gleam application the simulation is managed by the Gaudi algorithm G4Generator.

Figure 2 shows an event generated using Geant4 within the GLAST LAT experiment.

The next algorithm to be applied is the digitization which transforms the hits generated by the Monte Carlo simulations into the signal as read by the electronics. To implement a detailed digitization of the Tracker system a full simulation code has been developed[7]. Starting from Monte Carlo hits in the Silicon detectors, the current signals induced on each strip are evaluated and are converted into voltage signals using the transfer function associated to the detector electronics, taking

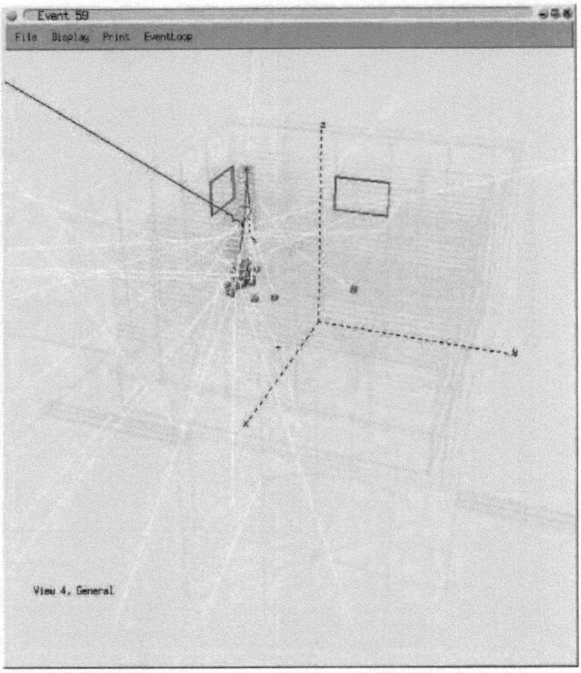

Figure 2. High energy gamma-ray interacting with the GLAST LAT detector

into account the detector noise as well as the noise associated to the electronics. The fired strips and the time over threshold are then determined.

The signals in the Tracker are then analysed by the reconstruction package. It generates a series of clusters, that are used to find and fit the best track candidates. This last procedure is done using alternative pattern recognition algorithms and a Kalman Filter based algorithm. Finally, using the best track found, another algorithm finds the best vertex candidate for gamma events.

Although it is not part of the simulation, the visualization package is essential for the use of the simulation itself. A new version [8, 9] of the event display based on the HepRep protocol was developed and integrated in the offline software. The figure 3 shows a recent FRED-based event display of GLAST.

3 Simulating the GLAST LAT High Energy Sky

Using Gleam it's possible to compute and store time, direction and energy of each incoming simulated and reconstructed gamma ray. The Data Challenge One (DC1), organized by the LAT Collaboration from December 2003 to February 2004, represented the first opportunity to test the complete simulation chain, and the first attempt to perform scientific analyses on simulated data. For DC1, only

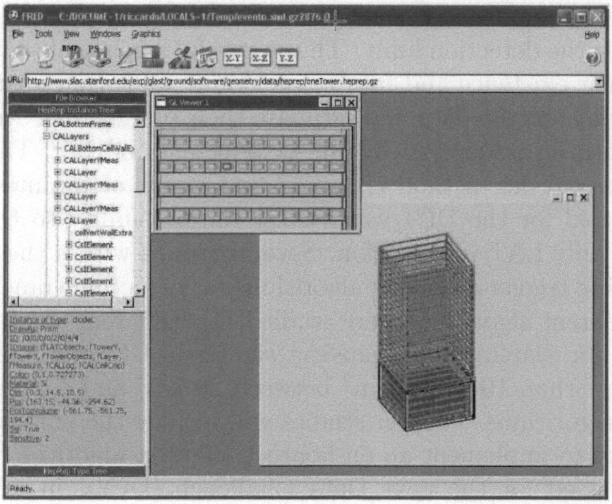

Figure 3. GLAST LAT event display based on FRED

the gamma-ray sky was simulated, while the cosmic-ray flux (about 10^4 times greater) was modeled separately for development of the background rejection algorithms. These algorithms were then applied to the simulated gamma-ray data. The gamma-ray sky simulated for DC1 was quite rich: a variety of sources was included. The simulation software takes into account the relative fluxes and computes from which source the next photon arrives. A map of the simulated sky is reported in fig. 4.

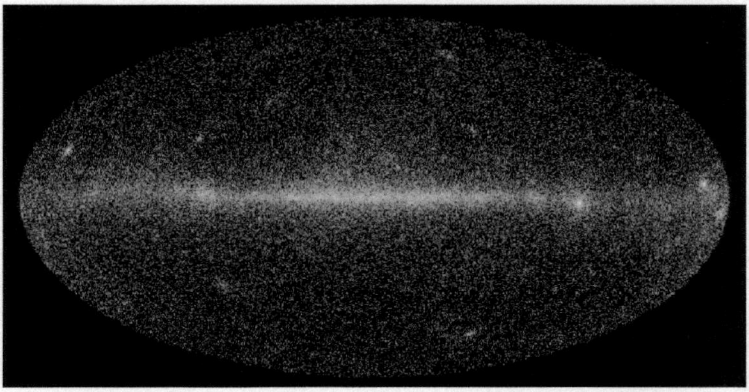

Figure 4. Full simulated gamma-ray sky seen by GLAST LAT

Based on the simulated data, several groups tried to develop the best algorithms to detect unknown gamma-ray sources. GLAST observations contain photons from astrophysical sources, convoluted with the spatial and spectral instrument response. Moreover in most astronomical gamma-ray images a large fraction of sources is near the detection limit. Thus a careful statistical treatment is needed to determine their existence and properties. Many tools (parametric methods) need a priori model to fit the data and estimate their parameters. No model or hypotesis on the data are requested by the wavelet method[10]. Through an iterative procedure the wavelet method allows a blind search of gamma-ray point sources.

Another goal for the DC1 was to test the possibility to trigger Gamma-Ray Bursts using only LAT information. Several groups within the LAT collaboration are prototyping trigger and alert algorithms for detecting transient signals in DC1 data, and different algorithms were studied[11]. Different algorithms were successfully applied for searching for transient signals in DC1 data. Bright bursts (with fluence greater than 10^{-5} erg/cm^2 between 20 keV and 1 MeV) can be detected with simple algorithms. Further studies will include the particle background, and the possibility to implement an on-board LAT alert algorithm. All of these items will be addressed for the next Data Challenge (DC2), in which one month of simulated data will be produced.

4 Conclusions

The *Gleam* simulation program has been developed in the last few years and now it's ready for simulating the full GLAST satellite and is being used for deriving the final instrumental parameters and for generating a full set of events for the developing of scientific analysis software.

References

[1] http://glast.gsfc.nasa.gov/

[2] GLAST Science Brochure (March 2001), http://glast.gsfc.nasa.gov/science/resources/

[3] P.Boinee et al. (2003), "Gleam: the GLAST Large Area Telescope Simulation Framework", in "Science with the New Generation of High Energy Gamma-ray Experiments. Between Astrophysics and Astroparticle Physics", eds. S. Ciprini, A. De Angelis, P. Lubrano, O. Mansutti, Forum Editrice Universitaria, Udine, arXiv:astro-ph/0308120

[4] Gaudi project, http://proj-gaudi.web.cern.ch/proj-gaudi/

[5] N. Omodei (2005), PhD thesis, University of Siena

[6] S. Agostinelli *et al.*, (2003), NIM-A, **506**, 250

[7] M. Brigida *et al.* (2002), LAT internal note, LAT-TD-1058

[8] M. Frailis and R. Giannitrapani, "The FRED Event Display: an Extensible HepRep Client for GLAST," arXiv:cs.gr/0306031. Talk given at 2003 Conference for Computing in High-Energy and Nuclear Physics (CHEP 03), La Jolla, California, 24-28 Mar 2003. Published in eConf C0303241:MOLT010,2003

[9] http://www.fisica.uniud.it/~riccardo/research/fred

[10] F. Marcucci (2004), PhD thesis, Uinversity of Perugia

[11] F. Longo et al., in press on the proceedings of the 4th workshop on "Gamma Ray Bursts in the Afterglow Era", (2005)

DARK MATTER DETECTION IN GAMMA ASTROPARTICLE EXPERIMENTS

ERICA BISESI [a], MOSÈ MARIOTTI [b],
VILLI SCALZOTTO [b]

[a] *Department of Physics, Udine University and INFN, Italy*
[b] *Department of Physics, Padua University and INFN, Italy*

Abstract

The content of matter in the Universe is estimated to be the 27% of its critical density. It is almost universally accepted that most ot this matter is non-baryonic. Constraints from primordial nucleosynthesis and cosmic background radiation measurements impose that the baryonic content of the Universe cannot exceed the 4% of the critical density, so the nature of the remaining 23% has yet to be identified. In this sense, one of the most promising candidates is represented by supersymmetric neutralinos. If they exist, they give rise to relic densities in the required range, and are very well motivated in the framework of theoretical extensions of the Standard Model of particle physics. In addition to direct neutralino searches and collider experiments, neutralino annihilation into gamma rays, neutrinos and synchrotron emission from the charged products represents a reliable way of detecting these intriguing particles. The strongest signals are expected to come from the Galactic Center and from the nearest dwarf spheroidals. Clumps of dark matter in galactic haloes are well predicted by high resolution cold dark matter numerical simulations. In this work we present our studies on the gamma-ray emission from the Galactic Center and from the Draco dwarf spheroidal. We investigate the effect of clumpiness on the detection of signals from neutalinos for different mass density profiles. One of the scientific goals of the MAGIC telescope are just searches for the stable lightest supersymmetric particle in the different physical scenarios in which they are produed. Assuming MAGIC specifications, we draw some conclusions about the potentialities of this telescope in such a kind of investigation.

B.G. Sidharth, F. Honsell and A. de Angelis. (eds.), Frontiers of Fundamental Physics, 315–320.

1 Introduction

One of the most promising candidates for halo dark matter are weakly interacting massive particles (WIMPs), although it is not excluded that particles of other kinds, not yet predicted by particle physics models, might represent the solution of this issue. These particles give a relic density which is of the right order of magnitude to explain the dark matter on all scales. Neutralino is the lightest stable supersymmetric particle in most models, so we focus on detection prospects of such a candidate, working either in the MSSM or in the mSUGRA frameworks.

High resolution numerical simulations of dark haloes formation suggest that the strongest signals are likely to come from the Galactic Center and from the nearest dwarf spheroidals. The persistence of substructures in these simulations induces to argue that at least a fraction of the dark matter in haloes is clustered in clumps.

Taking a phenomenological approach, we here discuss the implications of clumpiness on neutralino dark matter searches.

2 Particle physics models

Minimal supersymmetric model has many free parameters, but with some assumptions we are left with seven parameters in the MSSM model and with five parameters in the mSUGRA setup, namely the supersymmetric extension of the Standard Model defined in a supergravity inspired framework. For details about the parameters which fully define the action of MSSM and mSUGRA see Ref. [6, 7].

Table 1 shows the range of parameter values used in our scan of the MSSM space. Choosing the cosmologically interesting relic density range $0.094 < \Omega_\chi h^2 < 0.129$, we generate in this framework 500000 models and impose that they are not excluded by accelerators constraints.

We sample the 5-dimensional mSUGRA parameter space choosing a few values of $tg\,\beta$ and A_0, and slices along the $m_{1/2}, m_0$ plane for both $sign\,\mu$. We consider both the slepton and the stop coannihilation regions and calculate the relic density with all coannihilations. Visited benchmark points are indicates at the upper left of Fig. (3).

Table 1. Scans of the MSSm space

Parameter	μ	M_2	$tg\beta$	m_A	m_0	A_b/m_0	A_t/m_0
Unit	GeV	GeV	1	GeV	GeV	1	1
Min	10	10	1	10	50	-3	-3
Max	10000	10000	60	1000	10000	3	3

3 Dark matter distribution models

We focus on indirect searches of neutralinos in the Milky Way and in the Draco dwarf spheroidal. We model haloes of these galaxies using the following dark matter profiles:

- **Navarro–Frenk–White** cuspy model [4]: $\rho_{cusp}(r) = \frac{\rho_0}{(r/r_s)^\gamma (1+(r/r_s))^{3-\gamma}}$, $\gamma = 1$;
- **Moore & al.** cuspy model [3]: the same as above with $\gamma = 1.5$;
- a **milder cuspy** profile [2]: the same as above with $\gamma = 0.5$;
- **Burkert & al.** profile [5]: $\rho_{Burk}(r) = \frac{\rho_0}{(1+(r/r_s))(1+(r/r_s)^2)}$.

4 Gamma ray flux

Neutralino annihilation in the galactic halo produces both a gamma-ray flux with a continuum energy spectrum and monochromatic gamma-ray lines. Considering a detector with an angular acceptance $\Delta\Omega$ pointing in a direction of galactic longitude and latitude (l, b), the gamma-ray flux from neutralino annihilation at a given energy E is:

$$\Phi_\gamma(E, \Delta\Omega, l, b) = const. \sum_F \frac{< \sigma v_F >}{m_\chi^2} \frac{dN_\gamma^F}{dE} < J(l,b) > (\Delta\Omega) \text{ cm}^{-2} \text{ s}^{-1} \text{ sr}^{-1} \tag{1}$$

for the continuum gamma. If we assume a spherical dark matter halo in the form of a perfectly smooth distribution of neutralinos, $< J(l,b) > (\Delta\Omega)$ is equal to:

$$< J(l,b) > (\Delta\Omega) = const' \frac{1}{\Delta\Omega} \int_{\Delta\Omega} d\Omega' \int_{lineofsight} \rho(L, \psi')^2 dL, \tag{2}$$

where L is the distance from the detector along the line of sight, ψ is the angle between the direction of observation and that of the center of the galaxy and the integration over $d\Omega'$ is performed over the solid angle $\Delta\Omega$ centered on ψ.

4.1 EFFECT OF CLUMPINESS

To discuss the implications of clumpiness on neutralino dark matter searches, we follow the phenomenological approach of Bergström & al. (1999) [1]. This model is mainly focused on a many small clumps scenario, where substructures are light, with M_{cl} less than $10^4 - 10^6 M_\odot$. Postulating that a fraction f of the dark matter is concentrated in clumps, they find that the increase of the signal compared to a smooth halo is determined by the enhancement factor $f\,\delta$, where δ is the effective contrast between the dark matter density in clumps and the local halo density ρ_0.

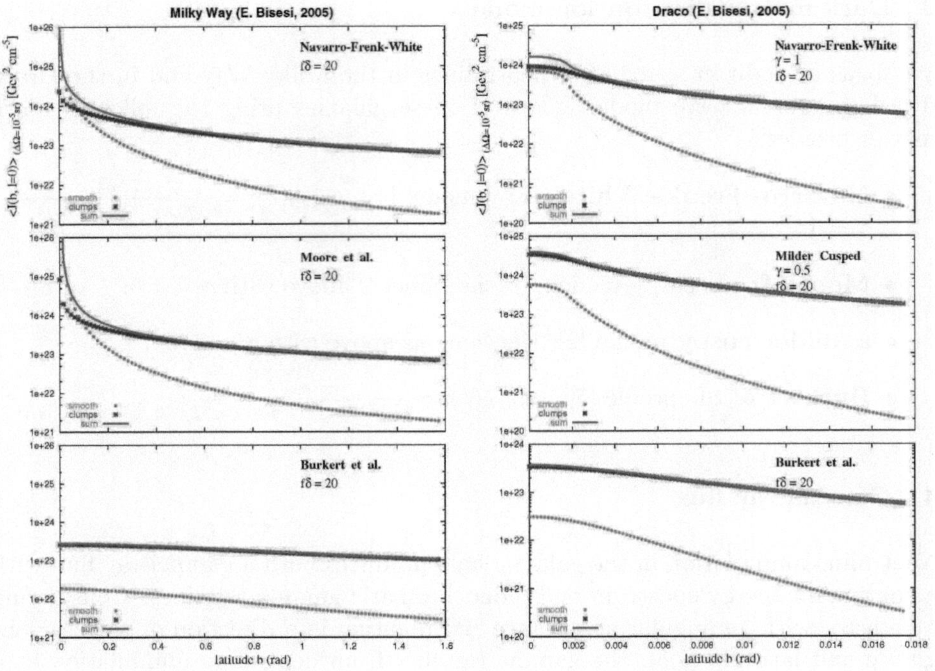

Figure 1. Values of $< J(l,b) > (\Delta\Omega)$ for different halo profiles for the Milky Way and Draco.

Assuming that the clumps can be regarded as pointlike sources, authors derive the anoalogous of Eq. (2) in the clumpy scenario:

$$< J(l,b) > (\Delta\Omega) = const' \frac{f\,\delta}{\Delta\Omega}\,\rho_0 \int_{\Delta\Omega} d\Omega' \int_{lineofsight} \rho(L,\psi')dL. \tag{3}$$

Fig. (1) illustrates the values of $< J(l,b) > (\Delta\Omega)$ for three of the halo profiles introduced above, giving the smooth and clumpy components in the cases of the Milky Way and Draco respectively. The clumpiness enhancement factor is taken reasonably equal to 20. Values of the scale lenght and local halo densities follow prescriptions of Ref. [1, 2].

5 Results and discussion

As the background follows a poissonian statistics, the minimum detectable flux of gamma rays from an ACT telescope is determined by the condition:

$$\frac{\Phi_\gamma\, A_{eff}\, t\, \Delta\Omega}{\sqrt{N_b}} \geq 5, \tag{4}$$

for a 5σ detection level. N_b is the number of background counts, hadrons and electrons, which is obtained on the ground of Ref. [1]. We assume the following

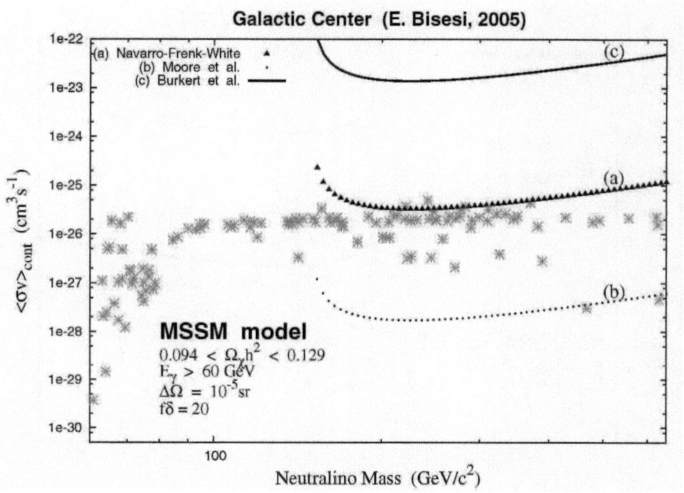

Figure 2. The minimum detectable $< \sigma v >_{cont}$ versus m_χ for the NFW, Moore & al. and Burkert & al. profiles. Dots are points of the parameter space of MSSM, lines represent the 5σ detection level for the MAGIC telescope. Only models corresponding to SUSY points above the curves yield a detectable signal.

MAGIC specifications: $E_{th} = 60$ GeV, $A_{eff} = 10^9$ cm^2, $t = 250$ h, $\Delta\Omega = 10^{-5}$ sr and an energy resolution of 25%. Plotting inequality (4) with the equality sign onto the SUSY parameter space, we divide it into the detectable (above the line) and undetectable (below the line) regions. Results for the Galactic Center are shown in Fig. (2) and Fig. (3), for the MSSM and mSUGRA scenarios respectively and for a clumpiness enhancement factor of 20. This factor is anyway uninfluential in this case.

As we can see from Fig. (2)−(3) plots, detectability of SUSY particles is very sensitive to the choice of the dark matter profile. We find that the scenario which gives the best opportunities for the MAGIC telescope is the Navarro-Frenk-White. Anyway, our model doesn't take into account distribution functions for substructures; an extension of our investigations at higher galactic latitudes does need this is to be modeled in detail, so we will address our future interests in this direction.

References

[1] L. Bergström, J. Edsjö, P. Gondolo & P. Ullio, *Phys Rev D* **59**, 043506 (1999).

[2] N.W. Evans, F. Ferrer & S. Sarkar, *Phys Rev D* **69**, 123501 (2004).

[3] B. Moore, F. Governato, T. Quinn, J. Stadel & G. Lake, *ApJ* **499**, L5 (1998).

[4] J.F. Navarro, C.S. Frenk & S.D. White, *ApJ* **490**, 493 (1997).

[5] A. Burkert, *ApJ* **447**, L25 (1995).

[6] J. Edsjö, M. Schelke, P. Ullio & P. Gondolo, arXiv:hep-ph/0301106 v2 (2003).

[7] J. Edsjö, M. Schelke & P. Ullio, arXiv:astro-ph/0405414 v1 (2004).

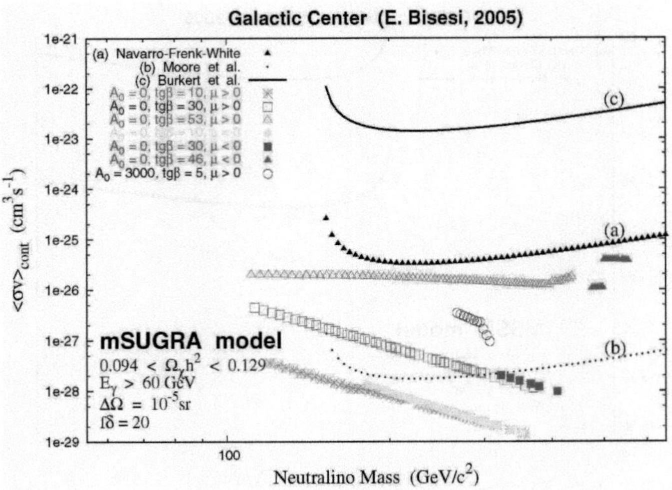

Figure 3. The minimum detectable $< \sigma v >_{cont}$ versus m_χ for the NFW, Moore & al. and Burkert & al. profiles. Dots are points of the parameter space of mSUGRA, lines represent the 5σ detection level for the MAGIC telescope. Only models corresponding to SUSY points above the curves yield a detectable signal.

[8] A. Tasitsiomi & A.V. Olinto, *Phys Rev D* **66**, 083006 (2002).

[9] L. Bergström, P. Ullio & J.H. Buckley, arXiv:astro-ph/9712318v1 (1997).

[10] http://hegra1.mppmu.mpg.de/MAGICWeb/

CONTRIBUTION OF PULSARS TO THE GAMMA-RAY BACKGROUND AND THEIR OBSERVATION WITH THE SPACE TELESCOPES GLAST AND AGILE

ERICA BISESI [a]

[a] *Department of Physics, Udine University, via delle Scienze 208, 33100 Udine, Italy*

Abstract

Luminosities and fluxes of the expected population of galactic gamma-ray pulsars become foreseeable if physical distributions at birth and evolutive history are assigned. In this work we estimate the contribution of pulsar fluxes to the gamma-ray background, which has been measured by the EGRET experiment on board of the CGRO. For pulsar luminosities we select some of the most important gamma-ray emission models, taking into account both polar cap and outer gap scenarios. We find that this contribution strongly depends upon controversial neutron star birth properties. A comparison between our simulation results and EGRET data is presented for each model, finding an average contribution of about 10%. In addition, we perform the calculation of the number of new gamma-ray pulsars detectable by GLAST and AGILE, showing a remarkable difference between the two classes of models. Finally, we suggest some improvements in the numerical code, including more sophisticated galactic m odels and different populations of pulsars like binaries, milliseconds, anomalous pulsars and magnetars.

1 Introduction

The estimation of the contribution of pulsars to the gamma-ray background is based on the comparison between theoretical predictions on the flux emitted by these sources and experimental observations. Each model for the emission of radiation by a stellar population is specified by the galactic distribution of sources and their individual luminosities.

B.G. Sidharth, F. Honsell and A. de Angelis. (eds.), Frontiers of Fundamental Physics, 321–326.

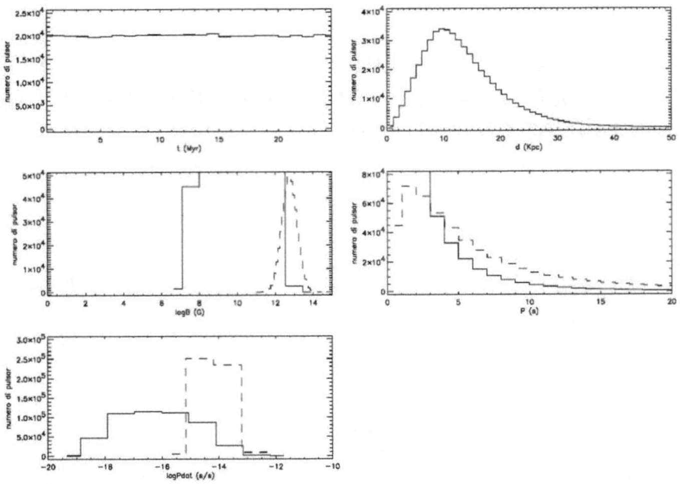

Figure 1. Distributions of pulsars ages (t), distances (d), magnetic fields (B_0), periods (P_0), period derivatives ($\dot{P}$). Dot lines represent a situation with constant magnetic field, solid lines are referred to the field decay case (E. Bisesi, 2002).

2 Galactic model

2.1 SPATIAL DISTRIBUTIONS

We here present a numerical simulation of the galactic population of radio pulsars giving all initial spatial and physical parameters in the framework of the galactic models of Paczyński (1990) and of Gonthier & al. (2002) [4]; we then make them evolve accordingly with the velocity model of Sturner & Dermer (1994) [10]. We show histograms of distributions of various pulsar properties in Fig. (1), assuming both constant magnetic field and field decay.

2.2 PULSAR LUMINOSITIES

In the commonly accepted hypotesis that every pulsar irradiates on the whole electromagnetic spectrum, we assign a gamma luminosity to each source. We consider a suitable choise of emission models.

POLAR CAP MODELS:

- **Harding (1981) [5]:** $L_\gamma(>100 \text{ MeV}) = 1.2 \times 10^{35} B_{12}^{0.95} P^{-1.7}$ ph s^{-1};
- **Zhang & Harding (2000) [12]:**

$$L_\gamma(I) = 5.87 \times 10^{35} {B_{12}}^{6/7} P^{-1/7} \text{ ph s}^{-1} \quad \text{for} \quad {B_{12}}^{1/7} P^{-11/28} > 6.0,$$

$$L_\gamma(II) = 1.0 \times 10^{35} B_{12} P^{-9/4} \text{ ph s}^{-1} \quad \text{for} \quad {B_{12}}^{1/7} P^{-11/28} < 6.0;$$

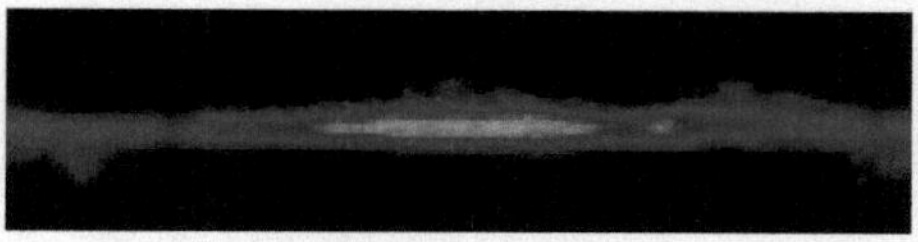

Figure 2. Map of the gamma-ray sky as seen by EGRET (E. Bisesi, 2002).

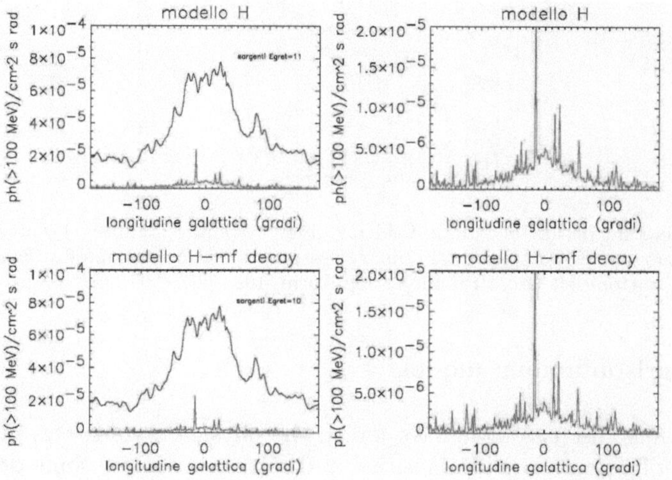

Figure 3. Comparison between the EGRET gamma-ray background (in blue) and predictions for the polar cap model of Harding (1981) (in violet). We also give the simulated number of pulsars above the EGRET detection threshold for this model. Plots on the right side are enlargements of the simulated profiles (E. Bisesi, 2002).

- **Sturner & Dermer (1994) [10]:** $L_\gamma = 6.25 \times 10^{35} {B_{12}}^{3/2} P^{-2}$ ph s^{-1}.

OUTER GAP MODELS with the *death line* $5 \log B - 12 \log P \leq 72$:

- **Romani & Yadigaroglu (1995) [9]:** $L_\gamma = 1.56 \times 10^{36} {B_{12}}^{0.48} P^{-2.48}$ ph s^{-1};
- **Cheng & Zhang (1996) [2]:** $L_\gamma = 3.93 \times 10^{37} {B_{12}}^{0.3} P^{-0.3}$ ph s^{-1}.

We calculate the integrated flux on the galactic latitude $-10° < b < 10°$ as a function of the longitude l; results are plotted together with the experimental points as they have been measured by the EGRET experiment.

Fig. (2) and (3) respectively show the map of the gamma-ray sky as seen by EGRET and our results for the polar cap model of Harding (1981).

The contribution of pulsars to the gamma-ray background is less than 10% for this model; the foreseen number of sources above the EGRET detection threshold of 10^{-7} ph cm^{-2} s^{-1} in not far from the observed value.

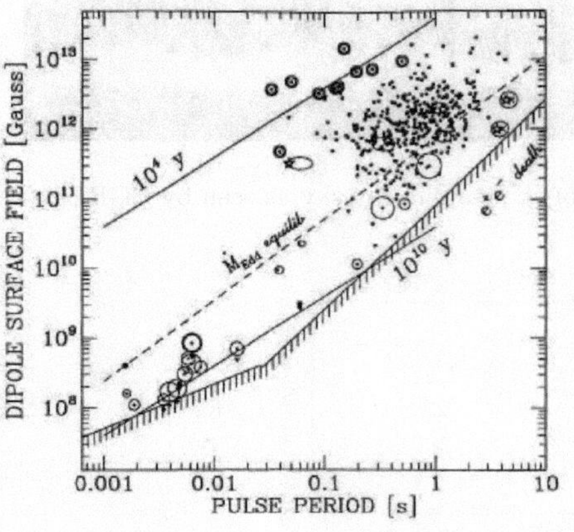

Figure 4. Diagram period–magnetic field for different pulsar families. Points indicate young isolated pulsars, circle–dots pulsars in binary systems. The population of millisecond pulsars is visible at the bottom left (from Phinney & Kulkarni, 1994 [8]).

3 Comparison among models

In the previous discussion we have forced the physical parameters involved in the calculation of gamma ray luminositied and fluxes to assume some definite values. In order to give a meaningful comparison among different emission models an extension to a more general parameter space is required.

So we define a parameter space whose extremes are corresponding to the limit conditions for pulsating gamma neutron stars; we then follow the evolution of a representative point inside this space. The most relevant physical parameters are the rotation period, the magnetic field and the velocity fron the Galactic Centre.

3.1 PULSAR POPULATIONS

Different families of pulsars follow distinct evolutive histories, and this has a direct impact on their physical properties. Furthemore, physical mechanisms ruling over the gamma-ray emission are peculiar of each class. Fig. (4) shows a diagram period–magnetic field for young isolated, binaries and millisecond pulsars. Magnetars are characterized by very high magnetic fields, typically $10^{14} - 10^{16}$ G.

We bind the parametric space in this way:

- $P_{0min} = 0.01$ s (when the centrifugal force equals molecular bonds of the star, yielding its survival);
- $P_{0max} = 1$ s (fixed from observations);
- $B_{0min} = 10^{11}$ G (the lowermost value for young isolated pulsars);
- $B_{0max} = 4.413 \times 10^{13}$ G (condition for the magnetar regime);

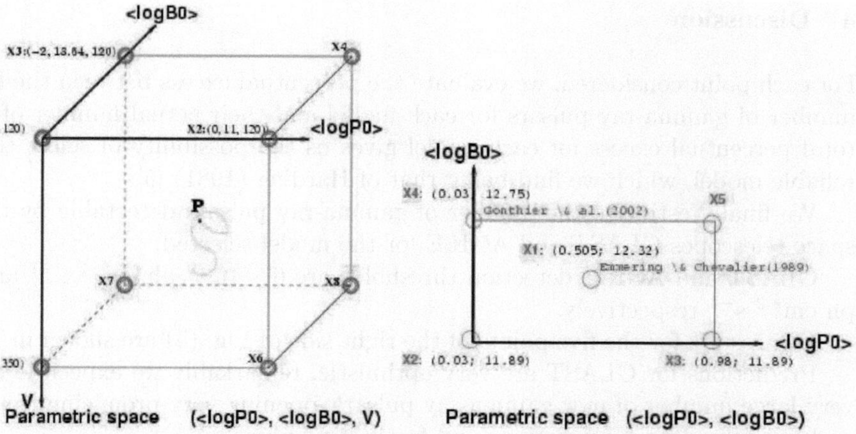

Figure 5. Comparison among models (E. Bisesi, 2002).

- $v_{0min} = 120$ km/s (Gonthier & al., 2002 [4]);
- $v_{0max} = 350$ km/s (Lyne & Lorimer, 1994).

We show a tridimensional representation of the parametric space $\langle \log P_0 \rangle$, $\langle \log B_0 \rangle$, V in the left side of Fig. (5). The bidimensional portion on the right shows the positions of two models of interest [4, 3].

We compare predictions for the five models introduced above for a set of points of the parametric space. As physical mechanisms for the gamma-ray emission differ from one pulsar family to another, we would give luminosity function expressions for each of them. We purpose to improve our work in such a way in future; in this context we restrict our analysis to young isolated pulsars. An example relevant to the point X_3 is shown in Fig. (6).

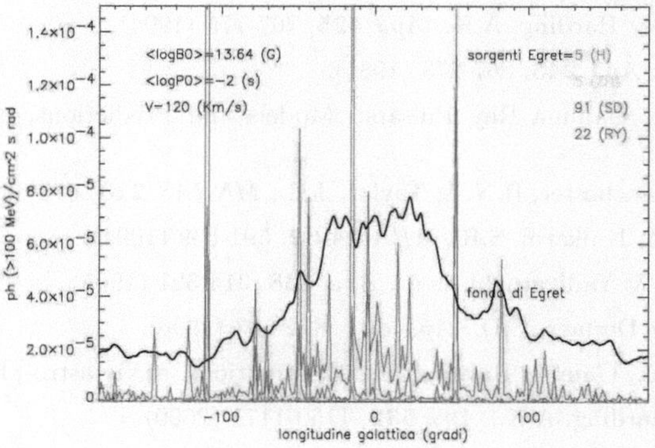

Figure 6. Comparison between EGRET observations and simulation results for the point X_3 of the parametric space (E. Bisesi, 2002).

4 Discussion

For each point considered, we evaluate the percentual excess between the foreseen number of gamma-ray pulsars for each model and their actual number of 7. The total percentual excess for each model gives us the possibility of select the most reliable model, which we find being that of Harding (1981) [5].

We finally estimate the number of gamma-ray pulsars detectable by the next space telescopes GLAST and AGILE for the model selected.

GLAST and AGILE detection thresholds are 6×10^{-9} ph cm^{-2} s^{-1} and 10^{-7} ph cm^{-2} s^{-1} respectively.

Our results for the five points of the right side of Fig. (5) are shown in Table 1.

Predictions for GLAST are very optimistic, remarkably we expect to detect a very large number of new gamma-ray pulsars, opening very promising frontiers in understanding these mysterious and fascinating objects.

Table 1. Total percentual excess for each model on the whole parametric space for young pulsars.

Model	$\lvert$Total percentual excess$\rvert = \sum_{i=1}^{8} \lvert$Percentual excess$(X_i)\rvert$
H	27
ZH	220
SD	174
RY	30506
CZ	2277

References

[1] Bailes, M. & Kniffen, D.A., *ApJ* **391**, 659-666 (1992).

[2] Cheng, K.S. & Zhang, J.L., *ApJ* **463**, 271-283 (1996).

[3] Emmering, R.T. & Chevalier, R.A., *ApJ* **345**, 931-938 (1989).

[4] Gonthier, P.L. & Harding, A.K., *ApJ* **425**, 767-775 (1994).

[5] Harding, A.K., *ApJ* **245**, 267-273 (1981).

[6] Harding, A.K., Gamma Ray Pulsars: Models and Predictions, arXiv:astro-ph/0012268 (2000).

[7] Lyne, A.G., Manchester, R.N. & Taylor, J.H., *MNRAS* **213**, 613-639 (1985).

[8] Phinney, E.S. & Kulkarni, S.R., *ARA&A* **32**, 591-639 (1994).

[9] Romani, R.W. & Yadigaroglu, I.-A., *ApJ* **438**, 314-321 (1995).

[10] Sturner, J.S. & Dermer, C.D., *ApJ* **461**, 872-883 (1996).

[11] Thompson, D.J., Gamma Ray Pulsars: Observations, arXiv:astro-ph/101039 (2001).

[12] Zhang, B. & Harding, A.K., *ApJ* **532**, 1150-1171 (2000).

[13] http://glast.gsfc.nasa.gov/

[14] http://agile.ifctr.mi.cnr.it/Homepage/collaboration.shtml

DATA MINING IN GAMMA ASTROPHYSICS EXPERIMENTS

MARCO FRAILIS [a], ALESSANDRO DE ANGELIS [a],
VITO ROBERTO [b]

[a] *INFN and Dipartimento di Fisica, Università degli Studi di Udine, Italy*
[b] *Dipartimento di Matematica e Informatica, Università degli Studi di Udine, Italy*

Abstract

Data mining techniques, including clustering and classification tasks, for the automatic information extraction from large datasets are increasingly demanded in several scientific fields. In particular, in the astrophysical field, large archives and digital sky surveys with dimensions of 10^{12} bytes currently exist, while in the near future they will reach sizes of the order of 10^{15}. In this work we propose a multidimensional indexing method to efficiently query and mine large astrophysical datasets. A novelty detection algorithm, based on the Support Vector Clustering and using density and neighborhood information stored in the index structure, is proposed to find regions of interest in data characterized by isotropic noise. We show an application of this method for the detection of point sources from a gamma-ray photon list.

1 Characterization of the astrophysical datasets

At present, several projects for the multi-wavelength observation of the Universe are underway, for example SDSS, GALEX, POSS2, DENIS, etc. In the next years, new spatial missions will be launched (e.g. GLAST, Swift), surveying the wall sky at different wavelengths (gamma-ray, X-ray, optical). In the Astroparticle and Astrophysical fields, data is mostly characterized by multidimensional arrays. For instance, in X-ray and Gamma-ray astronomy, the data gathered by detectors

B.G. Sidharth, F. Honsell and A. de Angelis. (eds.), Frontiers of Fundamental Physics, 327–331.

are lists of detected photons whose properties include position (RA, DEC), arrival time, energy, error measures both for the position and the energy estimates, quality measures of the events . Source catalogs, produced by the analysis of the raw data, are lists of point and extended sources characterized by coordinates, magnitude, spectral indexes, flux, etc.

1.1 MINING MULTIDIMENSIONAL DATA

Data mining applied to multidimensional data analyzes the relationships between the attributes of a multidimensional object stored into the database and the attributes of the neighboring ones. Several queries are required by this kind of analysis:

- *point queries*, to find all objects overlapping the query point;
- *range queries*, to find all objects having at least one common point with a query window;
- *nearest-neighbor queries*, to find all objects that have a minimum distance from the query object.

Another important operation is the *spatial join*, which in the astrophysical field is needed to search multiple source catalogs and cross-identify sources from different wavebands. This multidimensional (spatial) data tend to be large (sky maps can reach sizes of Terabytes) requiring the integration of the secondary storage, and there is no total ordering on spatial objects preserving spatial proximity. This characteristic makes it di cult to use traditional indexing methods, like B+-trees or linear hashing.

2 An optimized R-tree structure

The **R-tree** is a data structure meant to efficiently index multidimensional point data or objects with a spatial extent. The structure of an R-tree is the following:

- an **inner node** of the R-tree has entries of the form (cp, MBB), where cp is the address of a child node and MBB is the n-dimensional Minimum Bounding Box of all entries in that child node;
- a **leaf node** has entries of the form (*cp, MBB*), where *cp* refers to a record describing a particular object and *MBB* is its minimum bounding box, or (*Point, Attributes*), where *Point* is a coordinate in the n-dimensional space and *Attributes* are data associated to that point.

An optimized index, in terms of construction time, memory occupied and query performances, can be built using a priori information on the dataset by means of bulk loading algorithms. We have followed a top-down construction method called VAMSplit algorithm to build and optimized R-Tree. This method preserves

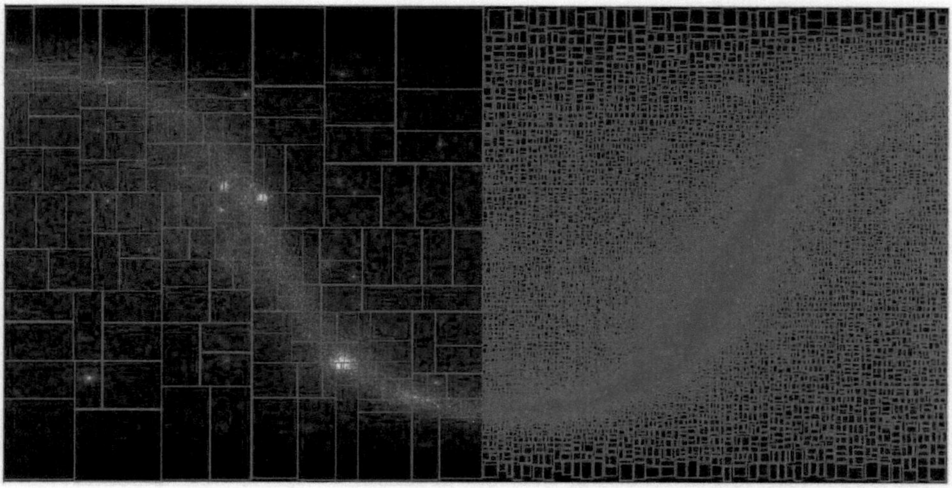

Figure 1. The structure of the optimized R-tree built a photon dataset

the spatial proximity between sibling nodes, resulting in a partition of the dataset with no overlapping between MBBs. Moreover, the volume of the data space covered by each node (at a particular level) is variable and dependent on data density. The main idea of this method is to recursively split the dataset on a near median element along the dimension with maximum variance. In particular, at each recursive step, child subtrees capacity is calculated by:

$$cscap = B \cdot F^{\left\lceil \log_F \left\lceil \frac{N}{B} \right\rceil \right\rceil - 1}$$

where B is the page capacity, F is the internal fanout and N is the number of elements to index in the current step. The near median element is computed by:

$$med = cscap \cdot \left\lfloor \frac{1}{2} \cdot \left\lceil \frac{N}{cscap} \right\rceil \right\rfloor$$

Our implementation uses a sampling strategy to find a good pivot value in the partition step and reduce the number of I/O operations; a caching strategy has been adopted to partition the data into the secondary storage. The total construction time is $O(\frac{N}{B} \log_{\frac{M}{B}} \frac{N}{B})$.

3 A scalable novelty detection algorithm

The structure of the optimized R-tree can help exploring the data and finding regions of interest. For this purpose, other information can be added to each node: the total number of data points covered by the node, their mean and variance, other statistical moments. In this work we propose a novelty detection algorithm based on the *Support Vector Clustering* (SVC).

3.1 THE SVC ALGORITHM

The SVC algorithm estimates the *support* of a high dimensional distribution. It computes the hypersphere with minimal radius which encloses the data points when mapped to a high dimensional feature space. Given a set of points $\mathbf{x}_1, \ldots, \mathbf{x}_N$, with $\mathbf{x}_i \in X \subseteq \mathbb{R}^d$, it finds the hypersphere $(\mathbf{c}, r)$ that solves the optimization problem:

$$\min_{\mathbf{c},r,\rho} \; r^2 + C\sum_{i=1}^{N} \rho_i$$

$$\text{s.t.} \;\; \|\phi(\mathbf{x}_i) - \mathbf{c}\| \leq r^2 + \rho_i$$

$$\rho_i \geq 0, \quad i = 1, \ldots, N$$

where ρ_i are slack variables and C is a positive constant. Defining the Lagrangian, the solution is obtained solving the dual problem:

$$\max_{\alpha} \; \sum_{i=1}^{N} \alpha_i k(\mathbf{x}_i, \mathbf{x}_i) - \sum_{i,j=1}^{N} \alpha_i \alpha_j k(\mathbf{x}_i, \mathbf{x}_j)$$

$$\text{s.t.} \;\; 0 \leq \alpha_i \leq C, \quad i = 1, \ldots, N$$

where α_i are the lagrange multipliers and $k(\mathbf{x}_i, \mathbf{x}_j) = \langle \phi(\mathbf{x}_i), \phi(\mathbf{x}_j) \rangle$.

3.2 FEATURES FROM THE R-TREE

The partition generated at a given level of the optimized R-tree is used as the input space of the novelty detection algorithm. For each node of the partition, the input parameters include the *center* $\mathbf{c}$ of its bounding box and the *density* (the ratio between the number n of elements covered by the node and the volume V of its bounding box). These features are not orthogonal. Therefore, we first apply the PCA method to find the directions along which the variance is higher and project the features on the corresponding eigenvectors. The projected features are passed to the SVC algorithm.

3.3 GAMMA SOURCE DETECTION

Point sources are mostly characterized by a stronger flux, with respect to the surrounding, focused on a small angular region. The area covered by a point source depends also on the instrument point spread function. An optimized R-tree index can be built on a dataset including photons gathered in a certain range of time (we are using, for the analysis, a minimum interval of 6 days). To find static or strong variable sources (e.g. gamma-ray bursts) we use only a bidimensional indexing on the RA and DEC values. The algorithm estimates the support of the diffuse background. The output of the SVC algorithm is filtered to single out the nodes with lowest density. Point sources are considered as the remaining outliers.

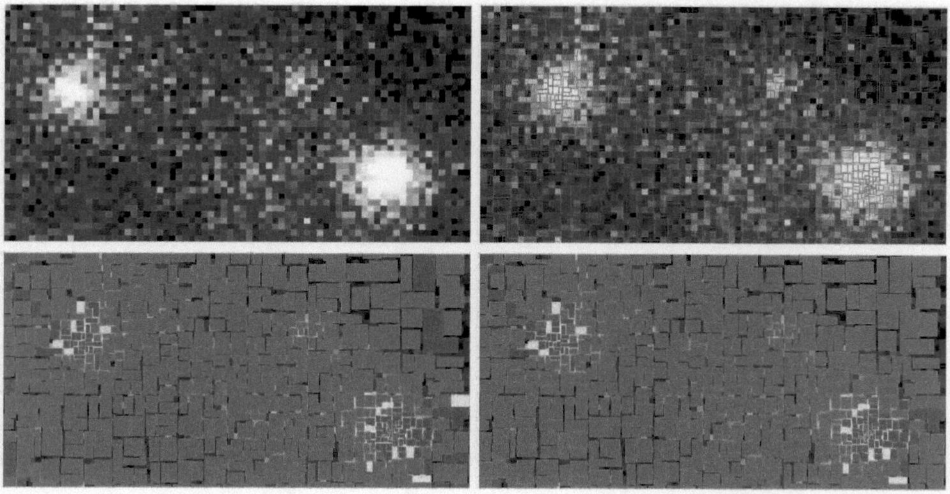

Figure 2. The novelty detection algorithm applied to the anticenter

Figure 2 shows the application of our method to the anticenter region. Green boxes represent the background (support) while yellow boxes are *support vectors* and the red ones are the outliers. In particular,the three major sources of the anticenter are highlighted as novelty.

4 Conclusions

In this work we have realized a multidimensional indexing method to efficiently access and mine large multidimensional astrophysical data. The index adapts the VAMSRtree to large datasets. The partition generated by the optimized R-tree is used to scale the SVC algorithm and find regions of interest where a more accurate analysis can be performed.

References

[1] D. A. White, R. Jain, *Similarity Indexing: Algorithms and Performance*, Proc. SPIE, 2670:62-73, San Diego, USA (1996)

[2] C. Böhm, H.-P. Kriegel, *Efficient Bulk Loading of Large High-Dimensional Indexes*, DaWaK 99, 31:251-260 (1999)

[3] A. Ben-Hur et al., *Support Vector Clustering*, Journal of Machine Learning Research 2(Dec), 125-137 (2001)

[4] J. Shawe-Taylor, N. Cristianini, *Kernel Methods for Pattern Analysis*, Cambridge (2004)

GRID SERVICES FOR THE MAGIC EXPERIMENT

A. FORTI [a], S.R. BAVIKADI [a], C. BIGONGIARI [b], G. CABRAS [a], A. DE ANGELIS [a], B. DE LOTTO [a], M. FRAILIS [a], M. HARDT [c], H. KORNMAYER [c], M. KUNZE [c], M. PIRACCINI [a] AND THE MAGIC COLLABORATION

[a] *INFN and Dipartimento di Fisica, Universitá degli Studi di Udine, Italy*
[b] *INFN and Dipartimento di Fisica Galileo Galilei, Padova, Italy*
[c] *Institut für wissenschaftliches Rechnen, Forschungszentrum Karlsruhe, Germany*

Abstract

Exploring signals from the outer space has become an observational science under fast expansion. On the basis of its advanced technology the MAGIC telescope is the natural building block for the first large scale ground based high energy γ-ray observatory. The low energy threshold for γ-rays together with different background sources leads to a considerable amount of data. The analysis will be done in different institutes spread over Europe. Therefore MAGIC offers the opportunity to use the Grid technology to setup a distributed computational and data intensive analysis system with the nowadays available technology. Benefits of Grid computing for the MAGIC telescope are presented.

1 MAGIC

The MAGIC telescope has been designed to search the sky to discover or observe high energy γ-rays sources and address a large number of physics questions [1]. Located at the Instituto Astrophysico de Canarias on the island La Palma, Spain,

B.G. Sidharth, F. Honsell and A. de Angelis. (eds.), Frontiers of Fundamental Physics, 333–337.

at 28° N and 18° W, at altitude 2300m asl, it is the largest γ-ray telescope in the world. MAGIC is operating since October 2003, data are taken regularly since February 2004 and signals from Crab and Markarian 421 was seen. The main characteristics of the telescope are summarized below:

- A 17m diameter (f/d=1) tessellated mirror mounted on an extremely light carbon-fiber frame (< 10 tons), with active mirror control. The reflecting surface of mirrors is 240 m^2; reflectivity is larger than 85% (300 - 650nm).
- Elaborate computer-driven control mechanism.
- Fast slewing capability (the telescope moves 180° in both axes in 22s).
- A high-efficiency, high-resolution camera composed by an array of 577 fast photomultipliers (PMTs), with a 3.9° field of view.
- Digitalization of the analogue signals performed by 300 MHz FlashADCs and a high data acquisition rate of up to 1 KHz.
- MAGIC is the lowest threshold ($\approx$ 30 GeV) IACT operating in the world.

Gamma-rays observation in the energy range from a few tenth of GeV upwards, in overlap with satellite observations and with substantial improvement in sensitivity, energy and angular resolution, leads to search behind the physics that has been predicted and new avenues will open. AGNs, GRBs, SNRs, Pulsars, diffuse photon background, unidentified EGRET sources, particle physics, darkmatter, quantum gravity and cosmological γ-ray horizon are some of the physics goals that can be addressed with the MAGIC telescope.

2 Grid

The idea of computational and data grids dates back to the first half of the 90's. The vision behind them is often explained using the electric power grid metaphor. The electric power grid delivers electric power in a pervasive and standardized way. You can use any device that requires standard voltage and has a standard plug if you are able to connect it to the electric power grid through a standard socket. When you use electricity you don't worry were it is produced and how it is delivered, you just plug your device into the wall socket and use it. Currently we have millions of computing and storage systems all over the planet connected through the Internet. What we need is an infrastructure and standard interfaces capable of providing transparent access to all this computing power and storage space in a uniform way. This is the concept behind Grid. More precisely, Grid is a kind of parallel and distributed system that enables the sharing, selection and aggregation of services of heterogeneous resources distributed across multiple administrative domains based on their availability, capability, performance, cost, and users quality-of-service requirements [2]. As network performance has outpaced computational power and storage capacity, this new paradigm has evolved to enable the sharing and coordinated use of geographically distributed resources.

2.1 VIRTUAL ORGANIZATION

The Virtual organization is an important Grid concept. Grid allows a pool of heterogeneous resources both within and outside of an organization to be virtualized and form a large, virtual computer. This virtual computer can be used by a collection of users and/or organizations in collaboration to solve their problems. The rules governing the participants providing the resources and the consumers using the resources, as well as the conditions for sharing, dictate the nature of the virtual organization. Hence, a virtual organization groups people and resources without worry about their physical location or institute boundaries. For security reasons, the use of the resources is constrained by an authentication process.

3 Benefits of Grid computing for Magic

The collaborators of the MAGIC telescope are mainly spread over Europe, 18 institutions from 9 countries, with the main contributors (90% of the total) located in Germany (Max-Planck-Institute for Physics, Munich and University of Wuerzburg), Spain (Barcelona and Madrid), Italy (INFN and Universities of Padova, Udine and Siena).

The geographical distribution of the resources makes the management of the experiment harder. This is a typical situation for which Grid computing can be of great help, because it allows researchers to access all the resources in a uniform, transparent and easy way. The telescope is in operation during moonless nights. The average amount of raw FADC data recorded is about 500-600 GB/night. Additional data from the telescope control system or information from a weather station are also recorded. All these information have to be taken into account in the data analysis.

The MAGIC community can leverage from Grid facilities in areas like file sharing, Monte Carlo data production and analysis. In a Grid scenario the system can be accessed through a web browser based interface with single sign-on authentication method. We can briefly summarize the main benefits given by the adoption of Grid technology for the MAGIC experiment:

- Presently, users analyzing data must know where to find the required files and explicitly download them. In a Grid perspective, instead, users don't care about data location and files replication policies improve access time and fault tolerance.

- Grid workflow tools can manage the MAGIC Monte Carlo simulation. The resources from all the members of the MAGIC community can be put together and exploited by the Grid. Easy access to data production for every user, or accordingly to the VO policies.

- Analysis tools can be installed on the Grid. They are thus shared and available for all the users (no need for single installations), moreover, they can

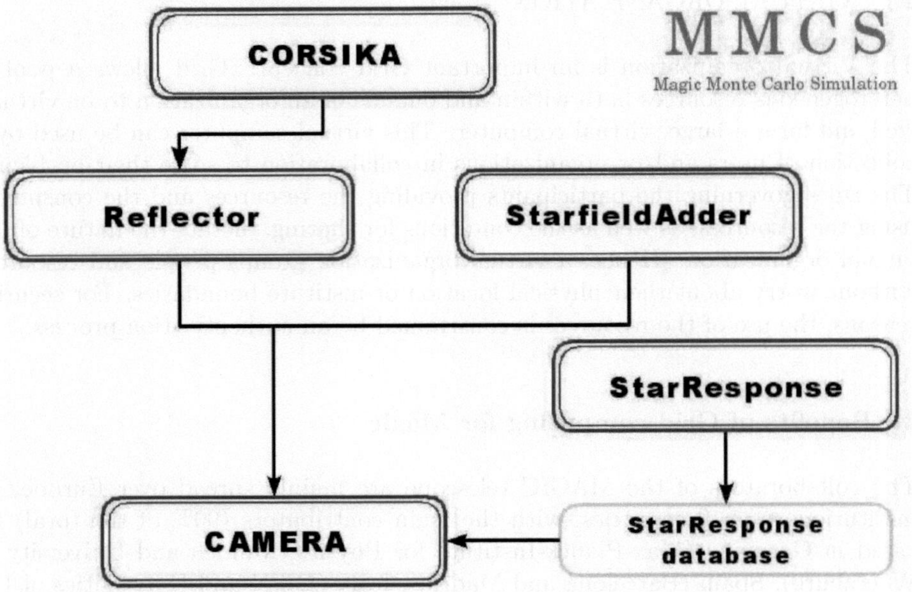

Figure 1. MAGIC Monte Carlo Simulation workflow

exploit the facilities of a distributed system.

3.1 MMCS

The MAGIC Monte Carlo simulation workflow is a series of programs which simulate the properties of different physics processes and detector parts (figure 1):

- *CORSIKA*: air shower and hadronic background simulation. The output contains information about the particles and the Cherenkov photons reaching the ground around the telescope.
- *Reflector*: simulates the propagation of Cherenkov photons through the atmosphere and their reflection in the mirror up to the camera plane. The input for the Reflector program is the output of CORSIKA.
- *StarfieldAdder*: simulation of the field of view. It adds light from the non-diffuse part of the night sky background, or the effect of light from stars, to images taken by the telescope.
- *StarResponse*: simulation of NSB (night sky background) response.
- *Camera*: simulate the behavior of the photomultipliers and of the electronic of the MAGIC camera. It also allows to introduce the NSB (optionally), from the stars and/or the diffuse NSB.

3.2 THE PRESENT

The EGEE (Enabling Grids for E-science in Europe) project brings together experts from 70 organizations and 27 countries with the common aim of building on recent advances in Grid technology and developing a service grid infrastructure in Europe which is available to scientists 24 hours-a-day. Recently MAGIC has became part of the EGEE project. This is the first step for enabling MAGIC on the Grid. This process migration is a big effort and must be divided in smaller steps. The first step was chosen to be the migration of the MAGIC Monte Carlo simulation workflow and it is now working [3] thanks to the effort of the Udine, Padova and CNAF (Bologna) groups.

4 Conclusions

Grid technologies promise to change the way organizations tackle their complex computational and data-intensive problems. The vision of large scale resource sharing is nowadays becoming a reality in many areas. However, it must be realized that Grid is an evolving field in computer science, where standards and technology are still being under development to enable this new paradigm. Presently, many efforts are being made to attract a wide range of new users to the grid. MAGIC has caught this big opportunity and now another step is made towards the realization of a wide project that wants to connect the compute and storage resources of the astroparticle institutions in order to collaborate across the institute borders as well as across the collaboration borders. This new challenge will be the ASTROPA Grid project [4].

References

[1] Lorenz, E., New Astron. Rev. vol. 48, 339-344 (2004).

[2] Foster, I., Kesselman, C. (eds.). The Grid: Blueprint for a New Computing Infrastructure. Morgan Kaufmann, 1999.

[3] M. Piraccini, tesi di Laurea in Fisica Computazionale, Udine 2004.

[4] ASTROPA Grid, proposal by the MAGIC Collaboration to EGEE, 2004.

Part IV

Complex Systems

TOWARDS A PHYSICAL MODEL FOR MEMORY

LEON COOPER [a]

[a] *Brown University, Providence, USA*

Abstract

The Interaction of theory with experiment?normal in physics since Galileo, but novel in neuroscience? has proven fruitful in neuroscience.

Theory has suggested experiments that have uncovered new phenomena:

a. Long Term Depression (LTD) and bi-directional synaptic modification dependent on the depolarization of the postsynaptic cell in agreement with the BCM synaptic modification function.

b. The sliding modification threshold. This has provided experimental verification of the postulates of the BCM theory of synaptic plasticity.

Theory has been able to relate diverse observations in different brain regions such as LTD/LTP in hippocampus to monocular deprivation and reverse suture results in visual cortex.

Theory has also shown that one underlying calcium dependent mechanism can account for the various methods of inducing synaptic plasticity. Cellular and molecular mechanisms that underlie this synaptic plasticity have been proposed, and some of these have been confirmed experimentally. The interaction of local and global modulatory signals, the probable basis for memory consolidation, as well as the implication of LTD mechanisms in such pathologies as the Fragile X Syndrome are being explored.

B.G. Sidharth, F. Honsell and A. de Angelis. (eds.), Frontiers of Fundamental Physics, 341–341.

TOWARDS A PHYSICAL MODEL FOR MEMORY

Abstract

PATTERNS AND DISSIPATIVE WAVES, INCLUDING SOLITONS, IN LATTICES AND AT INTERFACES

MANUEL G. VELARDE [a,b]

[a] *Instituto Pluridisciplinar, Universidad Complutense de Madrid (UCM), Paseo Juan XXIII 1, 28040 Madrid, Spain*
[b] *International Centre for Mechanical Sciences (CISM), Palazzo del Torso, Piazza Garibaldi 18, 33100 Udine, Italy*

Abstract

A succint account is given here of recently discovered salient features of selforganizing processes (disorder-order transitions) leading to cellular convective patterns and dissipative (and solitonic) waves in fluid layers with an interface and in active lattices. Emphasis is given to the recently predicted transition from (Drude) Ohm's linear electric conduction to a form of (high-temperature) superconduction mediated by the formation of electron-soliton (solectron) dynamic bound states in a nonlinear lattice, a phenomenon reminiscent of surfing.

1 Introduction

The apparently unrelated phenomena of cellular convective patterns and waves occurring at fluid interfaces and in discrete lattices can be explained using a common ideology, if not a theory. I refer to the onset of cooperative phenomena driven by a constraint that, on the one hand, leads to a (non-equilibrium) phase transition to a more ordered state and, on the other hand, provides a steady (dynamic) balance between energy input and dissipation. Beyond threshold the form and evolution of the newly formed "dissipative structures" [1] depend on the level of that energy balance sustaining the opportunistic mode (pattern or wave) dominating the others. This enslaving mode or "order parameter" in the supercritical regime is

B.G. Sidharth, F. Honsell and A. de Angelis. (eds.), Frontiers of Fundamental Physics, 343–354.

the "master/conductor" in the selforganizing, synergetic process occurring in the system [2]. Lack of space permits me to just itemize a few salient features with emphasis, in particular, on a recently predicted curious phenomenon triggered by dissipative solitons in a lattice operating as an electric circuit: the transition from (Drude) Ohm's linear electric conduction to an apparently purely classical form of electric (high-temperature) superconduction (the more ordered phase), an intriguing result indeed [3, 4].

2 Cellular patterns at interfaces

The onset and development beyond threshold of (steady) cellular convection driven by interfacial agents, e.g. surface tension gradients, is a typical case of selforganization and cooperative processes in a fluid layer subjected to a thermal constraint, like heating or ad/absorption and (mass) diffusion of a (surface active agent) surfactant. The engine providing the (steady) energy balance that allows the disorder-order transition is the Marangoni effect [5]-[8]. Here a variation of temperature and/or (excess) surfactant concentration at the interface creates shear stress and flow through the coupling between the "thermal" field and the velocity. The essential character of this process is that the engine operates not in the bulk of the fluid but rather at an open surface or interface between two fluids (e.g. an air-liquid interface). The enslaving of the velocity field by the "thermal" one feeds forward on the bulk equations where the "energy" (Fourier or, alternatively, Fick) equation couples to the (Navier-Stokes) equations of motion. Literally, a strict enslaving of the velocity by e.g. the temperature occurs with heated layers of high Prandtl number fluids like silicone oils. Fig. 1 illustrates the onset of cellular convection in one such liquid. It shows a disorder-order transition from a homogeneous, diffusion, motionless state to a finite-wavenumber, cellular, and hence more ordered convective state. This is not the only possibility offered by the dynamics or observed in experiment. Indeed, a long wavelength (zero-wavenumber) convective mode is also possible, and may even coexist with the above mentioned case. The control, or bifurcation, parameter of the dynamics is the temperature difference or the surfactant concentration difference or gradient (e.g. across the layer) creating the surface tension gradient that leads to and sustains the "ordered" flow.

To study the (nonlinear) evolution of the cellular convection illustrated in Fig. 1 both the (Newton/Navier-Stokes) equations of motion and the energy equation are needed. However, this poses a formidable task. A drastic albeit significant and universal approximation to the dynamics valid near and slightly above threshold is possible. Generic arguments permit one to reduce the problem to a set of Landau-Ginzburg equations for the amplitudes of the velocity and the thermal field [9, 10]. This is indeed the natural extension of Landau's approach to disorder-order equilibrium phase transitions where now the amplitudes of the cellular flow are the enslaving modes or order parameters in the evolution. A notable phenomenon is the (unavoidable) appearance of defects in such cellular structures. The defect not only contains significant information in itself but, moreover, contains the whole

Figure 1. Benard cells. A shallow (mm depth) layer of silicone oil open to the ambient air is heated from the bottom side. As the temperature difference across the layer increases there is the transition from the motionless, heat diffusion phase to the cellular convective one. Once the critical temperature difference for the onset of convection is achieved the sequence proceeds with a constant value of this constraint. Velocities are about a hundredth of a centimeter per second. [Details can be found in Refs. 6, 8 and 11].

dynamics as it is known in crystal growth processes. The evolution of a defect, e.g. the screw dislocation is enough to grow the full crystal, perfect or otherwise. The reader is referred to several recent publications where in-depth analyses are provided and due comparison with experimental observations given [8]-[11].

3 Patterns in (active, nonlinear) lattices

Cellular patterns arising from a disorder-order transition with a supporting (steady) input-output energy balance also appear in (active) lattices. Here (think about action potential transmission and pattern dynamics in brain activity or about arrays of lasers or collections of linked mechanical or electronic oscillators or the discrete version of reaction-diffusion systems) the equations of motion may be Newton's equations (in a generalized sense) for the units considered as (nearest-neighbor, n.n. in short) particles interacting with a given force (again in a generalized sense). The "activity" of the units and the role played by the bonds (springs, resistors, etc.) coupling the units necessitates the energy equation. Diffusion, noise, etc. provide the disorder when we consider not a purely deterministic case but a stochastic one. The latter is easy to visualize if we take the units to be immersed in a "heat" bath where the noise level (via Einstein's fluctuation-dissipation theorem) also provides a temperature. Then the disorder-order transition could depend on the temperature as well as on the control or bifurcation parameter of the dynamics that drives the change from one mode of operation to another in the time and/or space evolution of the system. Such a bifurcation parameter may affect only the "activity" of the units or the "interaction" between units or both. By "activity" of a unit I mean that if left alone it has an intrinsic evolution independent of the others. These units may all be the same, e.g. electronic oscillators with a rich panoply of dynamic states, from the simplest motionless one to bistability of steady states or oscillations, excitability or even time-chaotic states. Through cou-

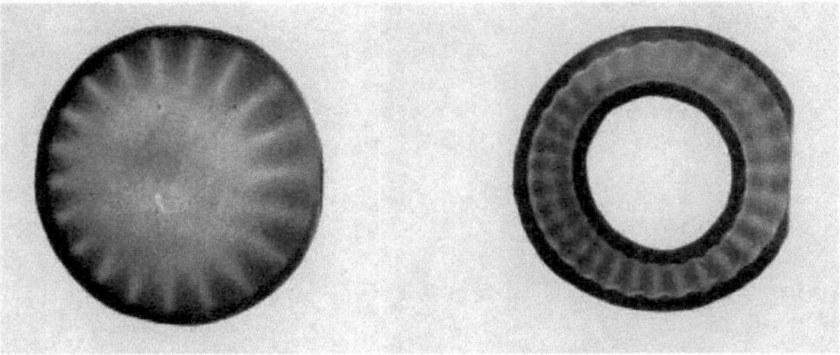

Figure 2. Typical replication of a pattern (a) using raw material (b) following the mutual interaction of two "active" lattices (130×130 units). The two (identical) off-springs, (c) and (d), almost perfectly reproduce the pattern but this need not be the case in general; combination/competition of patterns is also possible. The outcome of "evolution" depends on parameter values controling the "activity" of units and bonds/interactions, here between n.n. [Details can be found in Refs. 12 and 13].

pling and suitable synchronization patterns may appear in the lattice. Note that in a 2D, N×N lattice, if a unit possesses two accessible states, diffusive coupling, a resistor in an electric circuit/lattice, there may be as many as 2^{N^2} patterns, an extraordinary wealth indeed. Furthermore we can replicate a given pattern in a (father) lattice by dynamically "imprinting" it on a disordered (mother) lattice coupled to the former; the disordered state is also a possible steady pattern of the system [12, 13].

Fig. 2 illustrates two steps, initial and final, of one such replication process with two identical offsprings, almost cloning the father but this need not to be so. It is a dynamic process and not the mere action of a photocopying machine. We see again how significant the role of defects may be in this replication process. Replicating a given pattern is in general a difficult problem in the transfer process. Hence for purposes of information only it may be enough to replicate a defect in the original and nothing more. As a lattice and its dynamics with active units may be considered as a discrete version of a (continuous) reaction-diffusion system, the evolution of the "order parameters" (patterns) in many such problems can again be described, in approximate way, by a set of generic Landau-Ginzburg equations, much as in the case of cellular fluid flows described above.

4 Dissipative waves and solitons in fluid layers

Earlier we illustrated the onset of (steady) cellular convective patterns occurring at the open surface of a heated liquid layer. As indicated in the caption of Fig. 1 the heating is done from the liquid side, i.e., from the solid support below the liquid, while the open atmosphere is (mechanically passive) air. [N.B. The dynamic or shear viscosity of air is much lower than that of most liquids, hence air exerts negligible traction at the open surface]. In fact, if the ambient atmosphere is another liquid, cellular patterns may appear for either way of heating albeit depending on the values of the ratios of the corresponding (heat or mass) diffusivities and kinematic viscosities [14]. [N.B. This is a significant statement because the latter quantities control the diffusion of vorticity while the dynamic or shear viscosity comes in the stress balance at the interface and accounts for the relative traction of one liquid over the other (hence controlling momentum transfer)]. Restricting consideration to the case of a liquid layer open to air, what is the expected outcome of heating from the ambient atmosphere or cooling from the solid support? Theory predicts and experiments have confirmed that various types of waves, with and without significant surface deformation, transverse or longitudinal waves (and even internal waves in the otherwise stably stratified layer) can be excited and, eventually, sustained past threshold, again by the Marangoni "engine" operating at the open surface. The argument applies verbatim to the case of ad/absorption of a "light" surfactant that lowers the surface tension of the liquid in the layer. A surprising result is that some of the waves appearing at the open surface have solitonic features [11, 15]. For instance, as solitary waves or periodic wave trains (series of periodic wave peaks) they travel undeformed for long times, they cross/collide with each other with no appreciable change of form, they reflect at solid surfaces exhibiting Mach-Russell's stems, or not, according to the incident angle (a phenomenon also occurring in wide/obtuse angle collisions). These solitonic waves are very much like surf waves or hydraulic jumps (bores in English and mascarets in French) moving (upstream) in certain rivers and shallow channels and straits. [N.B. Solitary waves were found in canals. They were beautifully described by Russell, around the mid-XIXth Century. He also proposed, in 1840, that solitary waves could be used as a clever way to traverse canals (like surfing) in boats whose keel lengths might be greater than the depth of the canal. However, solitons were first described in the evolution of nonlinear lattices and their (quasi) continuum approximations around mid-XXth Century. Pioneering work by Fermi, Pasta and Ulam was followed by that of Zabusky and Kruskal (who coined the word soliton) and others].

Fig. 3 illustrates dissipative solitonic wave trains [16]. Here, the threshold for instability occurs in the long wavelength limit (zero wavenumber). In the energy balance the input of energy comes in this long wave range while dissipation (due to viscosity) occurs in the short wave range. In the zero-wavenumber case the approximation to the equations of motion and the energy equation cannot be a set of Landau-Ginzburg equations because there is a Goldstone mode (the conservation of liquid does not permit lifting the surface all at once in a homogenous way). The natural approach is that opened by Boussinesq and, subsequently, taken by others leading to a wave equation for the corresponding order parameters (e.g., for a one-sided propagation the Boussinesq-Korteweg-de Vries equation), augmented by the energy balance [11]. Mathematically, the original hyperbolic equations are then modified by (super) parabolic terms (e.g., the dissipation modified BKdV equation). Hence their solutions have been called dissipative solitons [17] as they are the generalization of the soliton concept introduced in conservative (hyperbolic) systems (for fluids those obeying the Euler equations and not the Navier-Stokes equations). As in the earlier cases discussed here there is a (non-equilibrium) transition from a disordered state, the motionless case or even a flow regime, to

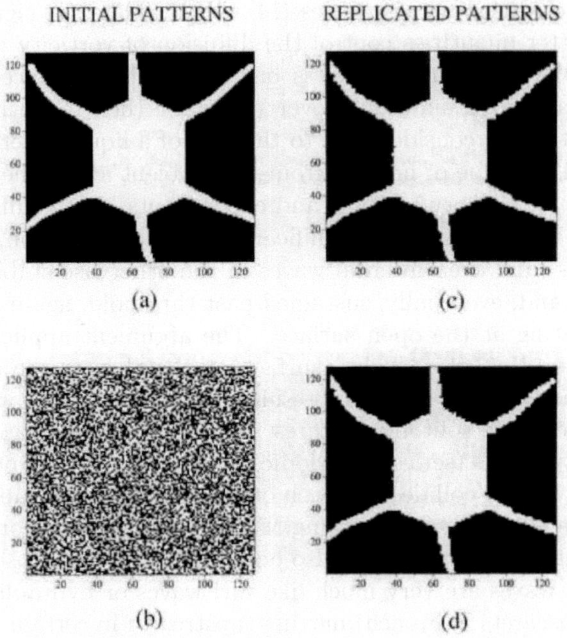

Figure 3. Typical dissipative periodic wave trains (like cnoidal waves) whose wave peaks/crests have been shown to exhibit solitonic features. The experiment corresponds to heating the liquid layer from the ambient air side or ad/absorbing a "light" surfactant, e.g., pentane vapor ad/absorbed by liquid toluene or hexane. Velocities are about a centimeter per second, hence two orders of magnitude higher than the velocities in Fig. 1 corresponding to steady convection. [Details can be found in Refs. 11 and 16]

a solitonic wave regime, the more ordered state, sustained by the steady input-output energy balance.

5 Dissipative solitons in (active) lattices and superconducting circuits

In the final section of this short text let me describe what can occur in a lattice of "active" units if we take ions as the units, electrons as "boats", and an electric field as the "wind" and/or powerful horses, all immersed in a heat bath. Can an electron surf on a nonlinear lattice of strongly interacting (positive) ions? What follows is a rough account of work done with W. Ebeling and A. P. Chetverikov [3, 4].

Toda showed that a 1D lattice with a suitable exponential (hence nonlinear) interaction between n.n. units of the form [18]

$$U^T(r) = \frac{a}{b}\left(\bar{e}^{br} - 1\right) + ar; \; a, b > 0, \tag{1}$$

has an exact solution in the form of solitonic (cnoidal or solitary) waves. The force corresponding to (1) is

$$F(r) = a\left(\bar{e}^{br} - 1\right) \text{ or else } F_n = a\left(\bar{e}^{br_n} - 1\right). \tag{2}$$

Thus for positive r the force is attractive (and constant) while for negative r it is repulsive (and very much like hard rods when b is large and $a \to 0$; this is the case of strong compression of the spring defined by (2)). [N.B. A more realistic interaction is the Morse potential, a combination of two exponentials that treats adequately the attractive component and fits well the Lennard-Jones interaction. However, we have no exact solutions of the corresponding equations of motion. Both the Toda and Morse potentials are easy to implement electronically]. The Taylor expansion of (1) yields the (cubic and quartic) anharmonic interactions used by Fermi, Pasta and Ulam in their pioneering computational physics work (with the earliest computer available at the Los Alamos Labratory) and of current use in solid state theory to explain equipartition, thermal expansion and heat transfer in a lattice (for b small the compression-expansion obeys Hooke's law and the motions are harmonic vibrations or phonons in the lattice). [N.B. The pure Toda lattice is transparent to heat but here we use more than the pure Toda interaction]. The solution found (constructed) by Toda is (using the force)

$$\bar{e}^{br_n} - 1 = \sinh^2 \kappa \operatorname{sec h}^2 (\kappa n - \beta t + \delta), \tag{3}$$

hence defining a solitary wave propagating supersonicaly with velocity

$$c = \frac{|\beta|}{\kappa} y = y\left(\frac{ab}{m}\right)^{1/2} \frac{\sinh \kappa}{\kappa} > \left(\frac{ab}{m}\right)^{1/2}, \tag{4}$$

where y accounts for the mean interparticle distance between the units (of mass, m).

For a 1D lattice of length L, with periodic b.c. $x_{i+N} = x_i + L$ we have

$$U = \sum_{i=1}^{N} U_i^T(r_i), \tag{5}$$

with U_i^T given by (1) with $r \to r_i - \sigma$, $r_i = x_{i+1} - x_i$, $\sigma > 0$. Then the (Newton) equations of motion of the units are

$$m\frac{dv_k}{dt} + \frac{\partial U}{\partial x_k} = 0. \tag{6}$$

To these deterministic and Hamiltonian equations we shall add an energy equation after having also added the following: (i) the consideration of the units as Brownian particles in a heat bath (white Gaussian noise, delta correlated in space -discrete units- and time), (ii) as Brownian particles they experience dissipation/friction and fluctuations and, eventually, they are able to pump energy, first internal and, subsequently, kinetic energy/motion from the (disordered) noise, (iii) the units are charged particles, e.g., ions with positive charge ($+e$), and (iv) the lattice/system is subjected to an external electric field, E. Thus the equations of motion (6) become stochastic (Langevin) equations,

$$m\frac{dv_k}{dt} + \frac{\partial U}{\partial x_k} = e_k E + F(v_k) + \sqrt{2D}\xi_k(t), \tag{7}$$

where the force F accounts for two terms. We set $F(v_k) = -m\gamma(v_k)v_k$, with $\gamma = \gamma_0 + \gamma_1$. Here γ_0 is the standard (fluctuating) friction coefficient obeying Einstein's fluctuation-dissipation theorem. The other term, γ_1, is taken to be nonfluctuating and with values at our "will" in view of the energy balance to be satisfied in the evolution of the system.

Let us introduce now a particular form of the energy balance [19]. Each Brownian particle is considered as "active", capable of pumping internal energy from the bath (noise), dissipating internally and transforming internal energy into motion, hence feed forwarding the equations of motion. [N.B. We are now exploring the possibility of direct energy pumping by "passive" ($\gamma_1 = 0$) particles from the electric field alone taken indeed as "wind"]. Let $\epsilon(t)$ be the energy, then we may write

$$\frac{d\epsilon}{dt} = q(t) - c\epsilon(t) - d(v)\epsilon(t), \tag{8}$$

which is reminiscent of metabolic processes in biosystems. The quantity $q(t)$ accounts for the energy flux from the bath to the unit's "internal" depot. The internal dissipation is assumed proportional to the actual energy level with a constant rate of energy loss, c. The third term describes how internal energy is converted into motion, with a rate depending on the velocity value, $d(v)$, and indeed proportional to the actual energy level. For simplicity, we assume a steady state, $q = q_0$, and $d(v) = dv^2$, with d constant and positive. Hence if at the initial time $\epsilon(0) = \epsilon_0$ we have

$$\epsilon_0 = q_0 / \left(c + dv^2\right). \tag{9}$$

Then we should write $\gamma_1 = -q_0 d / \left(c + dv^2\right)$. However, to further simplify we shall use $\gamma_1 \approx \gamma_1^R \approx \gamma_0 v^2$, which is a function introduced by Lord Rayleigh [20] that approximates γ_1 well for negative values, when γ_1 acts opposite to γ_0, as an energy pumping (negative friction) mechanism. The two quantities γ_1 and γ_1^R differ drastically for positive values where γ_1 saturates while γ_1^R diverges when v is large. Hence we write

$$F(v) = m\gamma_0 v \left[\mu - \frac{v^2}{v_1^2}\right], \tag{10}$$

where for convenience we have rewritten γ_1 in terms of two new parameters μ and v_1. The former, μ, is now the key parameter of the problem. Indeed for $\mu = 0$ the montionless state is the only stable phase of the system that for μ positive yields a new state with motion (expected to be the more ordered phase), $v = \pm v_0 = v_1\sqrt{\mu}$.

To the 1D lattice of ions we now couple a 1D lattice of electrons. For simplicity we take only one electron (-e) whose mass is a thousand times smaller than that of the ions (hence $m_e\gamma_{0e} << m_i\gamma_{0i}$; e denotes electron and i, ion). The electrostatic

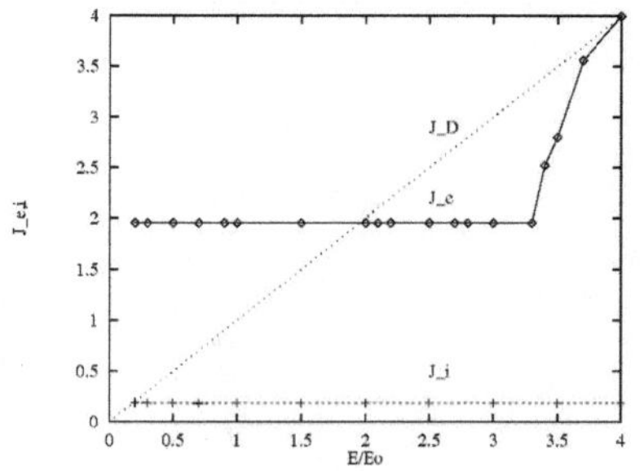

Figure 4. Typical current-voltage/field characteristics of the electrically conductive Toda lattice described in the main text. The dotted diagonal line corresponds to (Drude) Ohm's law, the broken line at the bottom is the ionic current (can be set to zero or a reference value by changing the frame of reference), and the solid line (with dots) is the solectronic current, which significantly deviates from Ohm's law as we lower the field intensity. [Details can be found in Refs. 4 and 5]

electron-ion interaction is taken as the pseudopotential

$$U_e\left(y_j,\, x_k\right) = \frac{(-e)e_k}{\left[\left(y_j - x_k\right)^2 + h^2\right]}, \tag{11}$$

(h denotes a cut-off, $h \approx \sigma/2$, with σ the interion mean distance; note that to rule out unnecessary difficulties the interaction (11) is taken in 3D). Hence to the equations of motion of the ions we add the equation of motion for the electron

$$\frac{m_e d^2 y_j}{dt^2} + \frac{\partial U_e}{\partial y_j} = -eE - m_e \gamma_{eo} v_j + \sqrt{2D_e}\xi_j(t). \tag{12}$$

Ions and electrons with a field E create currents. The current density (per unit length) a la Drude is

$$j_D = \left(e^2/m_e\gamma_{e0}\right)E, \tag{13}$$

while the total current is

$$j = e_i \left\langle v_i \right\rangle - e \left\langle v_e \right\rangle. \tag{14}$$

The brackets indicate average values taken over numerous computer simulations of the motions. Figs. 4 and 5 illustrate results found for $\sigma = 1$, $a = 13.69$, $b = 1$, $h^2 = 0.08$, $m_i = 10^3 m_e$, $m_e = 1$, $\gamma_{e0} = \gamma i0 = 1$, $\mu = 1$, and $v_1 = 1$. Upon compressing two nearby ions, thus forming an electrostatic valley, the electron tends to seat on its minimum. [N.B. By plotting (11) for one electron between two positive ions it can be seen that given any compression (respectively, h) there is always a value of h (respectively, compression) providing such a valley].

The electric field breaks the symmetry pushing the electron off-center/off-minimum. Then the solitonic wave catches the electron and displaces it to the nearest ion-pair (some kind of promiscuity of the electron) and so on. This is the soliton-electron ("solectron") dynamic bound state that, thanks to the energy balance, sustains the ordered state with higher current. Note that the actual value

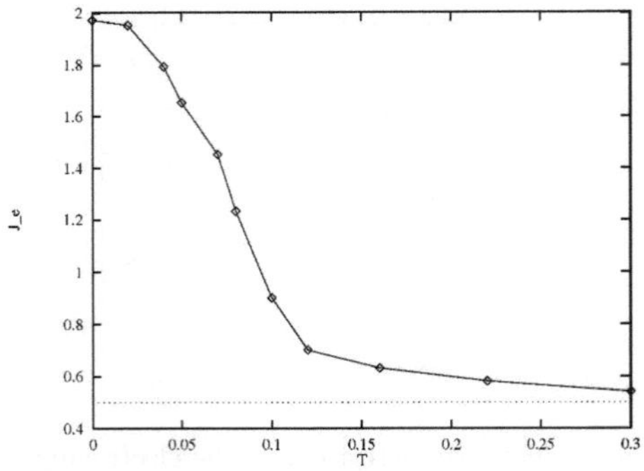

Figure 5. Typical current-temperature characteristics for the lattice of Fig. 4 when we lower the noise level, and hence the temperature, showing a significant increase in the value of the solectronic current (solid line with dots) relative to Ohm's law (dotted line at bottom). [Details can be found in Ref. 4]

of the field intensity is immaterial. What really matters is its symmetry breaking role. As the phenomenon exhibits a transition from (Drude) Ohm's linear current to a kind of supercurrent, the latter is indeed the more ordered state (recall that this phase transition occurs at $\mu = 0$). The results found with such (an apparently) simple lattice model are intriguing for they are reminiscent of experimental curves obtained with high-T superconductors [21, 22].

6 Concluding remarks

The sketchy account here given, as a summary of my presentation at the Sixth International Symposium on Frontiers of Fundamental and Computational Physics, hopefully offers an overview of fascinating phenomena occurring in driven, non-equilibrium, dissipative, systems when nonlinearities dominate their evolution. I have focused on work I have been doing in the recent past, years and weeks, and hence there is indeed much more in the field that I have not described. Reference to publications other than mine is expected to help the reader proceeding on his/her own.

In all cases described we have seen how disorder-order transitions occur when the equations of motion, being neither Hamiltonian and integrable nor conservative, are augmented with an appropriately coupled energy equation providing, e.g., a steady input-output energy balance. Steady patterns or dissipative waves (including solitons) are typical cases of "selforganization" and "ordering" in fluid layers (with an interface) or lattices (discrete version of, e.g., reaction-diffusion systems). I have emphasized an intriguing result found in a 1D lattice with nonlinear (anharmonic) interactions: the transition from the standard linear electric conduction (Drude-Ohm) to a form of (an apparently purely classical high-T) superconduction mediated by the formation of electron-soliton (solectron) dynamic bound states. Is it not fun to see how an electron could "surf" on a nonlinear

(Toda) lattice?

Despite the fact that I have been talking about a driven-dissipative system, one should be aware of one train hiding another and also of pictures that look alike yet refer to different objects! Linking the predictions made using the drastically simplified 1D lattice model studied here with real data about high-T superconductors is beyond the scope of this discussion. However, should the model (or variations on it like using the Morse potential or removing, one way or another, the "activity" of the particles, etc.) be derived from first principles (quantum mechanics) by generalizing Bloch's theorem (a non trivial problem since we have nonlinearity and time-dependence from the beginning), should (relatively strong) spring compressions be possible in a conductor/metal, should the major predictions in 1D hold for 2D lattices (with a Fermi-like level), etc. then it might be that the present results already provide a key to understanding "room" temperature superconductivity: the formation of "solectrons" and the (nonlinear) cooperativity leading to the disorder-order transition.

References

[1] G. Nicolis, I. Prigogine, *Self-Organization in Non-Equilibrium Systems. From Dissipative Structures to Order through Fluctuations*, Wiley, New York (1977)

[2] H. Haken, (a) *Synergetics. An Introduction* (3rd Ed.), Springer, Berlin (1983); (b) *Advanced Synergetics. Instability Hierarchies of Self-Organizing Systems and Devices*, Springer, Berlin (1983)

[3] M. G. Velarde, W. Ebeling, A. P. Chetverikov, *Int. J. Bifurcation Chaos* (to appear)

[4] A. P. Chetverikov, W. Ebeling, M. G. Velarde, *Phys. Rev. Lett* (submitted)

[5] M. G. Velarde, Ch. Normand, *Sci. American* 243 (1), 92-108 (1980)

[6] E. L. Koschmieder, *Benard Cells and Taylor Vortices*, Cambridge Univ. Press, Cambridge (1993)

[7] Ch. Normand, Y. Pomeau, M. G. Velarde, *Rev. Modern Phys.* 49, 581-624 (1977)

[8] P. Colinet, J. C. Legros, M. G. Velarde, *Nonlinear Dynamics of Surface-Tension-Driven Instabilities*, Wiley-VCH, New York (2001)

[9] M. C. Cross, P. C. Hohenberg, *Rev. Mod. Phys.* 65, 851-1112 (1993)

[10] I. S. Aranson, L. Kramer, *Rev. Mod. Phys.* 74, 99-143 (2002)

[11] A. A. Nepomnyashchy, M. G. Velarde, P. Colinet, *Interfacial Phenomena and Convection*, Chapman and Hall/CRC/Pitman, London (2002)

[12] M. G. Velarde, V. I. Nekorkin, V. B. Kazantsev, J. Ross, *Proc. Nat. Acad. Sci. USA* 94, 5024-5027 (1997)

[13] V. I. Nekorkin, M. G. Velarde, *Synergetic Phenomena in Active Lattices. Patterns, Waves, Solitons, Chaos*, Springer-Verlag, Berlin (2002)

[14] R. V. Birikh, V. A. Briskman, M. G. Velarde, J. C. Legros, *Liquid Interfacial Systems. Oscillations and Instabilities*, M. Dekker Inc., New York (2003) (and references therein)

[15] A. C. Scott, *Nonlinear Science*, Oxford Univ. Press, Oxford (1999)

[16] H. Linde, M. G. Velarde, A. Wierschem, W. Waldhelm, K. Loeschcke, A. Ye. Rednikov, *J. Colloid Interface Sci.* 188, 16-26 (1997). This is the first of a series of papers in the same journal where reference to further publications by the same authors can be found.

[17] C. I. Christov, M. G. Velarde, *Physica D* 86, 323-347 (1995)

[18] M. Toda, *Nonlinear Waves and Solitons*, Kluwer, Dordrecht (1983)

[19] W. Ebeling, I. M. Sokolov, *Statistical Thermodynamics and Stochastic Theory of Nonequilibrium Systems*, World Scientific, London (to appear)

[20] J. W. S. Rayleigh, *Theory of Sound*, Dover Edition, New York (1945)

[21] P. Chaudhari, J. Mannhart, D. Dimos, C. C. Tsuei, J. Chi, M. M. Oprysko, M. Scheuermann, *Phys. Rev. Lett.* 60, 1653-1656 (1988)

[22] J. Mannhart, P. Chaudhari, D. Dimos, C. C. Tsuei, T. R. McGuire, *Phys. Rev. Lett.* 61, 2476-2479 (1988)

LATTICE PROTEIN MODELS: A COMPUTATIONAL APPROACH TO FOLDING AND AGGREGATION PHENOMENA

MARCO CITOSSI [a] AND GILBERTO GIUGLIARELLI [a,b]

[a] *Dipartimento di Fisica, Università di Udine, I–33100 Udine, Italy*

[b] *INFM–Dipartimento di Fisica, Università di Padova, I–35131 Padova, Italy*

Abstract

Square lattice protein models are used to study the competition between folding and aggregation phenomena. The problem is approached by considering Metropolis Monte Carlo simulations of non-isolated lattice protein models; different protein molecules can interact each other and, in competition with folding, can aggregate by forming dimers. The calculations take in exam the behavior of three types of proteins: *a*) proteins with a very well designed sequence (good folders); *b*) proteins which folding kinetics present kinetic partitioning effects (intermediate folders); *c*) small proteins with native states having the geometry of pure secondary structure motives (like alpha helices or beta sheets). The results show that good folders very rarely form aggregates; on the contrary, in almost all considered cases, intermediate folders display high tendency to form dimers. Finally, alpha helices display a low tendency to aggregate in comparison to that found for beta-sheets. However, also for these systems, structural intermediates in the folding kinetics can strongly influence the aggregation tendency.

B.G. Sidharth, F. Honsell and A. de Angelis. (eds.), Frontiers of Fundamental Physics, 355–358.

1 Introduction

Protein molecules are found in all superior living systems and contribute at many levels to their functionality. The folding process of a protein molecule consists in its self–assembly into a specific conformation: the *native state.* Most of proteins found in nature are good folders and therefore self–assembly to the native conformation efficiently and reversibly: they have very small folding times [1]. In this respect, besides the intrinsic properties relied to a specific protein amino acid sequence, the efficiency of the folding process can be influenced by native state geometry (secondary structure motives), as well as, by the presence of structural intermediates in the folding path (*kinetic partitioning mechanism*) [2].

Failures in protein folding bring to misfolding events and possibly to aggregation phenomena [3]. A striking example if these phenomena is offered by prion proteins misfolding; although not yet completely clear, such events seem to trigger very unhealthy protein aggregations processes degenerating in disorders like bovine spongiform encephalopathy (BSE) [4].

In the present work we present some preliminary results of a study in which equilibrium and non–equilibrium properties of single and interacting protein molecules are analyzed with the purpose of getting insights into the competition between folding and aggregation mechanisms. The study, based on lattice protein models and Metropolis Monte Carlo simulations, reveal that, in most of the cases, protein molecules with tendency to aggregation are also affected by kinetic partitioning effects. Moreover, on a preliminary base, we observed some relations between these effects and the α–helix or β–sheet nature geometry of the protein native state.

2 Model and calculation methods

In our minimalist model a protein chain is seen as a self–avoiding walk in a 2D square lattice. We have considered chain lengths ranging between 20 and 36 and amino acid interactions have been simulated by a pair–wise interaction energy [5] between nearest neighbors (extra chain) beads.

Amino acid sequences used in the work have been selected to have near 45% of hydrophobic residues and, then, have been optimized (Z-score optimization) to have a preselected native states.

Equilibrium thermodynamics calculations has been obtained on the basis of several Metropolis Monte Carlo simulations [6] done at different temperatures; equilibrium averages were then computed by means of multi–histogram technique [7]. In the simulations concerning interacting molecules, two different chains were considered and, correspondingly, the Monte Carlo algorithm was modified in a way that elementary moves were accept with the additional condition to maintain the distance between the center of mass of the molecules, $D_{c,m}$, below a preselected value. Folding kinetics were based on several (up to 200-500) relaxation Monte Carlo trajectories to chain native state, after a temperature quenching.

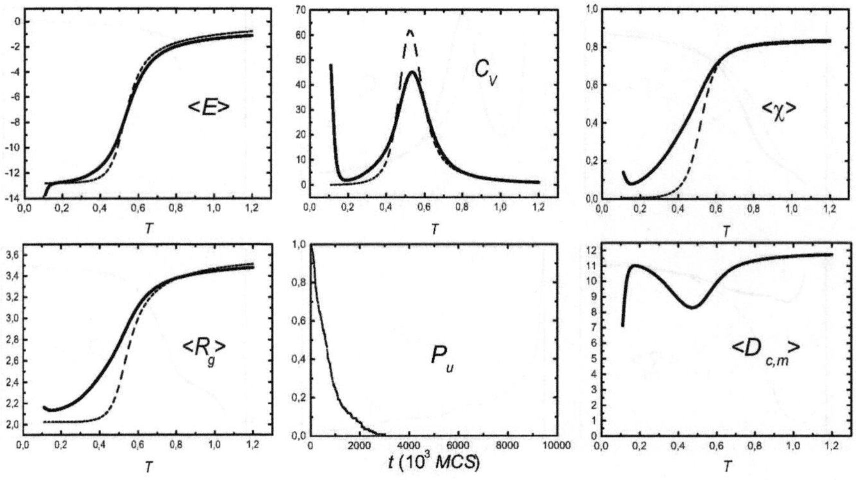

Figure 1. Typical behavior of a 25–beads good folder. Temperature dependence of the average values of energy ($\langle E \rangle$), heat capacity (C_V), χ factor ($\langle \chi \rangle$) and radius of gyration ($\langle R_g \rangle$) for single molecule (dashed lines) and two interacting molecules (continuous lines) simulations. Good folder character of the chain is stated by the pure exponential and fast decay of the fraction of unfolded molecules, P_u, (obtained over 500 trajectories). Note that the chain show no tendency to aggregation; the average distance between the centers of mass, $D_{c,m}$ is always well above the minimal distance between centers of mass of folded chains.

3 Results

We have considered a rather systematic analysis of thermodynamics and kinetics of both isolated and interacting protein chains. The series of calculations done for 2D protein models have revealed that:

- very good folders are usually not affected by aggregation phenomena. The good design of the sequence prevents the protein from aggregation also in very unfavorable conditions (for an example see Fig. 1).
- folders exhibiting kinetic intermediates are found to have a good propensity to form aggregates. In particular, models with β–sheet like native states seem to form aggregates easily than others (for an example see Fig. 2).
- folders with α–helix like native states are found more rarely to form aggregates (in particular with respect to models with β–sheet like native states). However, also in this case the presence of kinetic intermediate favor the formation of aggregates.

We are working to extend the calculations to 3D lattice models (also with higher coordination).

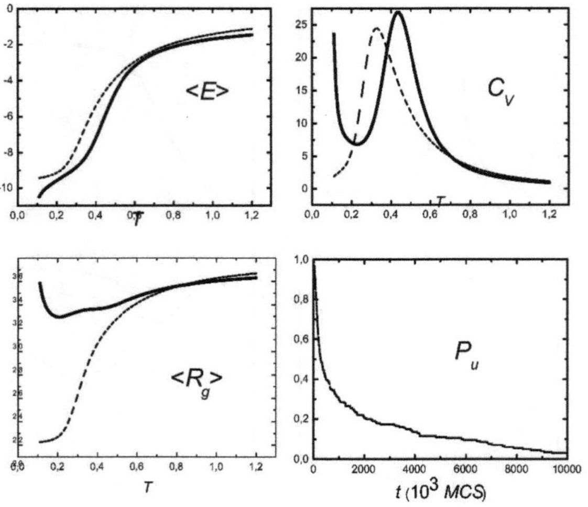

Figure 2. Behavior of a 25–beads β–sheet like folder exhibiting tendency to aggregation (as in Fig. 1 dashed and continuous lines refer to single and interacting molecules simulation, respectively). Note that now, As the temperature is decreased χ factor ($\langle\chi\rangle$) and radius of gyration ($\langle R_g \rangle$) do not goes to the native state prescribed values; C_V peak is now shifted a higher temperature. Fraction of unfolded molecules (average over 500 trajectories) display a double exponential decay; the analysis of the trajectories reveal the presence of kinetic partitioning effects and a structural intermediate. The chain show high tendency to aggregation; average $D_{c,m}$ reaches very small values and the resulting aggregate structure is very different from chain native state.

References

[1] D. K. Klimov and D. Thirumalai, J. Mol. Biol. **282**, 471 (1998); C.J. Camacho and D. Thirumalai, Proc. Natl. Acad. Sci. USA **90**, 6369 (1993); H. S. Chan and K.A. Dill, J. Chem. Phys. **100**, 9238 (1994).

[2] T. Kiefhaber, Proc. Natl. Acad. Sci. USA **92**, 9029 (1995); D. Thirumalai, D. K. Klimov and S. A. Woodson, Theor. Chem. Acc. **96**, 14 (1997); A. Matagne, S. E. Radford and C. M. Dobson, J. Mol. Biol. **267**, 1068 (1997); G. Giugliarelli, D. K. Klimov and D. Thirumalai, to be published.

[3] H. S.Chan and K. A.Dill, Proteins **24**, 335 (1996); G. Giugliarelli, C. Micheletti, J. R. Banavar and A.Maritan, J. Chem. Phys. **113**, 5072 (2000).

[4] S. B. Prusiner, Science **252**, 1515 (1991); S. B. Prusiner and S. J. DeArmond, Annu. Rev. Neurosci. **17**, 311 (1994); F. E. Cohen, K.–M. Pan, Z. Huang, M. Baldwin, R. Fletterick, and S. B. Prusiner, Science **264**, 530 (1994).

[5] A. Kolinski, A. Godzik and J. Skolnik, J. Chem. Phys. **98**, 7420 (1993).

[6] C. Micheletti, F. Seno, A. Maritan, and J. R. Banavar, Phys. Rev. Lett. **80**, 2237 (1998); C. Micheletti, A. Maritan and J. R. Banavar, J. Chem. Phys. **110**, 9730 (1999).

[7] A. M. Ferrenberg and R. H. Swendsen, Phys. Rev. Lett. **63**, 1195 (1989).

INTERFACE DEPINNING FROM WEDGES WITH A CENTRAL RIDGE

GILBERTO GIUGLIARELLI [a]

[a] *Dipartimento di Fisica, Università di Udine, I–33100 Udine, Italy and INFM–Dipartimento di Fisica, Università di Padova, I–35131 Padova, Italy*

Abstract

The depinning properties of a fluctuating interface near $2D$ and $3D$ wedges with a central ridge are studied by discrete models with short range interactions. The calculations demonstrate that, in both cases, depinning take place in two stages: *i*) a continuous filling-like transition in the pure wedge-like components of the system; *ii*) a final discontinuous jump from the central ridge. In $2D$ an exact transfer matrix approach shows that, in the thermodynamic limit, the threshold of the depinning from the central ridge coincides with the one corresponding to the continuous filling transition. In $3D$, on the contrary, accurate Metropolis Monte Carlo simulations show that the two transitions are separated by a finite gap. The mechanism at the basis of the phenomenon is studied in detail and, in $2D$, the whole interface phase diagram and free energy profiles are provided. The physical scenario emerging from these results is discussed also in relation with the problem of the wetting transition in the case of random rough walls.

1 Introduction

The wetting properties of the liquid film forming when an undersaturated vapor is put in contact with a solid inert substrate are, generally, determined by nature and range of intermolecular interactions (for a review on these phenomena see [1]). But these properties can be strongly influenced by substrate surface geometry. For example, adsorption isotherms of random rough and linearly sculpted substrates [2]

B.G. Sidharth, F. Honsell and A. de Angelis. (eds.), Frontiers of Fundamental Physics, 359–363.

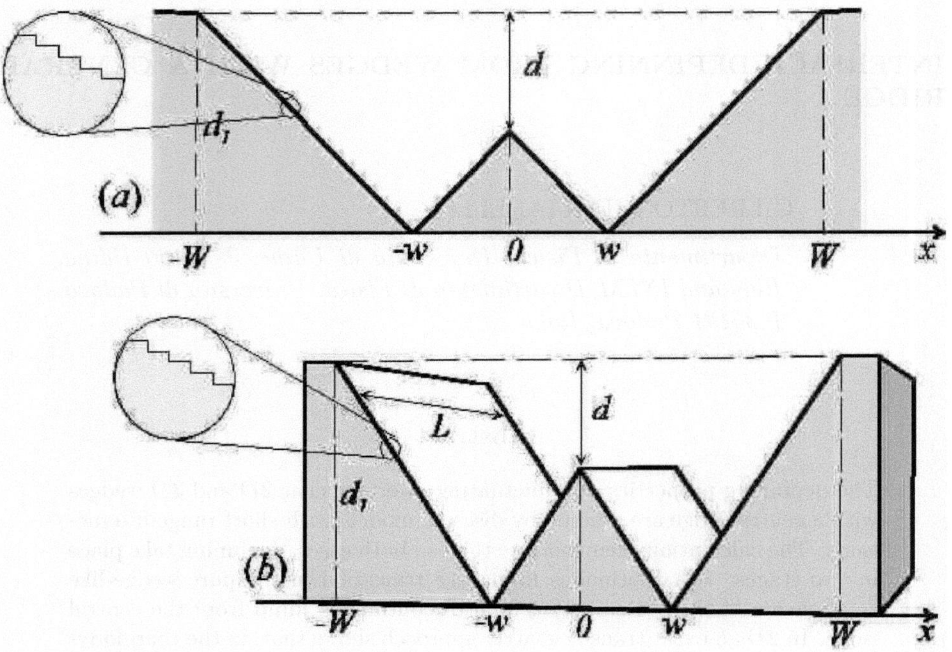

Figure 1. Sketch of the $2D$ (panel (a)) and $3D$ (panel (b)) SW geometries studied in this paper. The *magnifying glasses* show the staircase nature of the tilted smooth walls (with slope $|1/n|$) by which the SW are constructed

exhibit unusual geometry–determined exponents; while partial wetting prevents the growth of macroscopic films, in pure wedges, one finds continuous filling phenomena [3]. Finally, there are many indications that increasing surface roughness can drive wetting transition from second to first–order [4].

In this work we report the results of an accurate investigation about the unbinding properties of a thermally fluctuating interface from substrate geometries (see Fig. 1), we denoted *structured wedges* (SW). Such a surface structure is motivated by the fact in these systems coexist contrasting geometrical motives (like wedges and ridges) which are usually present in the geometry of rough surfaces, which combination determine peculiar effects. In the work we have considered both $2D$ (Fig. 1(a)) and $3D$ (Fig. 1(b)) SWs.

2 The Model

We construct a liquid–vapor interface in the framework of solid on solid (SOS) approach. Therefore, the interface corresponds to a lattice random walk (in $2D$) or to a random surface (in $3D$) in the vicinity of a fixed substrate boundary. Denoting with H_X the substrate boundary (integer) height at X, each interface configuration can be specified in terms of the local relative (integer) height variables z_X.

At coexistence the interface can be studied in terms of an Hamiltonian of the form

$$\mathcal{H} = \sum_{\langle X,X'\rangle} \mathcal{E}(1 + |h_X - h_{X'}|^{\gamma}) - \mathcal{U} \sum_X \delta_{z_X,0}, \tag{1}$$

where the first sum is done over all pairs of nearest neighbor columns, $\mathcal{E}$ and $-\mathcal{U}$ (with $\mathcal{E},\mathcal{U} > 0$) are the energy cost of any interface step (or plaquette in $3D$) and the energy gain of each interface contact (of horizontal step or plaquette) with the substrate, respectively, and $\gamma = 1$ or ∞ determines the SOS or restricted SOS (RSOS) character of the implemented walk model, respectively.

In $2D$ the problem can be fully treated by a transfer matrix approach based on the following definitions

$$[\hat{\mathbf{R}}_x]_{z,z'} = \omega^{|z'-z+H_{x+1}-H_x|^{\gamma}} k^{\delta_{z',0}}, \tag{2a}$$

$$[\hat{\mathbf{L}}_x]_{z,z'} = \omega^{|z'-z+H_{x-1}-H_x|^{\gamma}} k^{\delta_{z',0}}, \tag{2b}$$

where $\omega = \mathrm{e}^{-\mathcal{E}/k_BT} = \mathrm{e}^{-1/t}$ $(t = k_BT/\mathcal{E})$ and $k = \mathrm{e}^{U/k_BT} = \mathrm{e}^{u/t}$ $(u = U/\mathcal{E})$ correspond to step and wall fugacities

These matrices allows, by iterations, the direct calculation of the interface distance probability distribution functions (PDF) at any position along the boundary, P_{x_0}, in terms the corresponding ones, $P^{(R)}_{x_0}$ and $P^{(L)}_{x_0}$, for right and left travelling walks. That is

$$P_{x_0}(z) = P^{(R)}_{x_0}(z) P^{(L)}_{x_0}(z), \tag{3}$$

and the quantity

$$\Delta f_{x_0} = -\ln P_{x_0}(z) = -\left[\ln P^{(r)}_{x_0}(z) + \ln P^{(l)}_{x_0}(z)\right], \tag{4}$$

can be seen as the corresponding local excess interface free energy profile (more details about the treatment will be published elsewhere [5].

Specializing the method to our particular surface geometry, the wetting problem can be further reduced to an eigenvalue problem which, depending on the cases, can be exactly or numerically solved.

Transfer matrix approach is not applicable to the wetting problem in $3D$. In this case we have considered Metropolis Monte Carlo simulations of an RSOS fluctuating interface in a $3D$ SW geometry like the one sketched in Fig. 1(b). Simulations were done for squared system with size $L = 50, 100, 150$ (in the y axis direction) and $W = L/2$ (L and W are in lattice units) and periodical boundary conditions along both x and y axes. After equilibration, the calculation of equilibrium average parameters has been performed on the basis of very long simulations of up to $10^5 \div 10^6$ MCS (1 $MCS \equiv L^2$ Monte Carlo moves).

3 Results and Conclusions

The phase diagram of SW wetting in $2D$ is reported in Fig. 2(a). However, the nature of the unbinding transitions we found in $2D$ SWs can be extracted by

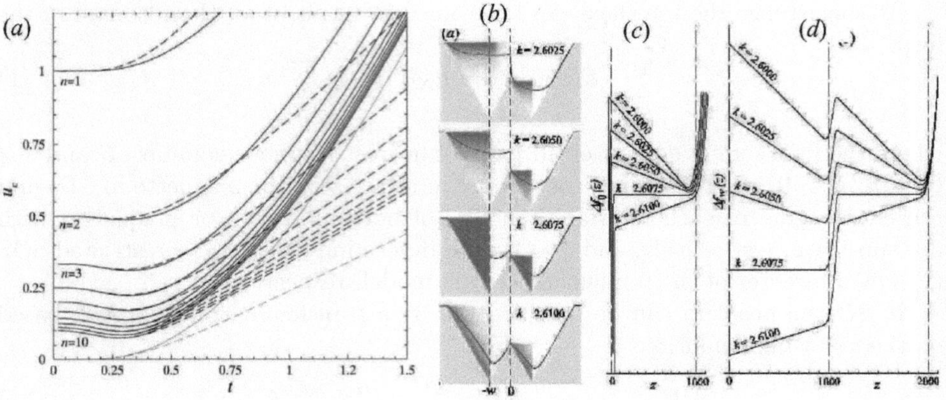

Figure 2. (*a*) Interface phase–line boundaries for $2D$ SWs (like in Fig. 1(*a*)) for SOS (continuous lines) and RSOS (dashed lines) models. Different lines correspond to tilted wall slope $|1/n|$ ($n = 1, 2, \ldots, 10$). Lower (light) curves are the interface phase–line boundary for a flat wall [6]. Interface profiles (panel (*b*)), and interface excess free energy profiles at $x = 0$ (panel (*c*)) and $x = \pm w$ (panel (*d*)). The free energy profiles have been obtained by a numerical calculation at $\omega = 0.2$ ($t = 0.6213\ldots$)

an analysis of free energy profiles shown in Fig.2 (*c*) and (*d*). In particular, at fixed temperature, as k is decreased from $k > k_c$ (the critical wall fugacity) the situation can be resumed as follows: *i*) a unique free energy minimum at $z = 0$ (for $k = 2.6100 > k_c$) which delocalizes into a wide one at $k = 2.6075 \simeq k_c$ (i.e. a continuous unbinding transition in the pure component wedges); *ii*) double minima profiles for $k = 2.6050, 2.6025 \lesssim k_c$ with the minima placed at $z = d_1 - d$ and $z = d_1$ (i.e. coexistence between the state localized at the SW central ridge height and the bulk unbound state); *iii*) dominance of bulk unbound state ($k = 2.6000 > k_c$). This scenario supports an interface unbinding in two steps, with a first–order unbinding from the SW central ridge.

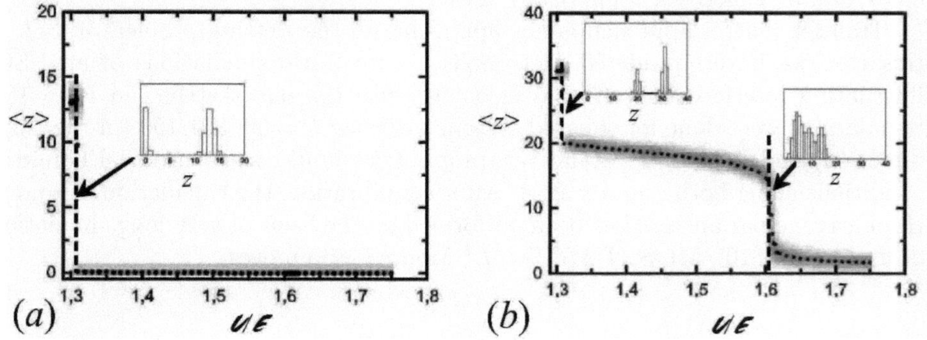

Figure 3. Average interface distances (black dots) from $3D$ SW central ridge (panel (*a*)) and bottoms (panel (*b*). The data come from Metropolis Monte Carlo simulations at constant $t = 2.0$ and system sizes $L = 100$, $W = 50$, wall unit slope and $d_1/d = 2$ (i.e. $d = 16$). In the insets is the detail of interface distance distributions at the u values marked by the dashed lines

Surprisingly, such a mechanism, as shown in Fig. 3, is fully confirmed by our Monte Carlo simulations for $3D$ SWs. In fact also in this case we have a discontinuous unbinding of the interface from the central ridge. The only difference is between $2D$ and $3D$ problems seems to be related to the fact that in $3D$ continuous and discontinuous unbindings seem to be separated by a finite gap.

Our investigation demonstrate that interface unbinding in SWs involves also a discontinuous transition. In this respect, this result opens new possible extensions to the wetting problem of random rough surfaces. In particular, our arguments give some more efforts on the key role of surface roughening exponent ζ_S in determining the nature of the wetting transition in self–affine rough substrates.

References

[1] S. Dietrich, in *Phase Transition and Critical Phenomena*, edited by C. Domb and J. L. Lebowitz (Academic Press, London, 1988), vol. 12, p. 1; G. Forgacs, R. Lipowsky, and T. M. Nieuwenhuizen, in *Phase Transition and Critical Phenomena*, edited by C. Domb and J. L. Lebowitz (Academic Press, London, 1991), vol. 14, p. 135.

[2] P. Pfeifer, Y. J. Wu, M. W. Cole, and J. Krim, Phys. Rev. Lett. **62**, 1997 (1989); M. Kardar and J. O.Indekeu, Europhys. Lett. **12**, 161 (1990); A. O. Parry, A. J. Wood, and C. Rascòn, J. Phys.: Condens. Matter **12**, 7671 (2000); C. Rascòn and A. O. Parry, Nature **207**, 986 (2000); L. Bruschi, E. Carlin, and G. Mistura, J. Phys.: Condens. Matter **15**, S315 (2003).

[3] A. O. Parry, C. Rascon, and A. J. Wood, Phys. Rev. Lett. **83**, 5535 (1999);

[4] G. Giugliarelli and A. L. Stella, Phys. Rev. E **53**, 5035 (1996); G. Giugliarelli and A. L. Stella, Physica A **239**, 467 (1997); A. L. Stella and G. Sartoni, Phys. Rev. E **58**, 2979 (1998); G. Giugliarelli and A. L. Stella, J. Phys.: Math. Gen. **32**, 5409 (1999).

[5] G. Giugliarelli, submitted to Phys. Rev. Lett. (2004); G. Giugliarelli, to be published;

[6] D. B. Abraham, Phys. Rev. Lett. **44**, 1165 (1980).

Part V

New Approaches to Physics Teaching

LEARNING PROBLEMS RELATED TO THE CONCEPT OF FIELD

FRANCESCA BRADAMANTE,[a] MARISA MICHELINI,[a]
ALBERTO STEFANEL[a]

Research Unit in Physics Education University of Udine, Italy

Abstract

The concept of field has formed itself through a process of elaboration of concepts both on the historical level, and on that of mathematic formalism (integral or differential approach, Lagrangian and Hamiltonian formalism), and in the relationship between mathematics and physics. In didactic tradition it is used as an example of the formalised analogical process. Its phenomenological aspects and its characteristics of time and space are only partially described for the various types of field, mixing the descriptive and interpretative levels. In this review we select some of the main teaching-learning problems and students' difficulties with learning the concept of field, that many researches have stressed, such as: the concept of field and its representation by lines of force, the difference between field and force, the connection between field and its sources, the concept of the field's superposition and the interpretation of mathematical formulas, the application of the third principle of dynamics, the motion of the field's characteristic particles in the field itself. We consider the field in static situations, because these learning problems are fundamental and basic for when we will pass to dynamic situations (electromagnetic field).

1 Introduction

The concept of field has formed itself through a process of elaboration of concepts both on the historical level and on that of mathematics formalism, and in the relationship between mathematics and physics. This concept is central in the

B.G. Sidharth, F. Honsell and A. de Angelis. (eds.), Frontiers of Fundamental Physics, 367–379.

description of the real world that physics gives nowadays. It is of fundamental importance when dealing with interaction processes, both in classic physics and in quantum physics, even if with different meanings. The extension of the Lagrangian and Hamiltonian formalisms in systems with a limited or definite number of levels of freedoms in dealing with continuous systems provides the formal tools for a unified treatment of classical fields and for their following quantization. The concept of field is no longer used only when dealing with interaction mediators, even in classical physics. It is more often used to describe physical quantities in relation to space and time within phenomenologies where local variations of physical dimensions are followed by global variations of the same quantities or of connected ones. The unified treatment of fields gives a unified view of the two different formal approaches that have been historically developed: the differential one, starting from the mathematic apparatus of fluid-dynamics and of physics of continuous systems based on the use of curl and divergence, and that of differential geometry, which uses integral-differential formalism. The first approach refers to many instruments which are typical of plane and geometry (curves and vectors), which initially facilitate the understanding for a broader public, but that have later a complex formal use in the forward developments and cause wrong local visions, inhibiting or strongly contrasting the formation of a global vision of the matter. The second approach is developed with a less common formalism and is therefore thought to be more complex, but it is actually more complete, general and elegant than the first. The concept of field can also be built without having to use other notions such as "force". Both approaches, in order to be fully developed, require mathematical instruments that are usually acquired during the first two years of a degree course in a scientific subject, but in all cases remain references for dydactic approaches of field concept also in secondary school. Historically [1, 2, 3] field lines and field concept, as "field of force", were introduced in 1845 by Faraday in the debate of action at a distance [4]. According to Faraday the propagation of forces takes place through a medium and not at a distance. Also, lines of force have a physical reality and represent the structure of material substances and their interactions. According to Maxwell *"Faraday ... saw lines of force traversing all space where the mathematicians saw centres of force, ... saw a medium where they saw nothing but distance ... "* [5].

Afterwards, Faraday's concept of field was adopted and changed by Thomson and Maxwell. The first gave a mathematical representation of Faraday's theory of magnetic field and developed a theory on the "ethereal continuum" as a physical expression of Faraday's "plenum of forces".

Maxwell, as well as completing the electromagnetic theory with his equations, formulated the first clear definition of a field, giving it the role of mediator of the interactions between systems. The field of forces was seen as a continuous or discrete system formed by interacting particles (the ether) or as a spatial distribution of forces. Maxwell also elaborated a "geometrical model" of field, in which Faraday's lines of force constituted a purely geometric representation of the field's structure. A confirmation of Maxwell's theory came from Hertz's experiments, which allowed to directly highlight the propagation of electromagnetic waves. Lorentz described the interaction between fields and matter, separating the concepts of ether and matter, finally setting up a theory that developed a universal physics based on electromagnetism.

With the ridding of the ether and the discovery of the quantum nature of the electromagnetic field (Einstein, 1905) it was possible to pass from classical physics to modern physics, based on the quantum theory, and reach a modern concept of field. The theory of field brought a redefinition not only of the physical model, but also of the meanings and the role given to the field, and it also drew attention on the need to acquire a unitary vision of the fields. The study on learning problems, on mental models and on the students' reasoning patterns is carried out considering these two problems: the historical point of view and the mathematical formalism.

As shown in literature, among the pupils' learning knots [6], there are often the ones[1] that scientists encountered in the process of building physics theories: they have a privileged position in offering different and partial representations, which are presented in an alternative and separate way to the physical meaning of the entity.

The concept of field is one of these. In mathematics it has an autonomous nature. In physics, it is re-interpreted with a role of synthesis and generalization, referring to different types of field and a complex nature in relation to various phenomenologies. It is a typical physical entity[2], object of interest also from the point of view of disciplinary foundations. Considered by physicists themselves only a formal (mathematical) object, an object/physics entity (for example the electromagnetic field), a property of a physical system (for example the field of the velocity in a fluid), a property of space or a structure or a tool for describing interaction [7, 8], it occupies a privileged position among the basic concepts for the building of physics culture.

In didactic tradition it is used as an example of the formalised analogical process. Its phenomenological aspects and its characteristics of time and space are only partially described for the various types of field, mixing the descriptive and interpretative levels[3]. The local characters are never discussed with the general ones and properties are presented depending on the type of field, without giving a role to the compared physical meaning of formal entities, which have the unifying power of the field concept (field lines, potential, flux, circuitation, etc.). As happens for other abstract and/or general concepts[4], which recall the human toil needed to acquire familiarity with their representative power, many ignore it

[1] The difficulties children meet obviously are not only the historical interpretative knots or those unifying concepts which single out new physical entities: there are also various cognitive aspects, which complicate the problem.

[2] The nature of the concept of field summarises physics' modalities in the interpretative processes.

[3] The confusion between the descriptive and interpretative levels is a typical problem of formal didactics and a lot of proof may be found in research on learning processes [22]

[4] The case of fractions in maths is famous: though handled with ease by primary school children, they are a complex learning problem at the ages of 14-16 [7].

or consider it in a reductive way in school practics, while even younger children have intuitions that allow to deal with the physical meaning at a high level of abstraction and symbolic representation [9].

Our research aims at building didactic proposals that precociously introduce the concept of field, giving a modern vision of physics, able to overcome learning problems or avoid the obstacles in learning that many researches have shown and that are encountered also by university students [10, 11, 12, 13, 14, 15, 16, 17, 18, 19, 20, 21]; for example those relating to the concept of field and its representation by lines of force, the difference between field and force, the connection between field and its sources, the concept of the field's superposition and the interpretation of mathematical formulas, the application of the third principle of dynamics, the motion of the field's characteristic particles in the field itself.

In this paper we present a critical overview of the learning problems related to the concept of field, which have been pointed out in literature. These problems, together with the analysis of the fundamental principles of the concept of field, both on the historical and on the formal level, allow us to consider the relationship between science and mathematics and to discuss the meaning of the concept of field, putting in correlation the knowledge problems with the conceptual aspects. Here we take into consideration only static fields; however, the learning difficulties and problems students encounter and that we analyse concerning the field concept foundations, are therefore important also when considering dynamic situations (electromagnetic field).

2 Learning problems related to the concept of field

In this work we select some of the main problems and difficulties students encounter when learning the concept of field, which have already been pointed out in literature. This review, concerning only static fields, is the base for future researches on the construction of learning proposals for teaching and learning the field as a unifying concept in static and dynamic contexts, that can be proposed from primary school to secondary and university level.

2.1 THE ROLE OF ACTION AT A DISTANCE IN OPPOSITION TO THE CONCEPT OF FIELD AS A MODIFICATION OF SPACE POINTS

The explanation of interaction as action at a distance has always been problematic, either on historical level [1, 2, 3] or common sense one [16, 23]. The action at a distance concept implies some presuppositions that causes learning difficulties in the comprehension of field concept. These presuppositions, as some researchers pointed out [23] in their studies with pupils of 9-18 age years old, are: the idea that the interaction is instantaneous, the need for a connection between the objects interacting at a distance (many pupils expressed a need for air as a conducting medium for magnetic and gravitational attraction [16, 23]), the need for a medium for the transmission of interaction, how various forces support and focus each other, and the idea of the field's action as a limited region of space. Bar [23, 24] analyzed children's conceptions about "gravity" and found that they considered the atmosphere as the limit of the Earth's gravitational attraction; while Borges' and Gilbert's studies [15] reveal one of the models of magnetism as a "cloud" or a "finite region of influence".

On the other hand a completely different lay-out is to consider the different phenomenologies, interpreted up to now in terms of action at a distance, in a unifying framework where the concept of field acts as a cognitive organizer. The analysis of difficulties of students in learning the concept of field, have been the bases for some proposals [17, 18, 25, 26] for the introduction of this concept in Secondary Education and High School. The purpose of these proposals is to overcome some learning problems, as:

1) the students' idea of field as a limited space region or a volume;
2) the students' identification of the field with the force and not with a physical entity and a modification of space which allow interactions;
3) the students' difficulty in understanding the differences between Newtonian conception of interactions and the conception of interaction described by fields.

2.2 GRAPHIC REPRESENTATIONS OF A FIELD AND THE ROLE OF FIELD LINES

Here are some learning problems related to field representation by field lines [13, 14, 27]:

- Interpretation of field representation by field lines [13, 15, 27]
- Objectualization of field lines [13, 15, 27]
- Difficulties to recognize differences and to integrate alternative representations of fields (field lines, equipotential surfaces...) [28, 29]
- Interpretation of the representation of objects in a field and situations involving fields.

The concept of field and field lines are sources of confusion among students even at university level. Tornkvist, Pettersson and Transtromer in a research on the electrostatic field [13] have shown that university students dont treat field lines using arguments in terms of mathematical concepts (uniqueness and continuity, proportionality and isomorphic mapping). In fact, students often consider field lines as isolated entities in the Euclidean space, not as a set of curves representing a vectorial property of that space; finally they don't fully understand the hierarchical sequence between the concepts, for instance: charge geometry, field line, force vector, velocity vector, trajectory (fig.1). The authors suggest that this confusion by representation is, at least partly, the cause of the well-known conceptions students have on the concept of force, reported by Johansson and Viennot [10].

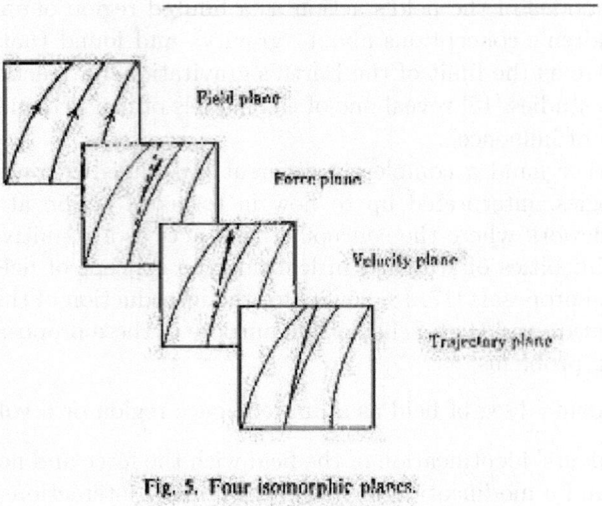

Figure 1. From [13]

Some other studies [29] on the graphical representation of electric and magnetic fields, suggest to not limit to field lines but also to draw the orthogonal equipotential surfaces. The authors believe that in such a field diagram it is possible not only to clearly distinguish the distribution of the "flux sources", but also that of the "circulation sources". Maxwell already used lines of force and "level surfaces" in all his graphical representations of fields, with and without circulation [5] (figure 2).

In the case of the gravitational field there are some studies that investigate the way pupils envisage the Earth, its shape and its relationship with the direction of the gravitational field [9, 23, 24, 30, 31, 32, 33, 34]. In the 80's Nussbaum [30, 31] made a first classification of children's ideas about Earth, finding three essential elements: the Earth's shape, the feature of the sky and the space, and the direction of falling objects from different locations of the Earth. We found this type of classification in Vosniadou's studies about conceptual change and teaching of science [32, 33], where she classified Mental models of the Earth. For the author (Vosniadou) the representation of Earth as flat is based on the assumption that the Earth is a physical body, and children must understand that Earth is an "astronomical" rather than a "physical object". In particular she observes that in the area of astronomy students understand the spherical shape of the Earth only after they have acquired an elementary notion of gravity; and that it seems particularly important to teach children something about gravity in order to understand how people can live on a spherical, rotating Earth.

On the other hand, Arnold [34] investigates the development of children's concept of the Earth and the direction of the gravitational field, and the relationship between the development sequence of children's drawings of the Earth's shape and

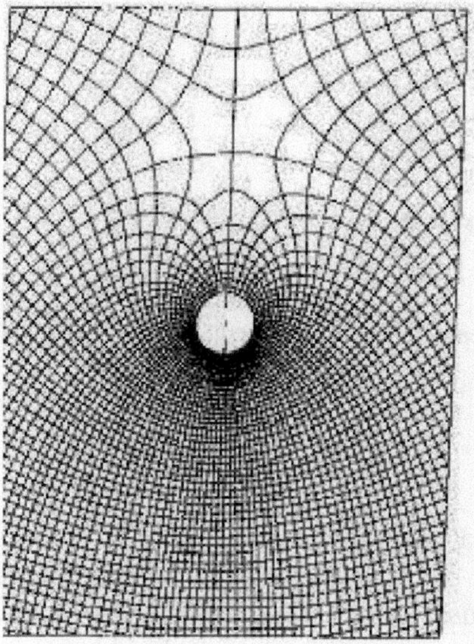

Figure 2. Maxwell's diagram of the superposition of the magnetic field of a linear current and a homogeneous magnetic field, in [5] quoted by [29]

the quality of their drawings of human figures.

2.3 THE FIELD AND ITS SOURCES: THE NATURE OF FIELD SOURCES; FIELD SUPERPOSITION

Even at university level, students don't identify correctly magnetic field sources, and most of them have difficulties in the distinction between electrostatic and magnetic interaction, i.e. they consider an electrostatic charge as a source of magnetic field [20, 35]. Students from 13 to 20 years old identify the magnet as a source of magnetic field, and motivate it considering its magnetic nature; but they think that the objects interact differently with it because of the difference of their electrostatic charge [35]. Borges Gilbert has investigated the ideas of students from 15 to 18 years old about the sources of the magnetic field and its action on objects [15]; they classify the answers in five categories: the models of magnetism.

Viennot and Rainson, while investigating university students' reasoning on electrostatics identify two main problems: "field only if mobility" and "cause in the formula" [11, 12]. The first one means that students have difficulty in accepting the existence of an electric field in a medium where charges are motionless. Students are not able to understand the different behaviours of conductors and insulators [11, 36] and to recognize the possible presence of an electric field in insulators. The second one ("cause in the formula") is a result of a questionnaire on Gauss's Theorem [12] and consists in ignoring the sources of the field that are not represented by their symbol in the formula which expresses the field. In these researches the authors describe a teaching sequence for the superposition of electric fields, which

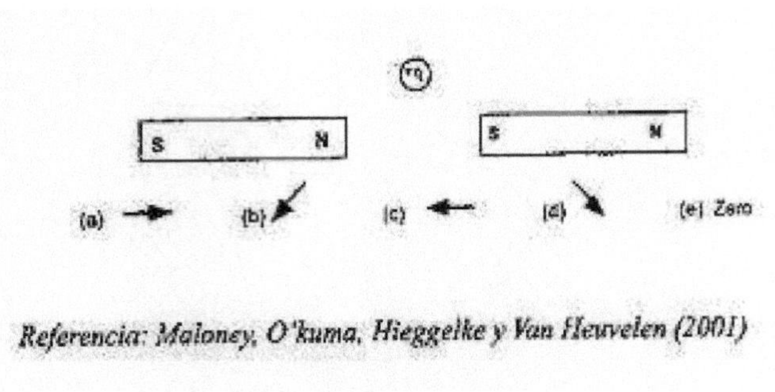

Figure 3. From reference [19, 20]

is a prerequisite for a basic understanding of electrostatics and for a unified view of electrostatics and electric circuits that can be fostered by teaching strategies, which emphasize causal aspects and transient phases. The authors suggest that the principle of field superposition is not obvious to students, and that it is useful to work on it in static situations before analysing electric circuits.

2.4 THE DYNAMICS OF INTERACTIONS MEDIATED BY A FIELD

Researches show that university students do not have a complete comprehension of the following:

- the objects on which a magnetic field can act, for example they think that a magnetic field can interact with electrostatic charges and can move them [19, 20] (Fig.3)
- the application of the third Dynamics Law and its applicability in contexts other than the mechanics one.

At every level from primary school [23, 24] to university [14, 19] students are not able to recognize that interaction is reciprocal in the different fields: the gravitational one [16, 23, 24, 37, 38], the electric one [18, 19, 25], the magnetic and electromagnetic one [14, 19, 20]. Most of them consider the force only in one direction: this shows the importance to teach force concept not as a cause that produces an effect but as an interaction (reciprocal). The motion of the field's characteristic particles in the field itself implies learning problems on:

- the recognition of the role of initial conditions when the particle is moving in a field,
- the distinction between trajectories of the particles and field lines [13, 14, 18, 20] (Fig.4)

Other studies [12, 20] reveal that at university level students have some learning problems related to the application of the magnetic force law: $\mathbf{F} = q\mathbf{v} \times \mathbf{B}$, due mostly, for the authors, to the incomprehension of the vectorial product.

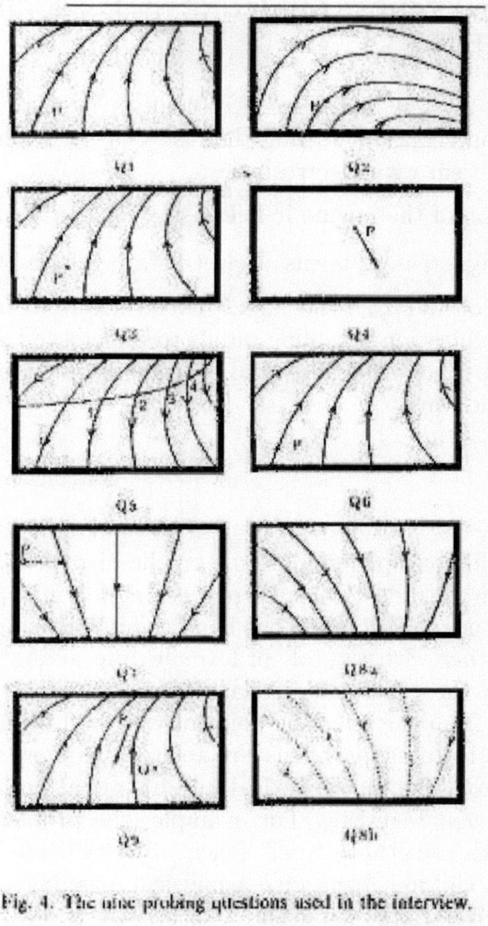

Figure 4. From reference [13]

2.5 CORRELATION BETWEEN GRAVITY AND MAGNETISM IN STUDENTS' CONCEPTIONS

Among the difficulties pointed out in research literature concerning these two types of interaction, in the last twenty years several studies [16, 23, 24, 34, 37, 38] have investigated children's commonsense ideas about gravity and its relationship with magnetism, and reveal that students (9-18 years old) identify gravitational effects with magnetic ones, considering the gravitational attraction as a magnetic one. They also connect gravity and magnetism on the Earth, considering gravity as necessary for a magnetic field on the Earth, a uniqueness of the Earth's system; in fact some children think that "the magnet cannot function without gravity", while others affirm that "electrostatic force cannot function without gravity", founding in that case a link between gravity and electrostatics.

2.6 THE STUDENTS' IDEAS OF MAGNETISM DESCRIBED BY ELECTROSTATIC EFFECTS

Many studies [15, 20, 21] reveal some difficulties in the comprehension of the magnetic field at university or during the last years of secondary school and the confusion of magnetism with electrostatics:

- the description of the magnetic field using electric field representations [20]
- models of magnetism in terms of electrostatic effects [15]

Some authors [23] consider that the "link between electricity and magnetism should be emphasized in education, enabling pupils to understand Ampere's explanation of magnetism. This relates magnetism to micro-currents within particles of matter and electromagnetic waves."

3 Conclusions

The analysis of field concept foundations, on the historical and formal level, and also the analysis, on the didactical level, of teaching-learning problems and schemes of reasoning, allow us to consider on one hand the relationship between science and mathematics, together with the role of formalism in giving interpretative models of physics, and on the other hand it lays down the problem of the meaning of this concept, both from the scientific point of view and from that of the students' conceptions. The learning problems are related to knowledge problems of the concept of field (from the mathematical and historical point of view), thus they have to be considered together. For example, the problem of identification of the field's lines with something "real" (their objectualisation) can be found both in students and in children, who give a physical meaning to the field lines, thus getting very near to Faraday's concept.

Can the field be considered as a property of space? Or should it be considered as a field of force momentum, or as a system? Or does it allow us to define the structured properties of a system? And finally, is it correct to observe all fields in only one way? From the point of view of mathematics, one could answer affirmatively, since scalar fields are a subclass of vectorial fields; however, from the physical point of view the answer is more difficult and questionable.

In light of the above, it is necessary to interpret the ways in which students and scientists consider this concept, and how they are associated to ideas and to the capacity to represent the concept dealt with. In view of a cognitive research for the development of a didactical proposal a question that could therefore be asked is: which are the representations that are triggered when the concept of field is explained by using an approach based on fluids' mathematics? Which are instead the ones when the approach adopted is based on differential geometry? What are the children's analogical representations according to the different approaches?

The problem here is that of knowledge and of the building of knowledge itself, together with the didactic problem. If we limited ourselves to the present requests of secondary school (applicative science conception), we could adopt a conceptual model of the nineteenth century. But if we want to give a unitary description of physical phenomena and a cognitive organisation of the different parts of physics, such as mechanics and electromagnetism, then it is necessary to adopt a unitary model (scientific competence conception) finding out new modalities to introduce in a simple way otherwise complex formalism.

Whatever choice is made from a didactical point of view, it is necessary to carry it on rigorous way, both with regard to the choice of the model and with regard to the formalization of the model itself, highlighting its qualities and limits.

This work has only taken into consideration the fundamental problems related to static fields. It is a preliminary work for future researches concerning the possible teaching proposals on field as a unitary and unifying concept, starting from elementary school and carrying on until secondary school and university.

References

[1] M.B. HESSE, *Forces and Fields, the Concept of Action at a Distance in the History of Physics*, Thomas Nelson and Sons Ltd., Edinburgh (1961).

[2] M. JAMMER, *Concepts of Force: A Study in the Foundations of Dynamics*, Harvard University Press edition (1957).

[3] E.J. DIJKTERHUIS, *Il meccanicismo e l'immagine del mondo*, pp 640-643, Feltrinelli (1980).

[4] M. FARADAY, *Experimental Researches in Electricity*, Richard Taylor and William Francis, Printers and Publishers to the University of London, Vol I, II, III. (1855).

[5] J.C. MAXWELL, *A Treatise on Electricity and Magnetism*, Dover Publications (1954).

[6] R. DUIT, *Constraint on knowledge acquisition and conceptual change: the case of physics*, Paper presented at the Symposium on Constraints on Knowledge Construction and Conceptual Change. Sixth European Conference for Learning and Instruction, Nijmegen (1995).

[7] P. GUIDONI, private communication (2003).

[8] R. GIANNITRAPANI, private communication (2004).

[9] F. BRADAMANTE, M. MICHELINI, *Children's ideas about gravitation, investigating a model of gravitational field*, Proceedings of Girep, Ostrava (2004).

[10] L. VIENNOT, *Spontaneous reasoning in elementary dynamics*, Eur. J. Sci. Educ. 1, pp 205-222 (1979).

[11] R. RAINSON, L. VIENNOT, *Design and evaluation of a research-based teaching sequence: the superposition of electric field*, Int. J. Sci. Educ. 21(1), pp 1-16 (1999).

[12] R. RAINSON, L. VIENNOT, *Students' reasoning about superposition of electric fields*, Int. J. Sci. Educ. 14(4), pp 475-487 (1992).

[13] S. TORNKWIST, K.A. PETTERSSON, G. TRANSTROMER, *Confusion by representation: On student's comprehension of electric field concept*, Am. J. Phys. 61(4), pp 335-338 (1993).

[14] I. GALILI, *Mechanics backgroud influences students' conceptions in electromagnetism*, Int. J. Sc. Educ. 17(3), pp 371-387 (1995).

[15] A.T. BORGES, J.K. GILBERT, *Models of Magnetism*, Int. J. Sc. Educ. 20(3), pp 361-378 (1998).

[16] I. GALILI, *Weight versus gravitational force: historical and educational perspectives*, Int. J. Sc. Educ. 22(10), pp 1073-1093 (2001).

[17] J. MARTIN, J. SOLBES, *Diseno y evaluacion de una propuesta para la ensenanza del concepto de campo en fisica*, Ensenanza de las Ciencias 19(3), pp 393-403 (2001).

[18] C. FURIO, J. GUISASOLA, *La ensenanza del cocepto de campo electrico basada en un modelo de aprendizaje como investigacion orientada*, Ensenanza de las Ciencias 19(2), pp 319-334 (2001).

[19] D.P. MALONEY, T.L. O'KUMA, C.J. HIEGGELKE, A. Van HEUVELEN, *Surveying students' conceptual knowledge of electricity and magnetism*, Phys. Educ. Res.,Am. J. Phys. Suppl. 69(7), pp S12-S23 (2001).

[20] J. GUISASOLA, J.M. ALMUDI, M. CEBERIO, *Concepciones alternatives sobre el campo magnetico estacionario. Seleccion de cuestiones elaboradas para su deteccion y tratamiento*, Ensenanza de las Ciencias 21(2), pp 281-293 (2003).

[21] J. GUISASOLA, J.M. ALMUDI, J. ZUBIMENDI, *Difficulties in Learning the Introductory Magnetic Field Theory in the First Years of University*, Wiley Periodicals, Inc. pp 443-464 (2004).

[22] S. BOSIO, V. CAPOCCHIANI, M. MICHELINI, J.S. PUGLIESE, C. SARTORI, M.L. SCILLIA, A. STEFANEL, *Playing, Experimenting, Thinking: Exploring Informal Learning Within An Exhibit Of Simple Experiments*, In New Way For Teaching, Girep Book, Ljubljana University Press (1997).

[23] V. BAR, B. ZINN, E. RUBIN, *Children's ideas about action at a distance*, Int. J. Sci. Educ. 19(10), pp 1137-1157 (1997).

[24] V. BAR, B. ZINN, GOLDMUNTZ, C. SNEIDER, *Children's concept about weight and free fall*, Science Education 78(2), pp 149-69 (1994).

[25] C. FURIO, J. GUISASOLA, *Concepciones alternatives y dificultades de aprendizaje en electrostatica. Seleccion de cuestiones elaboradas para su deteccion y tratamiento*, Ensenanza de las Ciencias 17(3), pp 441-452 (1999).

[26] C. FURIO, J. GUISASOLA, *The teaching of the concept of electric field: mountain or hill*, Proc. Esera Conf., Kiel (2001).

[27] E.R. SAVERLSBERG, T. DE JONG, M. FERGUSON-HASSLER, *Situational Knowledge in Physisc: the case of Electrodinamics*, J. Res. Sci. Teach. 39(10), pp 928-951 (2002).

[28] I.M. GRECA, M.A. MOREIRA, *The kinds of mental rapresentations-models, propositions and images-used by college physics students regarding the concept of field*, Int. J. Sci. Educ. 19(6), pp 711-724 (1997).

[29] F. HERRMANN, H. HAUPTMANN, M. SULEDER, *Representations of electric and magnetic fields*, Am. J. Phys. 68, pp 171-174 (2000).

[30] J. NUSSBAUM, J.D. NOVAK, *An assessment of children's concepts of earth utilizing structured interviews*, Science Education 60, pp 535-550 (1976).

[31] J. NUSSBAUM, *The Earth as a cosmic body Children's Ideas in Science*, Driver Gnesnes Tibershien (Eds) Open Univ Press Miltan Keynes, chapter 9 (1985).

[32] S. VOSNIADOU, W.F. BREWER,, *Mental Models of the Earth: A study of conceptual change in childhood*, Cognitive Psychology 24, pp 535-585 (1992).

[33] S. VOSNIADOU, *Conceptual change Research and Teaching of Science*, h. Behndt et al. (eds) Research in Science Education Past, Present, and Future, Kluwer Academic Publischers Printed in the Netherlands, pp 177-188 (2001).

[34] P. ARNOLD, A. SARGE, L. WORRALL, *Children's knowledge of the earth's shape and its gravitational field*, Int. J. Sci. Educ. 17(5), pp 635-641 (1995).

[35] A. MAAROUF, S. BENYAMNA, *La construction des sciences physiques par les representations et les erreurs: cas des phenomenes magnetiques*, Didaskalia, 11, pp 103-120 (1997).

[36] J. PARK, I. KIM, M. KIM, M. LEE, *Analysis of students' process of confirmation and falsification of their prior ideas about electrostatics*, Int. J. Sci. Educ. 23(12), pp 1219-1236 (2001).

[37] S. RUGGIERO, A. CARTELLI, F. DUPRE, M. VICENTINI-MISSONI, *Weight, gravity and air pressure: Mental representations by Italian Middle School pupils*, European Journal of Science Education 7(2), pp 181-194 (1985).

[38] D. PALMER, *Student's alternative conceptions and scientifically acceptable conceptions about gravity*, Int. Scie. Educ. 23(7), pp 691-706 (2001).

ELASTIC WAVES: MENTAL MODELS AND TEACHING/ LEARNING SEQUENCES

GIOVANNI TARANTINO [a]

[a] *Dipartimento di Fisica e Tecnologie Relative, Università di Palermo, viale delle Scienze Ed. 18, 90128 Palermo, Italy*

Abstract

In last years many research studies have pointed out relevant student difficulties in understanding the physics of mechanical waves. Moreover, it has been reported that these difficulties deal with some fundamental concepts as the role of the medium in wave propagation, the superposition principle and the mathematical description of waves involving the use of functions of two variables. In the context of pre-service courses for teacher preparation a teaching/learning (T/L) sequence based on using simple RTL experiments and interactive simulation environments aimed to show the effect of medium properties on the propagation speed of a wave pulse, has been experimented. Here, preliminary results of investigations carried out with a 120 trainee-teacher (TT) group are reported and discussed.

1 Introduction

According to experiences reported by many physics teachers, students are in trouble whenever they have to face up to elastic wave phenomena. Investigations [1] have shown that the difficulties students have with wave physics mainly involve the mathematical description of wave phenomena, requesting the use of two-variables functions, the concept and the application of the superposition principle, often mistaken by a mere overlap of waveforms, and the role that has to be assigned to the medium through which wave is propagating . This role is sometimes considered passive in the sense that the propagation is described as not directly affected by the elastic properties of the medium.

B.G. Sidharth, F. Honsell and A. de Angelis. (eds.), Frontiers of Fundamental Physics, 381–384.

Nevertheless, in real life, it is possible to find some examples which can be considered a clear proof that probably people perceive what a wave is. To produce the "Ola" (the typical wave in a stadium), for example, the fans know well when they have to stand up or sit down.

On the other hand, if we change the context, what happens is that students seem to apply inappropriate cognitive resources to describe the event propagation [2].

Consider, for example, a pulse wave travelling along a slinky. Many researches [1] have shown that a large group of students, even at university level, describe the pulse as if it were a moving body. Consequently, the medium is thought as playing a passive role and the propagation properties are wrongly considered as depending on the way the pulse has been generated. When we think of a propagating pulse as a whole (for example a pulse travelling along an elastic rope), we usually attribute to the pulse some object-like features (shape, width, speed etc.). But, while in the "Ola" case, the two levels of representation (on the whole and in terms of individual agents) are well distinguishable and people recognize that what is propagating is an event (standing up or sitting down), in the "slinky" case the switching between the two levels of representation is harder. Then, they are unable to recognize that they're dealing with the propagation of an event and use cognitive and perceptual resources which are adequate for describing the propagation of an object but inadequate to interpret the propagation of events.

Things get worse when we deal with sound waves because the difficulty of discerning between event and particle propagation is hampered by the imperceptibility of the medium: the perception of phenomenon is auditory and not visual and what is moving must be indirectly inferred.

In the next section, we discuss the results obtained in the preliminary phase of the research study carried out with a group of 120 trainee-teachers attending the teachers preparation school (SSIS) at University of Palermo concerning the understanding of the role of medium in elastic waves propagation.

2 The Pre-Test

By means of a questionnaire, students have been requested to describe three different situations dealing with propagation of pulses through elastic media. The situations were:

1. The free end of an horizontal elastic rope fixed on the wall is moved quickly up and down. A travelling pulse is generated. What happens if the hand generating the pulse is moved more rapidly?

2. The end of a two meters long metal bar is hit by an hammer. Two microphones are positioned at the same distance from the hit point but one is near the other end of the bar, while the second probe is far away from it. Which probe does the sound pulse reach before?

3. Consider a dust particle in front of a loudspeaker emitting a sound of a given frequency. Describe the motion of the particle.

The analysis of pre-test results has shown that the students answered to the questionnaire by using two different mental models.

The first model may be defined as the Active Medium Model (AMM). In this model the medium is thought as playing an active and essential role in the propagation mechanisms.

Those who used this model gave correlated interpretations of the three problematic situations of pre-test according to which the propagation speed is not affected by the way the pulse is generated (question 1), the sound pulse propagates faster along the metal bar because of the different elastic properties of metal and air (question 2) and the dust particle (question 3) moves back and forth.

The second mental model has been called the Passive Medium Model (PMM). Predictions made according this model have been based on the representation of the medium as a passive element in the propagation. While the pulse is travelling, it is subjected to the resistance exerted by the medium and depending on the medium density, just like a body moving in a viscous medium.

Those who applied this model provided some interpretations like:

- The pulse travels faster if the hand is moved faster
- Concerning question 2, the sound pulse reaches before the microphone placed far from the bar end because of resistance offered by the metal bar
- The dust particle in front of the loudspeaker moves forward or it keeps still.

All these wrong representations show that when the event-like nature of waves is not recognized, object-like properties are attributed to the wave pulse.

3 The Teaching/Learning Sequences

The teaching/learning sequences have been aimed to stimulating the use of the appropriate cognitive resources for description of wave phenomena and have been based on two different kinds of tools:

- RTL experiments
- Simulation environments

The experiments concerned essentially the measurement of the sound speed in different media (air and metals) with the aid of real time data logger.

Regarding the second kind of tools, we used the Interactive Physics environment to build up several simulations of linear chains of identical masses interacting through a linear nearest-neighborhood coupling. Students used such kind of tools in a constructivistic context stimulating the learning by means of work-sheets and adoption of a PEC (Prediction-Experiment-Comparison) cycle.

The analysis of materials produced by the students (work-sheets, interviews etc.) has shown the following two main results:

- The understanding of elastic wave phenomena is improved by choosing an approach based onto the representation of waves in terms of individual agents, rather than based "on the whole"
- The "pulse" approach is more effective than the "sinusoidal" one.

First result may be explained by considering that, when we observe a wave phenomenon, its wave-like nature is recognizable not until we think of it in terms of the behavior of the individual agents. In terms of conceptual resources, we find that the appropriate resources for description of wave phenomena are mostly stimulated by an approach based on individual agents representation.

On the other hand, the approach to waves based on the properties of sinusoidal and/or other periodic waves focus on the periodicity of a wave and not on its propagation. For this reason we think that, in order to explain the role of medium in the propagation process, a pulse approach be more effective. We found that students often associate the wave-like properties of a system to its shape. The term wave evokes the waves propagating through the surface of liquids (the sea waves, for example) and the sinusoidal approach fosters this kind of misunderstanding.

4 Conclusions

The starting point of the research study here described is based on the measured difficulties the students show to have in interpreting elastic wave phenomena.

By now, we are still analyzing the students work-sheets together with other materials and improving them on the basis of the feed-back.

However, the tutorials have met with measurable success and preliminary results obtained by examination of questionnaires, tests and interviews have shown an improvement in understanding and in the ability to describe the correct physics.

References

[1] M.C. Wittmann, R.N. Steinberg, E.F. Redish, *Making Sense of how Students Make Sense of Waves*, The Physics Teacher, 37 15-21 (1999)

[2] D. Hammer, *Student Resources for Learning Introductory Physics*, Phys Educ. Res., Am. J. Phys., Suppl. 68 (7) (2000)

INDEX OF AUTHORS

MIX
Papier aus verantwortungsvollen Quellen
Paper from responsible sources
FSC® C105338

If you have any concerns about our products,
you can contact us on
ProductSafety@springernature.com

In case Publisher is established outside the EU,
the EU authorized representative is:
Springer Nature Customer Service Center GmbH
Europaplatz 3, 69115 Heidelberg, Germany

Printed by Libri Plureos GmbH
in Hamburg, Germany